AF777944

INTEGRATED WEED AND SOIL MANAGEMENT

Edited by

J.L. Hatfield
D.D. Buhler
B.A. Stewart

Library of Congress Cataloging-in-Publication Data

Integrated weed and soil management / edited by J.L. Hatfield, D.D. Buhler, B.A. Stewart.
p. cm.
Based on papers from a workshop sponsored by the National Soil Tilth Laboratory.
Includes bibliographical references and index.
ISBN 1-57504-069-7
1. Weeds--Integrated control--Congresses. 2. Soil management--Congresses. I. Hatfield, Jerry L. II. Buhler, D.D. (Douglas D.) III. Stewart, B.A. (Bobby Alton), 1932–
SB610.2.I68 1998
632′.5--dc21 97-30972
CIP

ISBN 1-57504-069-7

COPYRIGHT 1998 by Sleeping Bear Press
ALL RIGHTS RESERVED

This book represents information obtained from authentic and highly regarded sources. Reprinted material is quoted with permission, and sources are indicated. A wide variety of references are listed. Every reasonable effort has been made to give reliable data and information, but the author and publisher cannot assume responsibility for the validity of all materials or for the consequences of their use.

Neither this book nor any part may be reproduced or transmitted in any form or by any means, electronic or mechanical, including photocopying, microfilming, and recording, or by any information storage or retrieval system, without permission in writing from the publisher.

No claim to original U.S. Government works.

ANN ARBOR PRESS
121 South Main Street, Chelsea, Michigan 48118
Ann Arbor Press is an imprint of Sleeping Bear Press

PRINTED IN THE UNITED STATES OF AMERICA
10 9 8 7 6 5 4 3 2 1

Preface

Weeds are considered to be one of the largest problems facing farmers each year. Weeds include a wide array of plants, ranging from cultivated plants growing in the wrong field, e.g., corn in a soybean field or oats in a rye field, to noxious plants that can cause injury or even death. Farmers spend large amounts of resources in controlling or at least attempting to control weeds in cultivated fields and pastures. Even gardeners and homeowners cannot escape weeds, and much effort is placed in trying to reduce or eliminate weeds from lawns and gardens. Weeds have been with humankind ever since cultivation was developed, and the Bible refers to weeds as choking out the crop in one of the parables. If weeds have been around for this length of time, why hasn't man with his ingenuity and skill figured out a way to eliminate these pests?

Many methods have been proposed to control weeds, but modern agriculture has come to rely on herbicides as the major means of controlling weed growth. However, this has come at a price and many critics believe that modern agriculture is responsible for the direct contamination of the drinking water supplies in the nation. Herbicides replaced tillage starting in the 1950s, and there is an increasing reliance on the use of manmade chemicals for weed control. The use of herbicides has not reduced weed problems, and some weeds have become resistant to chemicals, requiring different herbicides to be developed. This paradigm has become prevalent in the 1980s and early 1990s and suggests that we need to focus more on management of weeds. We decided to take a different approach and begin to examine the linkage between weed and soil management. This was the focus of the Fifth Long-Term Soil Management Workshop sponsored by the National Soil Tilth Laboratory, upon which these chapters are based.

Weed management implies a shift away from strict reliance on control of existing weed problems and places greater emphasis on prevention of propagule production, reduction of weed emergence in a crop, and minimizing weed competition with the crop. Weed management emphasizes integration of techniques to anticipate and manage problems rather than reacting to them after they are present. The goal is to maximize crop production where appropriate, optimize grower profit, and increase production efficiency by integrating preventative techniques, scientific knowledge, and management skills.

Weed seed is placed onto or into the soil; however, little is known about the interactions of weed seed with soil physical, chemical, or biological conditions. There are five major topic areas that link weed and soil management: surface residue, tillage, and weed and soil management; integration of soil and weed management to reduce environmental degradation; modeling weed emergence, interference, and management; new technology and weed management; and development of the next generation of weed management systems. Throughout this volume we have replaced weed control terminology with weed management terminology in an effort to shift the paradigm from one of control to better management. In order to manage, we need to understand the components of the system rather than simply striving to kill weeds.

The chapters prepared for this volume are written by experts dealing with weed and soil management. Many of these individuals are just beginning to develop research programs that connect our understanding of the soil and weed management, so the papers within this volume are the latest in our understanding of the processes involved. We also asked the authors to be bold in their concepts in an effort to approach the problem from as many different perspectives as possible. It is through our bringing pieces of the puzzle together in new and different ways that we can make progress toward this new paradigm. Chapters included in this volume address weed management from a weed population viewpoint and interactions among cover crops, smother crops, allelopathy, and tillage on the weed populations. Nutrients are supplied to foster more efficient crop growth but little attention is given to the effect on the weed growth and populations. Soil microbial populations affect weed seed and could also be used as a potential management tool. Environmental concerns of current weed management strategies and producers' attitudes toward weeds affect the decision-making process, and we have tried to show how these link to producers' attitudes about weeds and weed management. Decision-making relative to weed management will be based on improved models and economic value of this information, and we brought together individuals who have developed and are testing these approaches in the field. Weed management will require new technology, and chapters are included that describe the application of precision farming, new herbicide application technology, and approaches to reduce herbicide runoff, leaching, and volatilization. All of these chapters form a strong base for the development of the weed management paradigm.

Adoption of new technology by the producer and acceptance that our current and new technology does not pose an environmental risk can only be based on strong principles and insights into the problem. Weeds have been a problem for all of recorded humankind and we need to develop new and imaginative ways of weed management. It is our goal to begin this process with this volume, and we thank the authors of each chapter for so richly sharing their ideas and concepts with the world.

J.L. Hatfield
D.D. Buhler
B.A. Stewart

Contents

About the Editors:

Dr. J.L. Hatfield has been the Laboratory Director of the United States Department of Agriculture Agricultural Research Service, National Soil Tilth Laboratory in Ames, Iowa since 1989. He has been with the USDA-ARS since 1983, previously as the research leader of the Plant Stress and Water Conservation Unit in Lubbock, Texas. After receiving his Ph.D., Dr. Hatfield served on the faculty at the University of California, Davis, from 1975 through 1983. Dr. Hayfield received his Ph.D. from Iowa State University in 1975, a M.S. from the University of Kentucky in 1972, and a B.S. from Kansas State University in 1971. He is a Fellow in the American Society of Agronomy, Crop Science Society of America, and Soil Science Society of America. He served as editor of the *Agronomy Journal* from 1989 through 1995. Dr. Hatfield is the author or co-author of more than 225 articles and book chapters. He is the co-editor of *Biometeorology and Integrated Pest Management* and five volumes of *Advances in Soil Science*. He began the Long-Term Soil Management Workshops in 1991, of which this volume and other volumes of Advances in Soil Science are derived, as a means of evaluating the current state of knowledge regarding soil management and basic soil processes. He has an active research program in soil-plant-atmosphere interactions with emphasis on the energy exchanges as the soil surface under different tillage and crop residue management methods and the estimation of the evapotranspiration.

Dr. D.D. Buhler is a Research Agronomist at the USDA-Agricultural Research Service National Soil Tilth Laboratory in Ames, IA. His research includes the effects of crop production practices on weed biology and ecology and the development of alternative weed management systems. He has a B.S. degree from the University of Wisconsin-Platteville and M.S. and Ph.D. degrees from the University of Nebraska. Dr. Buhler is author or coauthor of more than 70 papers in refereed journals, has served as an associate editor for the journals of *Weed Science* and *Weed Technology*, and is a consulting editor for the Journal of Crop Production. He has received the Distinguished Young Scientist Award from the North Central Weed Science Society of America, and Midwest Area Early Career Scientist of the Year and Edminster Awards from USDA-Agricultural Research Service.

Dr. B.A. Stewart is a Distinguished Professor of Soil Science, and Director of the Dryland Agriculture Institute at West Texas A&M University, Canyon, Texas. Prior to joining West Texas A&M University in 1993, he was Director of the USDA Conservation and Production Research Laboratory, Bushland, Texas. Dr. Stewart is past president of the Soil Science Society of America, and was a member of the 1990-1993 Committee of Long Range Soil and Water Policy, National Research Council, National Academy of Sciences. He is a Fellow of the Soil Science Society of America, American Society of Agronomy, Soil and Water Conservation Society, a recipient of the USDA Superior Service Award, and a recipient of the Hugh Hammond Bennett Award by the Soil and Water Conservation Society.

Contributors

J.L. Baker, Agricultural and Biosystems Engineering Department, Iowa State University, Ames, IA 50011, USA

Roger L. Becker, Department of Agronomy and Plant Genetics, University of Minnesota, St. Paul, MN 55108, USA

Douglas D. Buhler, U.S. Department of Agriculture, Agricultural Research Service, National Soil Tilth Laboratory, Ames, IA 50011, USA

W.G. Crumpton, Botany Department, Iowa State University, Ames, IA 50011, USA

J. Anita Dieleman, Department of Agronomy, University of Nebraska, Lincoln, NE 68583-0915, USA

Michael Duffy, Department of Economics, Iowa State University, Ames, IA 50011, USA

T.J. Gish, U.S. Department of Agriculture, Agricultural Research Service, Hydrology Laboratory, Beltsville, MD 20705, USA

J.L. Hatfield, U.S. Department of Agriculture, Agricultural Research Service, National Soil Tilth Laboratory, Ames, IA 50011, USA

C.S. Helling, U.S. Department of Agriculture, Agricultural Research Service, Weed Science Laboratory, Beltsville, MD 20705, USA

Gregg A. Johnson, University of Minnesota, Southern Experiment Station, Waseca, MN 56093, USA

Robert P. King, Department of Applied Economics, University of Minnesota, St. Paul, MN 55108, USA

R.J. Kremer, U.S. Department of Agriculture, Agricultural Research Service, University of Missouri, Columbia, MO 65211, USA

K.-J.S. Kung, Department of Soil Science, University of Wisconsin, Madison, WI 53706, USA

Matt Liebman, Department of Applied Ecology and Environmental Sciences, University of Maine, Orono, ME 04469-5722, USA

John L. Lindquist, Department of Agronomy, University of Nebraska, Lincoln, NE 68583-0915, USA

Birl Lowery, Department of Soil Science, University of Wisconsin, Madison, WI 53706, USA

Donald W. Lybecker, Department of Agricultural Economics, Colorado State University, Fort Collins, CO 80523, USA

Alex R. Martin, Department of Agronomy, University of Nebraska, Lincoln, NE 68583-0915, USA

S.K. Mickelson, Agricultural and Biosystems Engineering Department, Iowa State University, Ames, IA 50011, USA

David A. Mortensen, Department of Agronomy, University of Nebraska, Lincoln, NE 68583-0915, USA

Tsutomu Ohno, Department of Applied Ecology and Environmental Sciences, University of Maine, Orono, ME 04469-5722, USA

Caleb A. Oriade, Department of Applied Economics, University of Minnesota, St. Paul, MN 55108, USA

Micheal D.K. Owen, Department of Agronomy, Iowa State University, Ames, IA 50011, USA

W.J. Rawls, U.S. Department of Agriculture, Agricultural Research Service, Hydrology Laboratory, Beltsville, MD 20705, USA

Edward E. Schweizer, U.S. Department of Agriculture, Agricultural Research Service, AERC, Fort Collins, CO 80523, USA

A. Shirmohammadi, Department of Agricultural Engineering, University of Maryland, College Park, MD 20742, USA

David E. Stoltenberg, Department of Agronomy, University of Wisconsin, Madison, WI 53706, USA

K.A. Sudduth, U.S. Department of Agriculture, Agricultural Research Service, Cropping Systems and Water Quality Research Unit, Agricultural Engineering Department, University of Missouri, Columbia, MO 65211, USA

Scott M. Swinton, Department of Agricultural Economics, Michigan State University, East Lansing, MI 48824, USA

John R. Teasdale, U.S. Department of Agriculture, Agricultural Research Service, Weed Science Laboratory, Beltsville, MD 20705, USA

B.J. Wienhold, U.S. Department of Agriculture, Agricultural Research Service, Northern Great Plains Research Laboratory, Mandan, ND 58554, USA

Lori J. Wiles, U.S. Department of Agriculture, Agricultural Research Service, Great Plains Systems Research, Fort Collins, CO 80522, USA

Important Biological Information Needed for Bioeconomic Weed Management Models

Edward E. Schweizer, Donald W. Lybecker, and Lori J. Wiles

I. Introduction

Farmers have relied on herbicides since the early 1950s as the primary method for controlling weeds in their crops because they are cheap, convenient, and more effective than most alternative methods. Annual loss from weeds for 46 major commodities in the United States was estimated to be $4.1 billion in 1991 with current control practices, and $19.6 billion if herbicides were not available (Bridges, 1992).

Public concern about food safety and the environment has caused considerable debate about the impact of pesticides, particularly herbicides, on our agroecosystems. U.S. agriculture is one of the most pervasive contributors to nonpoint source water pollution (Anonymous, 1992; U.S. Congress, 1990). Thus, the potential impact of herbicides on the quality and safety of our food and water supplies has prompted weed scientists to search for new weed control strategies that would balance herbicide use, crop productivity, food safety, profit, and water quality.

Excessive or inappropriate herbicide use could be avoided if applications are limited to fields, or portions of fields, for which weed control is justified economically, and if control is matched to the weed spectrum in each field. Thus, the development of computer software that farmers, crop consultants, extension personnel, and other decision makers could use to identify appropriate weed management strategies is desirable.

The number of computer models for making weed control decisions has increased significantly since the mid-1980s (Mortensen and Coble, 1991). The complexity of model-generated recommendations ranges from those based only on maximum weed control (herbicide efficacy aids) to those based on optimum economic weed control (population-based models). The latter group of models can be grouped into two broad categories. Category I models help decision makers select soil-applied and postemergence tactics while Category II models help decision makers select only postemergence tactics (Schweizer et al., 1993). This paper will focus on the biological aspects of population-based models. It is not meant to be an exhaustive review of the many theoretical and practical models of weed population dynamics and management. Our intent is to present our ideas (and biases) on important information gaps revealed by our experiences in developing bioeconomic weed management models.

II. Biological Aspects of Population-Based Models

All weed management decision aids must model the same basic biological processes. Yet, the development of realistic models is hindered by gaps in what we know about weed biology. Better information is needed not only for model developers, but for all weed scientists, on the following biological aspects of population-based models.

A. Composition of the Weed Population

An input for all current weed management models is an estimate of the composition of the weed population. This is usually an average seedbank or seedling count. Future models may also require a description of the spatial distribution of weed seed and seedlings as well as information on the vertical distribution of weed seed in the soil profile, seed viability, secondary dormancy, and mortality.

1. Seedbank

Information on seedbank dynamics is vitally important for simulating output with current soil-applied Category I research models. Simulations can only be run when we know the number of weed seed, by species, in the seedbank and the proportion of seed of each species that will emerge.

Of the biological factors discussed in this paper, weed seedbanks seem to be the most controversial among scientists, potential users, and administrators. Why is this, since quantitative information on seedbanks is such a critical input factor for soil-applied decision simulations? Critics say it's because characterizing weed seedbanks is costly. It requires intensive soil sampling, and commercial laboratories are not available to process soil samples and identify weed seed. These are valid criticisms. Either we have to overcome these problems or come up with a different approach. Seedbank data are currently the only basis of model recommendations as to whether we should not apply a soil-applied herbicide or employ some other tactic before weeds emerge.

We know that cultivated fields must be sampled intensively to characterize weed seedbanks because the composition and density of seedbanks are influenced by cropping history, tillage practices, and geographical location (Wilson, 1988). Seed densities can range from <200 to over >137,000 seed m^2 in agricultural soils (Forcella et al., 1993; Wilson, 1988).

The number of soil cores required to estimate the mean number of seed in a field also increases dramatically for those weed species whose seed are found at low frequencies versus a species whose seed are found at high frequencies (Wilson et al., 1985). For example, it took 66 times more soil cores to estimate the mean number of seed that is within 10% of the mean 95% of the time than it took to estimate the mean number of seed that is within 50% of the mean 80% of the time. The purpose of sampling the seedbank is to obtain data for decision model simulation to determine whether the seed density exceeds an economic threshold. If it does, we employ a management tactic. If it does not, we do nothing. Thus, there is no rule of thumb that will tell us how many soil cores we need to take. The number of samples required will depend on the species and how uniformly weed seed are distributed in the field. Since nonuniform spatial distribution of weeds complicates sampling, modeling, and management of weed populations, scientists are now combining classical and spatial statistics to study spatial distribution of seedlings (Cardina et al., 1995). This same technology should be employed to study the nonuniform spatial distribution of seed.

Commercial laboratories currently do not process soil samples for weed seedbank analyses, but would these laboratories process soil samples if the need existed? Probably! A case in point is there were no commercial laboratories available to process soil samples for the sugarbeet cyst nematode (*Heterodera schachtii*) in the 1960s and 1970s, but today several laboratories do. Based on the cost to process soil samples for nutrients and nematodes, an estimated cost would be $15 to $20 per sample for seedbank analysis. The initial cost to thoroughly map a field for the weed seedbank undoubtedly will be high. However, most Category I models simulate the number of weeds that escape control and these models could be modified to predict seed production by escapes and seedbank mortality. Then fields may have to be sampled only every fourth or fifth year, which will reduce sampling costs considerably.

In addition to how many soil samples should be taken per field, we still need to know the best way to describe the spatial distribution of weed seed in soil. Are spatial statistics, such as geostatistics, the answer? Mapping the spatial distribution of a weed population will require more intensive sampling than sampling to estimate mean density.

Practical application of models for soil-applied herbicide decisions will not be readily accepted until efficient methods of seed extraction, identification, and counting are developed, and seedling recruitment from the seedbank is better understood (Mortensen and Coble, 1991). Currently, two basic techniques are employed to determine the seedbank: the greenhouse tray method (Post, 1984) and the soil extraction method of Malone (1967) or a modification thereof (Schweizer and Zimdahl, 1984). The advantage of the greenhouse method is that seed do not have to be extracted from soil, identified, and counted. Counting and identifying seedlings is easier. The disadvantages are that it takes 3 to 5 cycles of wetting, drying, and cold stratification over several months to complete the test, and considerable bench space is needed for a large number of samples.

The primary advantage of the soil extraction method is that seed from a large number of soil cores can be extracted in a short period of time. Disadvantages are personnel must be trained to identify seed by species and no distinction is made between viable and nonviable seed. Thus, an assumption has to be made as to how many nondormant seed are likely to emerge. It is generally assumed that 2 to 10% of the nondormant seed will emerge, but predicting the number of weed seedlings that will emerge annually from the seedbank is not an exact science (Zimdahl, 1993).

Forcella (1992) reported that the greenhouse tray method was more reliable than the soil extraction method, with the latter method being more dependent on sample size and sampling date. Benoit et al. (1989) found that the greenhouse tray method was less labor intensive than the soil extraction method. Wiles et al. (1996) recently described a new soil sampler and elutriator that will make weed seedbank studies more feasible. These two methods, as well as any others that may be developed, need to be compared in terms of cost, time requirements, and reliability.

2. Seedling Populations

Information on seedling dynamics is vitally important for simulating output with current Category I and II postemergence research models. Computer simulations currently can only be run when we know the number of weed seedlings, by species, in each field. The procedure used to obtain these counts will influence their accuracy and ultimately, the quality of the recommendations generated by decision models (Ives and Moon, 1987). For practical application of these models, we must begin to consider how users will collect the required information about weed seedling populations.

A scouting plan outlines how to collect information about a pest population in a field in order to make a control decision. A plan is designed to achieve a balance between the cost of scouting and the value of the information obtained (Ives and

Moon, 1987; Southwood, 1976). The information is valuable from the perspective of preventing mistakes in selecting the most appropriate control treatment and thereby preventing unnecessary herbicide use or yield loss (Wiles et al., 1992c). The optimal design for a scouting plan is influenced by the cost of scouting, the sensitivity of the recommendation to the scouting information, and the characteristic spatial distribution of the pest (Ives and Moon, 1987; Southwood, 1976).

Scouting weed populations is anticipated to be almost as labor intensive as sampling seedbanks. For example, Marshall (1988) found that 18, 1-m^{-2} quadrats ha^{-1}, would have to be examined to estimate a mean grass density of about 2 inflorescences m^{-2} with an error of less than 30%. Considerable literature exists on sampling insect populations, but almost none on sampling weed populations. Unfortunately, scouting weeds involves problems that are not encountered with insects (Wiles et al., 1992b). Insect control decisions are usually a choice between one or two potential treatments for control of a single species. The spatial distribution of the insect may be fairly consistent and easy to describe. Postemergence weed control decisions commonly involve choosing among many potential treatments for control of a mix of weed species. The spatial distribution may also be difficult to characterize. These complexities will make it challenging to design robust, cost-effective weed scouting plans.

Those developing models must analyze trade-offs between the time needed to collect the information about the weed population and the value of that information. For example, what would be the impact if weeds were classified just as grasses and broadleaves, or as the predominant grass and broadleaf species, or as the predicted predominant grass and broadleaf species? Could the weed population be described by density levels rather than actual counts? After an analysis of scouting, the model delivered to users may be quite different than the original research model. It may be found that a simplified form, requiring less detailed inputs from the user, may be as valuable as the original research version. Or users may need to be educated on the value of information required by the model and guided on the most cost-effective way to collect the information.

In summary, we must sample seedbank and seedling populations to use decision models; however, this will be costly. We may need to implement new technology for practical application of decision models that use information on the composition of the weed seedbank or the spatial distribution of weeds. Identifying the seedbank may be economically feasible when the soil extraction method is combined with image analysis to count and identify seed (Benoit et al., 1992; Buhler and Maxwell, 1993). Mapping weed populations economically may require a combination of aerial photography and ground scouting using global positioning systems (GPS) equipment.

B. Weed Emergence

Weed management models are concerned with two aspects of weed emergence: the number of seed in the seedbank that will emerge and when they will emerge. Soil-applied weed management models are based upon or driven by an estimate of the number of annual weeds, by species, that will emerge. In some models, it is assumed that all weeds will emerge with the crop (Lybecker et al., 1994). In other models, the cumulative emergence is modeled over time (Swinton and King, 1994; Wiles et al., 1996). Modeling emergence over time can improve estimates of competitiveness, weed seed production, and the susceptibility of weeds to postemergence control. A weed population may be divided into cohorts, groups of individuals that have emerged at approximately the same time, that vary with respect to these characteristics.

Seedling emergence is controlled primarily by soil disturbance and temperature, water availability, and the vertical distribution of seed within the soil. Emergence is the result of two processes: release from dormancy and germination (Benech-Arnold and Sanchez, 1995). In a series of field studies conducted in the U.S. corn belt, researchers reported that the percentage of viable seed that emerged as seedlings ranged from < 1% for yellow rocket (*Barbarea vulgaris* R. Br.) to 30% for giant foxtail (*Setaria faberi* Herrm.) (Forcella et al., 1992). Redroot pigweed (*Amaranthus retroflexus* L.) and common lambsquarters (*Chenopodium album* L.) were the most frequently encountered species. Emergence percentages of these species were related inversely to rainfall or air temperatures in April or May, presumably because anoxia and/or high temperatures induced secondary dormancy in nondormant seed. This information can be employed by bioeconomic weed management models that currently use coarse estimates of emergence percentages to refine recommendations for weed control.

Emergence, release from dormancy, and germination have been modeled as a function of soil water availability and temperature to predict the pattern of emergence in the field. For example, Alm et al. (1993) developed a model for predicting the time pattern of seedling emergence based on soil water potential, average daily soil temperature, and the fluctuation in soil temperature. A seed population is divided into cohorts with individual germination and seedling elongation rates. The model was tested against the timing of emergence in the field for seeds planted 1, 2, 4, and 8 cm deep in natural or rain-excluded environments. The predictions were more accurate for all depths combined than for any particular depth. Sensitivity analysis of the model suggested the effect of soil temperature fluctuation could be ignored in future models (Alm et al., 1993).

Simpler models also have given reasonable predictions of cumulative emergence. Cumulative emergence of common lambsquarters from the top 5 cm of soil was predicted with germination rate as a function of average daily soil temperature (Harvey and Forcella, 1993). The number of emerged seedlings as well as the time pattern of emergence of velvetleaf (*Abutilon theophrasti* Medicus) have been modeled using data of average daily temperature and daily rainfall (Forcella, 1993). In the velvetleaf model, germination rate is a function of average daily soil temperature in the top 5 cm and is adjusted for soil water potential when daily water potential of the upper 5 cm of soil falls below a threshold. The model requires initialization of four seed zone parameters: soil water content at the start of the simulation period, soil water holding capacity, soil bulk density, and a water retention curve relating percent soil water to soil water potential. This model has been validated for field data collected in Illinois and Minnesota.

In some cases, prediction of seedling emergence may be more accurate if dormancy and germination are modeled as separate processes (Benech-Arnold and Sanchez, 1995). For example, total emergence and the time pattern of emergence for johnsongrass [*Sorghum halepense* (L.) Pers.] were different between bare and shaded soil even though recorded average daily temperatures were similar (Benech-Arnold et al., 1990). This result would be predicted with release from dormancy modeled separately from emergence and as a function of temperature fluctuation.

Weed management models should incorporate the effect of the tillage system and cultivation on weed emergence. So far, the focus has been on emergence in conventional tillage systems and the number and time pattern of emergence has been modeled independently of cultivation. There is extensive literature on emergence in response to tillage, but few attempts to model the response. This may be because the response is variable depending on species. One possible response is an increase in the number of seedlings that emerge, but no change in the seasonal pattern of emergence (Karssen, 1982).

Modeling emergence in response to soil water availability and temperature may partly capture the effect of tillage. Models that predict emergence from the distribution of seed within the soil and the effect of soil depth on emergence may be needed. Mohler (1993) has developed a simple, general model that describes changes in percent emergence with different types of tillage based on these two factors and the results are in general agreement with the literature.

We should incorporate emergence models like those described above into weed management models. Then we may be able to better predict emergence for different production practices, as well as study and account for the impact of year-to-year variation in emergence in response to weather on optimal decision-making. Also, more site-specific, real-time recommendations may be generated if emergence models are driven by readily available weather data.

C. Weed Competition

Yield loss predictions are an essential component of weed management models since recommending a treatment means evaluating the benefit of weed control. An obvious benefit is preventing yield loss. A difficulty in predicting yield losses is that populations consist of mixtures of species. Most Category I and II models predict crop yield loss from competition with mixed species populations using a hyperbolic relationship between weed density and yield loss (Cousens, 1986).

There have been two approaches for developing the relationship between density and yield loss for mixed populations. One approach is to directly estimate a set of competition coefficients from field data (Swinton et al., 1994b). This approach works well but requires extensive field data so that use of this approach has been limited to models with a limited number of weed species (Swinton and King, 1994). A second approach involves developing a relationship between yield loss and density for a single species and then ranking or rating other species for competitiveness as compared to this species. The ratings, called competitive indices, may be estimated from yield response data (Coble and Mortensen, 1992) or by experts

(Lybecker et al., 1994). The rankings are used to calculate a population of a single species that is as competitive as the mixed population. WEEDCAM estimates the degree of weed competition in corn (*Zea mays* L.) based on a subjective competitive index for 15 weed species (Lybecker et al., 1994). HERB estimates the degree of weed competition in soybean [*Glycine max* (L.) Merr.] based on a competitive index for over 75 weed species (Coble and Mortensen, 1992).

Additional research is still needed to establish the competitive indices for a number of weed species and to refine the use of competitive indices for mixtures of species. Also, recent research indicates that competition coefficients are not stable between years and location (Bauer et al., 1991; Knezevic et al., 1994; Oliver et al., 1991). The impact of variation in the competition coefficients must be established. In some cases, unique sets of competition coefficients may need to be established for geographic regions or particular production practices. Competition coefficients may also need to be established for the use of reduced rates of herbicides since sublethal doses of herbicides may suppress weed growth and their ability to compete with the crop (Pannell, 1990). This must be reflected in the competition coefficient in order to correctly calculate the benefit of a treatment.

Yield loss from weed competition also varies with the time of emergence of weed populations relative to the crop. Equations describing the relationship between yield loss and weed density have been modified to account for the impact of time of weed emergence (Cousens et al., 1987). However, with these equations, time of emergence of the entire population is a single day. Equations like these could be modified for multispecies populations and included in weed management models to improve yield loss estimates. When weed emergence is modeled over time, another approach is to divide the weed population into cohorts. Each cohort would have its own set of competitive indices. A simple implementation of this approach is to have two cohorts (weeds emerge either during or after the weed competitive period) and assume that weeds emerging after the competitive period do not compete with the crop (Wiles et al., 1996).

Weed leaf area or relative weed and crop leaf area may be more accurate predictors of crop yield loss than density (Ghersa and Martinez-Ghersa, 1991; Kropff and Lotz, 1992; Morin et al., 1993). Weed biomass or leaf area may be the measure of weed populations if all population processes in a model are based on this measure, or appropriate conversions are done where needed. For example, in a sugarbeet (*Beta vulgaris* L.) weed management model, weed population dynamics are based on density, but yield loss is predicted from weed biomass. A hyperbolic function is used to convert density to weed biomass (Schribbs et al., 1990).

It is often suggested that empirical yield loss equations of weed management models should be replaced with mechanistic models of weed-crop competition that respond to environmental conditions (Kropff et al., 1992; Dunan et al., 1994). Weed competition models have been developed for several major crops (Dunan et al., 1994; Kropff et al., 1993; Wilkerson et al., 1990). This substitution could overcome obstacles of the stability of competition coefficients for different regions, the effect of the time of emergence, and, in some cases, the impact of varying production practices. However, mechanistic models must be modified for mixed populations before this substitution is practical. Further, the extensive data on weed biology needed to parameterize these mechanistic models is not yet available for most species (Kropff et al., 1992).

D. Weed Seed Production

Another challenge for weed scientists is to develop more accurate predictions of seed production for weeds escaping control. This information is extremely important in developing multi-year models that simulate the effect of weed escapes on the weed potential in subsequent crops.

At least one Category I research model has the structure and equations to model seed production (Wiles et al., 1996). However, information on seed production is not available for all weed species or values are not realistic. A principal source for information on a number of weed species has been the work by Stevens (1932; 1957). Unfortunately, most of Stevens' data on seed production were from single, mature plants of average size that grew in the absence of crop competition. Thus, the number of seed reported per plant for some species may be too high, and for other species too low since seed values often included immature seed or seed lost through shattering.

More recently, researchers have reported that seed production for some weed species will be reduced, sometimes considerably, by crop competition (Evans et al., 1991; Senseman and Oliver, 1993; Lindquist et al., 1995), time of emergence relative to crop emergence (Knezevic et al., 1994; Lindquist et al., 1995; Mack and Pyke, 1983), intraspecific weed competition (Norris, 1992), timely foliar herbicide treatments (Khan and Donald, 1992; Willard and Griffin, 1993), or plant pathogens (Lindquist et al., 1995). Norris (1992) reported barnyardgrass [*Echinochloa crus-galli* (L.) Beauv.] seed production in California averaged 100,000 seed per plant, which was over 10-fold higher than the values reported by Stevens (1932). Seed production from individual velvetleaf plants also has been quite variable. Stevens (1957) reported a value of 4,300 seed produced per plant; whereas Munger et al. (1987) observed 236 seed per plant in 1 yr and 1,740 seed per plant in another year.

Seed production values for any given weed species can be quite variable from plant to plant and year to year. For species in which ripening extends over a considerable period and seed are shed as soon as they mature, collecting all seed produced is difficult. For species whose seed are retained on the plant until all mature, the odds are much higher that the total number of seed produced will be collected. Stevens (1957) estimated that the total yield for these species might be twice as much if some seed had dropped or had not matured, or four times as much if both of these factors prevailed.

Can we accurately assess seed production? Can appropriate field studies be conducted under different environmental and edaphic conditions that will provide realistic seed production values for weed escapes in different crops? What about immature weeds present in crops that are sprayed with desiccants or windrowed to promote drying, or are harvested green with hay crops or corn grown for silage? It may not be possible to obtain finite seed production values for every species, but more realistic values can be obtained by designing and conducting appropriate studies for different environs and production practices.

The bottom line is: the better job we do in predicting the number of escaped weeds, by species, in each crop and their seed production, the less frequently we will have to sample each field for its weed seedbank. Once the seedbank level has been determined initially via soil sampling, seed production for the next 3 to 5 yr could be estimated by model simulation.

E. Weed Seed Mortality

Most bioeconomic models have focused on the impact of weed control in a single year. Multiple year models are now needed to assess the impact of weed control on populations in future years. These models will have to predict seed survival. Seed are lost from the seedbank through emergence and mortality. Category I models predict emergence and must be modified to predict mortality.

Sources of mortality include *in situ* germination (seeds that germinate but do not emerge), decay, herbivory, and pathogens (Karssen, 1982; Lindquist et al., 1995; Wilson, 1988). Limited biological data are available on the sources or extent of mortality of weed seed in the seedbank as a whole or by species. In many studies, seed survival is measured and mortality can only be derived if emergence is also measured. We do know that mortality, like emergence, varies by species and tillage practices (Roberts, 1964; Roberts and Feast, 1972). Lindquist et al. (1995) reported that 70 ± 6% of all the velvetleaf seed in the seedbank that failed to emerge in the spring did not survive through August in soil. Furthermore, only 29 ± 3% of all velvetleaf seed in the soil in August survived through the next spring. Based on this study, we know that we will have to account for seed mortality when we sample the seedbank in the fall or several months before a crop is planted. Because there is so little data and mortality is inherently difficult to study directly, predicting mortality will be a major obstacle to developing multiple year weed management models.

F. Weed-Free Crop Yield

Knowledge of weed-free crop yield (WFCY), the expected crop yield without any weeds, is basic data for evaluating the benefit of weed control in most bioeconomic weed management models. If farmers and other weed management decision makers underestimate the WFCY, losses due to weeds will be undervalued (assuming losses are proportionate to expected WFCY) and overvalued if they overestimate WFCY.

Estimating the WFCY is not easy. This is demonstrated with data from 50 farmers producing irrigated corn in eastern Colorado who specified their WFCY in a 4-yr pilot test of WEEDCAM (D.W. Lybecker, 1995, personal communication). When actual yield was adjusted (increased) to account for the reduced yield due to weed escapes present at harvest, less than 50% of the farmers were within 10% of the WFCY, 75% within 20% of the WFCY, and 85% within 30% of the WFCY

Table 1. Percent of sites where actual yields (adjusted for weed escapes at harvest) were within 10%, 20%, and 30% of specified weed-free crop yield in WEEDCAM pilot tests

	Proportion of sites within stated percent of specified weed-free crop yield		
Year	±10%	±20%	±30%
1989	25.0%	57.1%	71.4%
1990	72.7%	90.9%	90.9%
1991	46.7%	76.7%	90.0%
1992	50.0%	81.8%	90.9%
All years	47.1%	75.5%	85.3%

(Table 1). Estimating the WFCY will be more difficult in rainfed areas because of precipitation variability.

Rather than relying on farmer recall or estimates, a preferable approach would be to review actual yields for the weed management site (field) for the past few years and use the average yield adjusted for weed competition. A refinement would be to consider yield changes over time and incorporate the annual trend. In the absence of yield history at a given site, yields of similar nearby sites or regional (county or multicounty) yield averages could be used. Figure 1 shows the yields for a 10-yr period starting in 1983 for northeast Colorado. The simple linear trend line shows an annual increase of 161 kg/ha. The 1993 expected yield would be 10,051 kg/ha (the 1983 yield of 8,441 plus 1,616 times 10, the number of years since 1983).

The number of years to include in computing the trend is important. Figure 1 used a 10-yr time period and showed a positive trend. If a 6-yr time period is used rather than 10 yr, the slope of the trend line is -15.3 kg/ha/yr (Figure 2).

A more complex yield function could be fitted and used to predict a future year's yield as alternative to a linear trend. A quadratic function was fitted for the 1983 to 1992 period for northeast Colorado irrigated corn yields in Figure 3. This function also shows declining yields for future years.

An approach developed by Swinton et al. (1994a,b) to estimate weed-crop interference parameters could be used to adjust given farm yield with known weed competition levels if weed-crop interference parameters for the region are known. This approach would provide good estimates of the WFCY in a given year but would still be subject to the problems of annual variation.

Regardless of the procedure used, careful review and analysis of the yields of previous years must be undertaken. Removing years when unusual events occur such as drought, flood, or early frost, and accounting for yield trend will provide a more accurate estimation of next year's expected WFCY. Areas with greater yield variation would be well served to use a longer, rather than a shorter time span for estimating their WFCY.

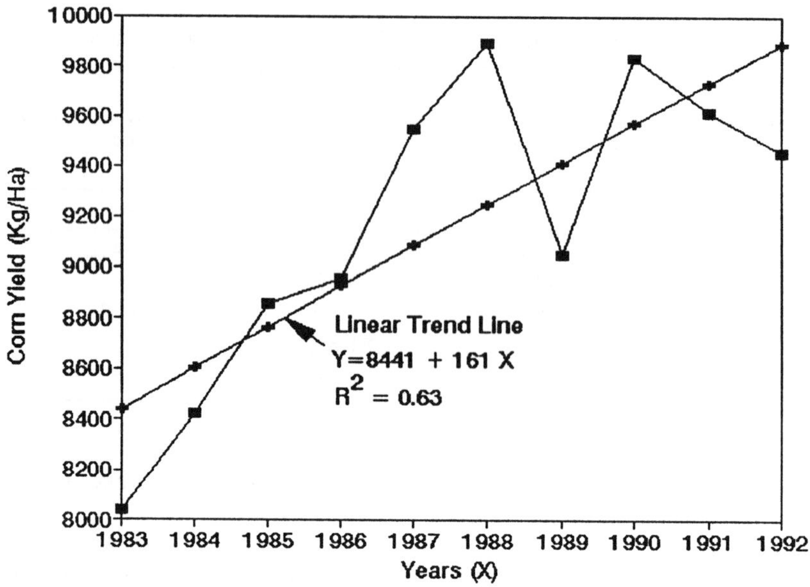

Figure 1. Northeast Colorado irrigated corn yields and linear trend, 1983-1992.

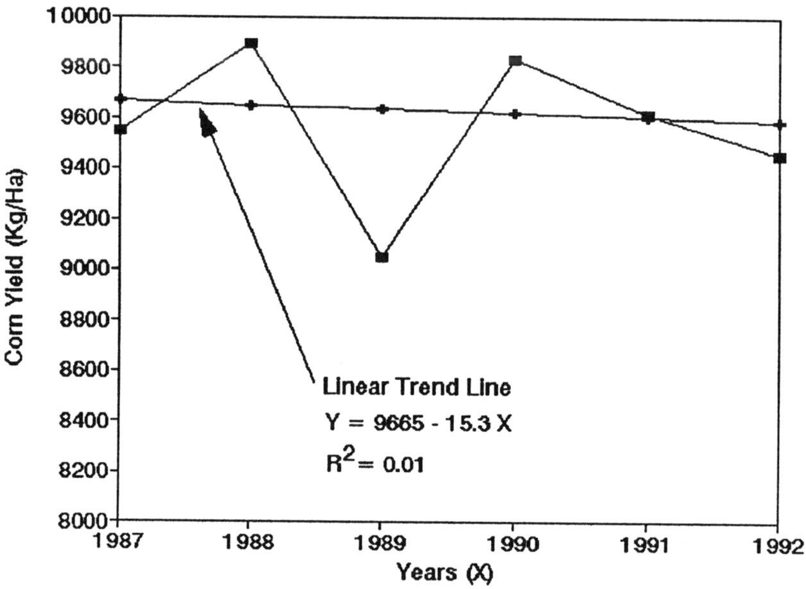

Figure 2. Northeast Colorado irrigated corn yields and linear trend, 1987-1992.

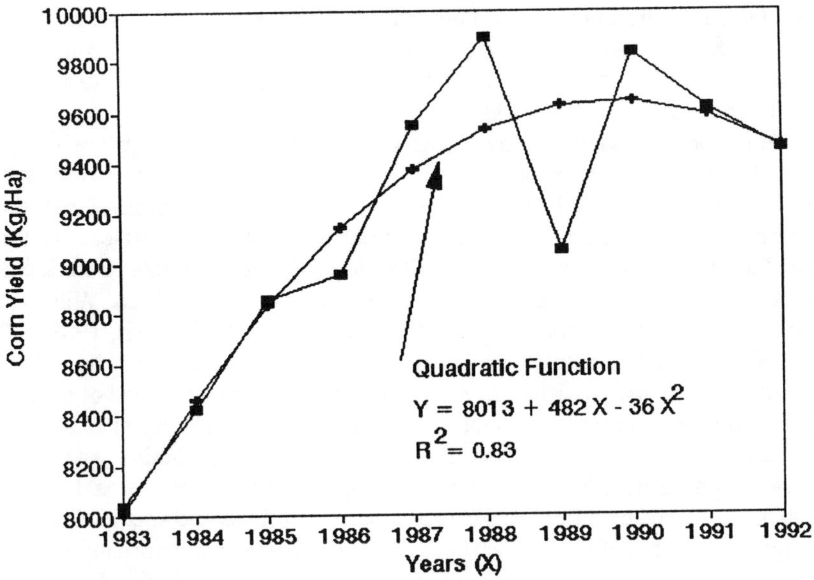

Figure 3. Northeast Colorado irrigated corn yields and quadratic function, 1983-1992.

Simulation studies with bioeconomic weed management models can determine the importance of the WFCY. The more sensitive a bioeconomic model is in terms of changing treatment recommendations, the more critical a good estimate of the WFCY is.

G. Weed Spatial Variability

For many years, researchers assumed that weeds were randomly distributed within fields and populations were not spatially structured. However, within the last 10 yr, several scientists have reported that weed populations were more likely to be aggregated or clumped rather than randomly dispersed (Benoit et al.,1989; Bigwood and Inouye, 1988; Cardina et al., 1995; Chauvel et al., 1989; Dessaint et al., 1991; Halstead et al., 1993; Mortensen et al., 1993; Thornton et al., 1990; Wiles et al., 1992a; Wilson and Brain, 1991). This is true for both seed and seedling populations. The aggregation and spatial structure of populations have been characterized by classical statistics (Cardina et al., 1995; Johnson et al., 1995b; Wiles et al., 1992a; Wilson and Brain, 1991), aggregation indices (Bigwood and Inouye, 1988; Cardina et al., 1995; Dessaint et al., 1991), and spatial statistics (Cardina et al., 1995; Halstead et al., 1993; Johnson, 1994).

As a result of competition between weeds, yield loss from a given weed population decreases as the degree of patchiness increases (Auld and Tisdell, 1988; Brain and Cousens, 1990; Hughes, 1989; Thornton et al., 1990). One implication of the clumped distribution of weed populations is that yield loss, when calculated from an average density of weeds in a field, may be overestimated. Models may recommend weed control when none is needed or more intensive control than is needed (Brain and Cousens, 1990; Wiles et al., 1992b).

Several models of yield loss have been developed that incorporate the effect of clumped distribution of weeds using classical statistics (Auld and Tisdell, 1988; Brain and Cousens, 1990; Wiles et al., 1992b). The approach is based on describing the weed population with a frequency distribution of counts for a particular quadrat size. The negative binomial is often appropriate (Benoit et al., 1989; Cardina et al., 1995; Johnson et al., 1995b; Wiles et al., 1992a; Wilson and Brain, 1991). A similar approach uses an aggregation index (Hughes, 1989). The value of these more complex yield loss models is not as clear. The value of accounting for the spatial distribution of weeds in a set of fields instead of assuming a regular distribution was estimated to average \$0.47 ha^{-1} (Wiles et al., 1992b). This value was calculated assuming the actual spatial distribution was known. However, spatial distribution, and the appropriate parameters for a frequency distribution of weed counts, are highly variable among fields and species and across time (Johnson et al., 1995b; Wiles et al., 1992a). This is not surprising when we consider that agronomic variables such as choice of crop, edaphic factors, soil fertility, crop rotation, and tillage practices, as well as previous field history all can have pronounced effects on weed species diversity and distribution (Andreasen et al., 1991; Wiese, 1985). An alternative to assuming values for the parameters of the distribution would be to estimate the values from scouting information (Johnson, 1994). However, this may be too time consuming and not economically feasible.

So far, most weed management models have been concerned with making one treatment or evaluation for an entire field. Models that consider or recommend spatially variable treatment within a field will be needed given the clumped distribution of weed populations. A study of weed distributions in 14 soybean fields in North Carolina found up to 29% of a field to be free of broadleaf weeds (Wiles et al., 1992a). In a study of 12 Nebraska soybean fields, where a preemergence herbicide was banded, on average 71% of the area sampled was free of broadleaf weeds and 94% was free of grasses. In the interrow area where no herbicide was applied, 30% was free of broadleaf weeds and 70% was free of grasses (Johnson et al., 1995a). In a study of eight irrigated corn fields in Colorado, 1% to 81% of a field was free of broadleaf weeds and 11% to 50% was free of grasses at the time of a postemergence decision (L.J. Wiles, 1995, personal communication). Weed management models should evaluate the possibility of varying the treatment across the field or not treating weed-free areas.

Classical statistics do not adequately describe the spatial structure of a weed population for spatially variable treatments (Cardina et al., 1995; Johnson, 1994). Instead, spatial statistics, such as geostatistics, are needed. While classical statistics may indicate what proportion of a field is weed-free, it does not describe the location of the weed-free areas in relation to each other. With spatial statistics, the location of the weed population can be specified.

Geostatistics has been the most commonly used spatial statistical method for describing the spatial distribution of weed populations. Geostatistics has been used to map intra- and interrow grass and broadleaf seedling populations in corn and soybean fields (Johnson, 1994), common lambsquarter seedling populations in no-tillage soybean fields (Cardina et al., 1995), a weed seedbank predominated by common lambsquarters and fall panicum (*Panicum dicotomiflorum* Mich.) (Halstead et al., 1993), and root and shoot density in a Canada thistle [*Cirsium arvense* (L.) Scop.] patch (Donald, 1994). Other mapping techniques also have been used (Benoit et al., 1992; Dessaint and Caussanel, 1994). These studies have confirmed the aggregation of weed populations and the value of spatially variable spraying, particularly, limiting spraying to weed patches. However, spatial statistics have not yet been incorporated into weed management models to improve estimates of yield loss.

Due to patchy distributions of weeds, future weed management models may be quite different than those today. Future models may not predict average weed population dynamics for a field. Instead, these models may predict weed population dynamics within patches and how the size of the patches change with time. This conceptual approach is illustrated by the theoretical model of Maxwell and Ghersa (1992) to assess the relative importance of weed competition and seed dispersal in determining long-term yield reductions due to weeds. The area of influence approach for modeling weed competition in SOYWEED is conceptually similar to that of Maxwell and Ghersa (1992). In simulations with this model, the field is divided into the area of influence of weeds and the area that is weed-free (Wilkerson et al., 1990). Weed competition is simulated only within the area of influence, and the size of the area of influence changes during the season as weeds grow. Modeling weed population dynamics within patches will require better information on the dispersal of weed propagules and changes in weed patches over time.

H. Weed Control Efficacy

Efficacy refers to the degree of weed control achieved by using one or more weed management tactics such as herbicides, secondary tillage, or biocontrol agents. Weed control efficacy databases in models can be state-specific or regional. Efficacy is modeled as the percent control by weed species. All current models have excellent databases on herbicide efficacy, limited databases on tillage, and nothing on biocontrol agents.

Information on herbicide efficacy for most weed species and crops is readily available in state weed control recommendation publications. Conversely, information on efficacy of mechanical tactics is limited for most weeds. Although efforts to look at alternative methods to control weeds have been minimal to date (Gunsolus, 1990; Schweizer et al., 1994), a few models include a limited database on the efficacy of rotary hoes, harrows, and cultivators. Additional information needs to be obtained with other types of secondary tillage equipment on their effectiveness to control weeds in different crops grown under conventional, reduced, and no-till practices.

Weed management models are concerned with three aspects of herbicide efficacy: control percentages for each weed species, period of herbicide activity, and growth stage of weeds. Once the period of herbicide activity or the growth stage of weeds is exceeded, the control percentages for each weed species defaults to zero in the current models. There is no decay of control. However, from a biological viewpoint this is unrealistic. For example, with soil-applied herbicides, and a few postemergence herbicides, their period of soil activity is usually a few days to several weeks. Once this time period is exceeded, the default for weed control becomes zero for all weed species. But, it is highly unlikely that if a soil-applied herbicide will control 80% of the weeds that germinate during the first 4 wk, that on the 29th day control drops to 0%.

We also need better information with postemergence herbicides on the percentage decline in weed control over time as affected by the stage of growth of each weed species. We know that small weeds are most susceptible to herbicides (Mathis and Oliver, 1980), and change in susceptibility is probably continuous, i.e., control declines as weeds become larger. Further, none of the current models deal with efficacy as a continuous phenomenon. However, HERB does break weed size down into three categories (Wilkerson et al., 1991). Most of the other weed management models have built in defaults once weeds exceed a certain size. For example, in WEEDCAM, the weed growth stage default is referred to as "postemergence triggers." These "triggers" are expressed as the number of leaves and/or height of each weed species as specified on each herbicide label. Should we assume that a postemergence herbicide that controls 85% of the weeds before they exceed 2.5 cm in height will not control any weeds when their average height exceeds 2.5 cm? No, we should not.

Thus, future models need to account for proportionally lower efficacy percentages as residual soil activity of herbicides and efficacy of postemergence herbicides on weeds declines over time. If we could override the present defaults with realistic efficacy percentages, then future models should provide users with better recommendations. Ideally, our future models need to have equations rather than discrete values to account for declining efficacy over time.

I. Herbicide Resistant Weed Species

The number of weed biotypes that have developed resistance to herbicides has continued to increase since the first report of triazine resistance in 1970 for common groundsel (*Senecio vulgaris* L.) in Washington (Ryan, 1970). By 1990, 107 resistant weed biotypes were documented throughout the world (LeBaron and McFarland, 1990). Resistance in the field becomes visible or detectable when it reaches about 30% of the population (Gessel and Segel, 1978). At least one current weed management model, WEEDCAM, reduces the herbicide efficacy percentage to zero for each weed species if resistance is indicated. However, in any field where resistance of a given weed species is indicated, the likelihood that the entire population of that species is resistant is slight. Thus, we need to know the population of each resistant species if the output from future weed management models is to be improved. Accurately modeling the population dynamics will be important to predict growth in the resistant population.

III. Issues for Modelers

Bioeconomic weed management models are needed to integrate vast databases of information on weed biology, ecology, and management practices generated from not only past weed research, but those studies that must be conducted in the future to address the information gaps identified in this paper. Before we conclude this paper, we want to share a few thoughts of Cousens (1995) for those weed scientists who are interested in conducting future weed biology studies that can contribute to modeling weed management. Central to all weed management models is a model of the population dynamics of weeds. Some of Cousens' thoughts on understanding and modeling the intrinsic dynamics of real plant populations were:

- If we understand the behavior of species in their natural environment, we should be able to predict what happens.
- Even if a realistic model can be formulated, its predictions can only tell us what might happen in a constant environment.
- The nature of the intrinsic population dynamics is expected to be masked by considerable environmental "noise."
- To ensure realism in a model, it needs realistic assumptions, a justifiable mathematical structure, and realistic parameter values.
- Qualitative predictions are model dependent.
- If we are to learn about real populations of a particular species, the parameter values used in a model need to be realistic. This surely means collecting them under realistic conditions. Extrapolation from the greenhouse into the field is renowned for its uncertainty throughout biology.
- Much more attention is given to describing seed production than to estimating germination success or seed longevity parameters. This is despite their critical importance in the behavior of models. We should be making more of an effort in collecting seed germination and survival data than we do.
- If we are to have confidence in our models, we not only have to believe that their assumptions are realistic but we must also be able to demonstrate that they are accurate in their predictions. Validation of our models is inherently difficult, perhaps almost impossible.

- Since there is uncertainty involved in the estimation of a parameter, there must also be uncertainty in the prediction from a model.

However:

- In applied science, some form of guess is better than no guess at all when a management decision has to be made.

IV. Potential Research Needs

Future computer bioeconomic weed management models will be much more realistic as scientists continue conducting appropriate research to address a number of the gaps in information in relation to weed biology, weed control, scouting, economics, and tillage. Potential research needs that should be considered and addressed are:

- How many soil cores need to be taken in a field to characterize the composition and density of the weed seedbanks?

- What is the best way to describe and map the spatial distribution of weed seed and weed populations in fields? Are spatial statistics, such as geostatistics, the answer?

- For practical application of computer bioeconomic weed management models, how should users collect the required information about weed seedbank and weed seedling populations? What would be the most cost-effective and reliable scouting method?

- Can weed seeds extracted from soil cores be counted and identified more economically with image analysis? What else can be done to make collecting and analyzing soil cores for the composition of the seedbank more economical?

- Can mapping weed populations in fields be done more economically by a combination of aerial photography and ground scouting using global positioning systems equipment?

- How can weed emergence models be improved to predict the cumulative emergence of weeds over time?

- How do we capture the effect of different tillage practices on weed emergence?

- Do we need competitive indices for individual weeds and mixtures of species in all crops? How can we improve on the current methodology used to determine the competitiveness of individual weed species and mixtures of species?

- Should empirical yield loss equations of weed management models be replaced with mechanistic models of weed-crop competition that respond to environmental conditions?

- How can we develop more accurate predictions of seed production for late-emerging weeds and weeds otherwise escaping control so this information can be used in multi-year models that simulate the effect of weed escapes on the potential in subsequent crops?

- How can we account for weed seed mortality when the seedbank is sampled in the fall or several months before a crop is planted? What is the pattern and extent of seedbank mortality so changes in the seedbank over years can be predicted?

- What new approaches need to be considered to predict expected crop yields in the absence of weeds?

- In terms of weed control efficacy, what is the effectiveness of tillage equipment to control weeds in crops grown under conventional, reduced, and no-till practices?

- How does weed control efficacy decay during the period of activity of a soil-applied herbicide? How do we estimate the percent control of each weed species once the period of herbicide activity or the growth stage of weeds, as stated on registration labels, is exceeded?

- In fields where herbicide resistant weed species are a problem, how can we better estimate the percentage of resistant and susceptible species?

V. Conclusions

Development of computer software that decision makers can use to identify appropriate weed management strategies is desirable to increase profits and avoid excessive or inappropriate herbicide use. Today, acceptance of these bioeconomic weed management models has been limited because the development of realistic models has been primarily hindered by gaps in information about weed biology. Modeling of weed biology and the economic impact of control decisions within models must be refined and a number of parameters need to be researched by weed scientists and economists. Ascertaining more biological information concerning emergence, competition, seed production, mortality, and composition of weed populations is critical. Other weaknesses of the models and our ability to use them that need to be dealt with are estimating weed-free crop yields, characterizing weed spatial variability, and maintaining and expanding our databases on weed control efficacy of herbicide-susceptible and resistant weed species. If weed scientists and economists can devote enough time and resources to obtain the necessary baseline database information, more realistic models will be developed that will have a greater acceptance by decision makers.

References

Alm, D.M., E.W. Stoller, and L.M. Wax. 1993. An index model for predicting seed germination and emergence rates. *Weed Technol.* 7:560-569.

Andreasen, C., J.C. Streibig, and H. Haas. 1991. Soil properties affecting the distribution of 37 weed species in Danish fields. *Weed Res.* 31:181-187.

Anonymous. 1992. Pesticides. p.13-20. In: *Agricultural resources: Situation and outlook report, AR-25.* USDA-ERS, Washington, DC.

Auld, B.A. and C.A. Tisdell. 1988. Influence of spatial distribution of weeds on crop yield loss. *Plant Prot. Q.* 3:81.

Bauer, T.A., D.A. Mortensen, G.A. Wicks, T.A. Hayden, and A.R. Martin. 1991. Environmental variability associated with economic thresholds for soybeans. *Weed Sci.* 39:564-569.

Benech-Arnold, R.L. and R.A. Sanchez. 1995. Modeling weed seed germination. p. 545-566. In: J. Kigel and G. Galil (ed)., *Seed Development and Germination.* Marcel Dekker, NY.

Benech-Arnold, R.L., C.M. Ghersa, R.A. Sanchez, and P. Insausti. 1990. Temperature effects on dormancy release and germination rate in Sorghum halepense (L.) Pers. seeds: a quantitative analysis. *Weed Res.* 30:81-89.

Benoit, D.L., N.C. Kenkel, and P.B. Cavers. 1989. Factors influencing the precision of soil seed bank estimates. *Can. J. Bot.* 67:2833-2840.

Benoit, D.L., N.C. Kenkel, and P.B. Cavers. 1992. Innovative approaches to seedbank studies. *Weed Sci.* 40:660-669.

Bigwood, D.W. and D.W. Inouye. 1988. Spatial pattern analysis of seed banks: An improved method and optimized sampling. *Ecology* 69:497-507.

Brain, P. and R. Cousens. 1990. The effect of weed distribution on predictions of yield loss. *J. Applied Ecol.* 27:735-742.

Bridges, D.C. 1992. *Crop losses due to weeds in the United States - 1992.* Weed Science Society of America, Champaign, IL.

Buhler, D.D. and B.D. Maxwell. 1993. Seed separation and enumeration from soil using K_2CO_3-centrifugation and image analysis. *Weed Sci.* 41:298-302.

Cardina, J., D.H. Sparrow, and E.L. McCoy. 1995. Analysis of spatial distribution of common lambsquarters (*Chenopodium album*) in no-till soybean (*Glycine max*). *Weed Sci.* 43:258-268.

Chauvel, B., J. Gasquez, and H. Darmency. 1989. Changes of weed seed bank parameters according to species, time and environment. *Weed Res.* 29:213-219.

Coble, H.D. and D.A. Mortensen. 1992. The threshold concept and its application to weed science. *Weed Technol.* 6:191-195.

Cousens, R. 1986. A simple model relating yield loss to weed density. *Ann. Appl. Biol.* 107:239-252.

Cousens, R. 1995. Can we determine the intrinsic dynamics of real plant populations? *Functional Ecol.* 9:15-20.

Cousens, R., P. Brain, J.T. O'Donovan, and P.A. Sullivan. 1987. The use of biologically realistic equations to describe the effects of weed density and relative time of emergence on crop yield. *Weed Sci.* 35:720-725.

Dessaint, F. and J.P. Caussanel. 1994. Trend surface analysis: a simple tool for modelling spatial patterns of weeds. *Crop Prot.* 13:433-438.

Dessaint, F., R. Chadoeuf, and G. Barralis. 1991. Spatial pattern analysis of weed seeds in the cultivated seed bank. *J. Appl. Ecol.* 28:721-730.

Donald, W.W. 1994. Geostatistics for mapping weeds, with a Canada thistle (*Cirsium arvense*) patch as a case study. *Weed Sci.* 42:648-657.

Dunan, C.M., F.D. Moore III, and P. Westra. 1994. A plant process-economic model for wild oats management decisions in irrigated barley. *Agric. Syst.* 45:355-368.

Evans, R.M., D.C. Thill, L. Tapia, B. Shafii, and J.M. Lish. 1991. Wild oat (*Avena fatua*) and spring barley (*Hordeum vulgare*) density affect spring barley grain yield. *Weed Technol.* 5:33-39.

Forcella, F. 1992. Prediction of weed seedling densities from buried seed reserves. *Weed Res.* 32:29-38.

Forcella, F. 1993. Seedling emergence model for velvetleaf. *Agron. J.* 85:929-933.

Forcella, F., K. Eradat-Oskoui, and S.W. Wagner. 1993. Application of weed seedbank ecology to low-input crop management. *Ecol. Appl.* 3:74-83.

Forcella, F., R.G. Wilson, K.A. Renner, J. Dekker, R.G. Harvey, D.M. Alm, D.D. Buhler, and J. Cardina. 1992. Weed seedbanks of the U.S. corn belt: magnitude, variation, emergence, and application. *Weed Sci.* 40:636-644.

Gessel, J. and L.A. Segel. 1978. The paucity of plants evolving genetic resistance to herbicides: possible reasons and implications. *J. Theor. Biol.* 75:349-371.

Ghersa, C.M. and M.A. Martinez-Ghersa. 1991. A field method for predicting yield losses in maize caused by johnsongrass (*Sorghum halepense*). *Weed Technol.* 5:279-285.

Gunsolus, J.L. 1990. Mechanical and cultural weed control in corn and soybeans. *Amer. J. Altern. Agric.* 5:114-119.

Halstead, S.J., K.L. Gross, and K.A. Renner. 1993. Geostatistical analysis of the weed seed bank. Weed Science Society of America abstracts 33:50.

Harvey, S.J. and F. Forcella. 1993. Vernal seedling emergence model for common lambsquarter (*Chenopodium album*). *Weed Sci.* 41:309-316.

Hughes, G. 1989. Spatial heterogeneity in yield-weed relationships for crop-loss assessment. *Crop Res.* 29:87-94.

Ives, P.M. and R.D. Moon. 1987. Sampling theory and protocol for insects. p. 49-75. In: P.S. Teng (Ed.), *Crop Loss Assessment and Pest Management.* APS Press, St. Paul, MN.

Johnson, G.A. 1994. Model parameterization, parametric sequential sampling, and geostatistical analysis of weed seedling populations. Ph.D. diss., Univ. Nebraska, Lincoln (Diss. Abstr. 94-25286).

Johnson, G.A., D.A. Mortensen, and A.R. Martin. 1995a. A simulation of herbicide use based on weed spatial distribution. *Weed Res.* 35:197-205.

Johnson, G.A., D.A. Mortensen, L.J. Young, and A.R. Martin. 1995b. The stability of weed seedling population models and parameters in eastern Nebraska corn and soybean fields. *Weed Sci.* (in press.)

Karssen, C.M. 1982. Seasonal patterns of dormancy of weed seeds. p. 243-270. In: A.A. Kahn (ed.), *the Physiology and Biochemistry of Seed Development.* Elsevier Biomedical Press, Amsterdam, Netherlands.

Khan, M. and W.W. Donald. 1992. Sulfonylurea herbicides reduce survival and seed production of green and yellow foxtails (*Setaria* spp.). *Weed Technol.* 6:284-290.

Knezevic, S.Z., S.F. Weise, and C.J. Swanton. 1994. Interference of redroot pigweed (*Amaranthus retroflexus*) in corn (*Zea mays*). *Weed Sci.* 42:568-573.

Kropff, M.J. and L.A.P. Lotz. 1992. System approaches to quantify crop-weed interactions and their application in weed management. *Agric. Syst.* 40:265-282.

Kropff, M.J., S.E. Weaver, and M.A. Smits. 1992. Use of ecophysiological models for crop-weed interference: relations amongst weed density, relative time of weed emergence, relative leaf area, and yield loss. *Weed Sci.* 40:296-301.

Kropff, M.J., S.E. Weaver, L.A.P. Lotz, J.L. Lindquist, W. Joenje, B.J. Schnieders, N.C. van Keulen, T.R. Migo, and F.F. Fajardo. 1993. Understanding crop-weed interaction in field situations. p. 105-135. In: M.J. Kropff and H.H. van Laar (eds.), *Modelling Crop-Weed Interactions*. CAB International, Wallingford, U.K.

LeBaron, H.M. and J. McFarland. 1990. Herbicide resistance in weeds and crops: An overview and prognosis. p. 336-352. In: M.B. Green, H.M. LeBaron and W.K. Moberg (eds.), *Managing Resistance to Agrochemicals: From Fundamental Research to Practical Strategies.* ACS Symposium Series No. 421. ACS Books, Washington, DC.

Lindquist, J.L., B.D. Maxwell, D.D. Buhler, and J.L. Gunsolus. 1995. Velvetleaf (*Abutilon theophrasti*) recruitment, survival, seed production, and interference in soybean (*Glycine max*). *Weed Sci.* 43:226-232.

Lybecker, D.W., E.E. Schweizer, and P. Westra. 1994. WEEDCAM Manual. Department of Agricultural and Resource Economics, Colorado State University, Ft. Collins, CO.

Mack, R.N. and D.A. Pyke. 1983. The demography of *Bromus Tectorum*: Variation in time and space. *J. Ecol.* 71:69-93.

Malone, C.R. 1967. A rapid method for enumeration of viable seeds in soil. *Weeds* 15:381-382.

Marshall, E.J.P. 1988. Field-scale estimates of grass weed populations in arable land. *Weed Res.* 28:191-198.

Mathis, W.D. and L.R. Oliver. 1980. Control of six morningglory (*Ipomoea*) in soybean (*Glycine max*). *Weed Sci.* 28:409-415.

Maxwell, B.D. and C. Ghersa. 1992. The influence of weed seed dispersion versus the effect of competition on crop yield. *Weed Technol.* 6:196-204.

Mohler, C.L. 1993. A model of the effects of tillage on emergence of weed seedlings. *Ecol. Appl.* 3:53-73.

Morin, C., D. Blanc, and H. Doorman. 1993. Limits of simple model to predict yield losses in maize. *Weed Res.* 33:261-268.

Mortensen, D.A. and H.D. Coble. 1991. Two approaches to weed control decision-aid software. *Weed Technol.* 5:445-452.

Mortensen, D.A., G.A. Johnson, and L.J. Young. 1993. Weed distribution in agricultural fields. p. 113-124. In: P.C. Robert, R.H. Rust, and W.E. Larson (Eds.), *Soil Specific Crop Management*. ASA, CSSA, and SSSA, Madison, WI.

Munger, P.H., J.M. Chandler, J.T. Cothren, and F.M. Hons. 1987. Soybean (*Glycine max*) - velvetleaf (*Abutilon theophrasti*) interspecific competition. *Weed Sci.* 35:647-653.

Norris, R.F. 1992. Case history for weed competition/population ecology: barnyardgrass (*Echinochloa crus-galli*) in sugarbeets (*Beta vulgaris*). *Weed Technol.* 6:220-227.

Oliver, L.R., J.M. Chandler, and G.A. Buchanan. 1991. Influence of geographic region on jimsonweed (*Datura stramonium*) interference in soybeans (*Glycine max*) and cotton (*Gossypium hisutum*). *Weed Sci.* 39:585-589.

Pannell, D.J. 1990. Model of wheat yield response to application of diclofop-methyl to control ryegrass (*Lolium rigidum*). *Crop Prot.* 9:422-428.

Post, B.J. 1984. Physical and chemical treatments for assessing the seed bank in soil samples. In: Int. Sym. Weed Biol. *Ecolog. Syst.* 7:71-79.

Roberts, H.A. 1964. Emergence and longevity in cultivated soil of seeds of some annual weeds. *Weed Res.* 4:296-307.

Roberts, H.A. and P.M. Feast. 1972. Fate of seeds of some annual weeds in different depths of cultivated and undisturbed soil. *Weed Res.* 12:316-324.

Ryan, G.F. 1970. Resistance of common groundsel to simazine and atrazine. *Weed Sci.* 18:614-616.

Schribbs, J.M., D.W. Lybecker, and E.E. Schweizer. 1990. Bioeconomic weed management models for sugarbeet (*Beta vulgaris*) production. *Weed Sci.* 38:436-444.

Schweizer, E.E. and R.L. Zimdahl. 1984. Weed seed decline in irrigated soil after six years of continuous corn (*Zea mays* L.). *Weed Sci.* 32:76-83.

Schweizer, E.E., P. Westra, and D.W. Lybecker. 1994. Controlling weeds in corn (*Zea mays*) rows with an in-row cultivator versus decisions made by a computer model. *Weed Sci.* 42:593-600.

Schweizer, E.E., D.W. Lybecker, L.J. Wiles, and P. Westra. 1993. Bioeconomic models in crop production. p. 103-107. In: D.R. Buxton, R. Shibles, R.A. Forsberg, B.L. Blad, K.H. Asay, G.M. Paulsen, and R.F. Wilson (eds.), *International Crop Science I.*, Crop Science Society of America, Inc., Madison, WI.

Senseman, S.A. and L.R. Oliver. 1993. Flowering patterns, seed production, and somatic polymorphism of three weed species. *Weed Sci.* 41:418-725.

Southwood, T.R.E. 1976. *Ecological Methods with Particular Reference to the Study of Insect Populations*. 2nd ed. John Wiley & Sons, NY.

Stevens, O.A. 1932. The number and weight of seeds produced by weeds. *Amer. J. Botany* 19:784-794.

Stevens, O.A. 1957. Weights of seeds and numbers per plant. *Weeds* 5:46-55.

Swinton, S.M. and R.P. King. 1994. A bioeconomic model for weed management in corn and soybean. *Agric. Syst.* 44:313-335.

Swinton, S.M., D.D. Buhler, F. Forcella, J.L. Gunsolus, and R.P. King. 1994a. Estimation of crop yield loss due to interference by multiple weed species. *Weed Sci.* 42:103-109.

Swinton, S.M., J.A. Sterns, K.A. Renner, and J.J. Kells. 1994b. Estimating weed-crop interference parameters for weed management models. Michigan Agric. Exp. Stn. Res. Rep. 538.

Thornton, P.K., R.H. Fawcett, J.B. Dent, and T.J. Perkins. 1990. Spatial weed distribution and economic thresholds for weed control. *Crop Prot.* 9:337-342.

U.S. Congress, Office of Technology. 1990. Beneath the Bottom Line: Agricultural Approaches to Reduce Agrichemical Contamination of Groundwater, OTA-F-418, Washington, DC, U.S. Gov. Print. Office, Washington, D.C.

Wiese, A.F. 1985. Weed control in limited-tillage systems. Monograph Ser. No. 2. Weed Sci. Soc. Am., Champaign, IL.

Wiles, L.J., G.W. Oliver, A.C. York, H.J. Gold, and G.G. Wilkerson. 1992a. Spatial distribution of broadleaf weeds in North Carolina soybean (*Glycine max*) fields. *Weed Sci.* 40:554-557.

Wiles, L.J., G.G. Wilkerson, and H.J. Gold. 1992b. Simulating weed scouting and weed control decision making to evaluate scouting plans. *Proc. 1992 Winter Simulation Conference* 25:1166-1171. Arlington, VA.

Wiles, L.J., G.G. Wilkerson, H.J. Gold, and H.D. Coble. 1992c. Modeling weed distribution for improved postemergence control decisions. *Weed Sci.* 40:546-553.

Wiles, L.J., D.H. Barlin, E.E. Schweizer, H.R. Duke, and D.E. Whitt. 1995. A new soil sampler and elutriator for collecting and extracting weed seeds from soil. *Weed Technol.* 10:35-41.

Wiles, L.J., R.P. King, E.E. Schweizer, D.W. Lybecker, and S.M. Swinton. 1996. GWM: General weed management model. *Agric. Syst.* 50:355-376.

Wilkerson, G.G., S.A. Modena, and H.D. Coble. 1991. HERB: Decision model for postemergence weed control in soybean. *Agron. J.* 83:413-417.

Wilkerson, G.G., J.W. Jones, H.D. Coble, and J.L. Gunsolus. 1990. SOYWEED: A simulation model of soybean and weed growth and competition. *Agron. J.* 83:1003-1010.

Willard, T.S. and J.L. Griffin. 1993. Growth response of wild poinsettia (*Euphorbia heterophylla*) following foliar herbicide applications. *Weed Technol.* 7:190-195.

Wilson, R.G. 1988. Biology of weed seeds in soil. p. 25-39. In: M.A. Altieri and M. Liebman (eds.), *Weed Management in Agroecosystems: Ecological Approaches*. CRC Press, Inc., Boca Raton, FL.

Wilson, B.J. and P. Brain. 1991. Long-term stability of distribution of *Alopecurus myosuroides* Huds. within cereal fields. *Weed Res.* 31:367-373.

Wilson, R.G., E.D. Kerr, and L.A. Nelson. 1985. Potential for using weed seed content in the soil to predict future weed problems. *Weed Sci.* 33:171-175.

Zimdahl, R.L. 1993. Weed biology: reproduction and dispersal. p. 59-89. In: R.L. Zimdahl, *Fundamentals of Weed Science*. Academic Press, NY.

The Economics of Weed Control and the Value of Weed Management Information

Robert P. King, Scott M. Swinton, Donald W. Lybecker, and Caleb A. Oriade

I. Introduction

Agricultural production management problems rarely respect disciplinary boundaries. This is certainly true of weed management problems. At the farm and watershed levels, weed management decisions are increasingly viewed from a systems perspective that integrates knowledge about biological and physical relationships, technology, economics, and social concerns. The papers in this volume illustrate the value of insights from disciplines ranging from plant breeding and ecology to soil physics and micrometeorology in understanding and evaluating strategies for managing weed populations.

This paper focuses on two important contributions economics has made to this systems view of the design and analysis of weed management strategies. The first of these contributions has been to focus attention on trade-offs between the benefits and the costs of alternative weed control actions. This shifts the emphasis in analysis and assessment from maximum yield or maximum control to maximization of economic benefit. The second contribution has been to draw attention to the role of information on situation-specific conditions in weed management and to the need to assess the value of that information. As in other aspects of agricultural production management, emphasis in weed management is shifting from the development of broadly applicable chemical, cultural, or mechanical technologies for weed control to the selection of situation-specific weed control actions from a diverse set of alternatives. Information plays a key role in this shift.

In the sections that follow, we first discuss economic frameworks for evaluating weed management strategies that allow for increasing sophistication in the assessment of trade-offs between the benefits and costs of alternative control actions. Next, we introduce general concepts related to assessing the value of information and review findings from studies that have measured the value of information in weed management systems. In the concluding section, we identify issues for future research on the economics of weed management.

II. Economic Frameworks for Evaluating Weed Management Strategies

The objective of an economic analysis of weed management decisions is to identify strategies that maximize utility. The definition of "utility" depends on the context of the analysis. In this section, we start with the simplest and most familiar framework for analysis — profit maximization in the absence of uncertainty. We then successively refine the analytical framework by adding new elements to the definition of utility. At each stage we explain how "standard" results may change as the analytical framework becomes more complex and more comprehensive.

A. Static Profit Maximization Under Certainty

In the simplest case, there is no uncertainty about crop yield, weed emergence, weed control efficacy, or market prices, and the objective is to maximize profit per hectare for a single growing season. This is the economic framework used in bioeconomic models such as HERB (Wilkerson et al., 1991) WEEDCAM (Lybecker et al., 1991b), WEEDSIM (Swinton and King, 1994a), and GWM (Wiles et al., 1994). Profit per hectare, π, is defined by the following expression:

π = revenue - cost of weed control inputs - cost of nonweed control inputs

$$= P_Y Y - \sum_{i=1}^{n} WCC_i CA_i - \sum_{j=1)}^{m} R_j X_j \qquad (1)$$

where P_y is the product price ($/kg), Y is crop yield (kg/ha), WCC_i is the cost for materials and application of weed control action i ($/ha), CA_i is a binary variable equal to one if treatment i is used and equal to zero if it is not, R_j is the cost per unit of nonweed control input j ($/unit), and X_j is the amount of input j used (units/ha). Crop yield is a function of weed density, which is, in turn, a function of conditions describing the potential for weed pressure and the efficacy of the control actions selected.

This formulation treats weed control actions as discrete entities. It can easily be modified, however, to allow for variable rates of herbicide applications. In general, the set of weed control actions includes mechanical as well as chemical treatments. It also includes the "no control" option, since all values of CA_i can be set to zero. When weed control actions are discrete entities, the set of actions that maximizes profit is identified by evaluating profit for all possible "strategies" or allowable sets of weed control actions.

This simplest of frameworks introduces the fundamental trade-off between the benefit from more effective weed control -- higher yield due to less weed pressure -- and the cost of more intensive weed control actions. In general, it is not optimal to control all weeds, and no weed control action will be used if the added revenue associated with higher yield is not greater than the cost of the weed control action. In settings with multiple weed species and/or multiple control actions, there will not necessarily be a single threshold density for weed control. Rather, there will be density zones within which one treatment will be preferred, as in the weed recommendation maps in Swinton and King (1994a, p. 328-329).

In this framework, the boundaries for weed control action zones will be affected, in general, by market prices. For example, an increase in crop price will increase the benefit of weed control and so will lower the weed density thresholds for weed control. Similarly, a reduction in the cost of a weed control action will generally expand the weed density zone over which it is preferred. Finally, lower costs for nonweed control inputs may provide incentives for increasing potential yield. If yield losses due to weeds are represented as a percentage of potential yield, this will lower the weed density levels at which more intensive and expensive weed control actions are taken.

B. Adding Uncertainty to the Static, Deterministic Framework

Adding uncertainty to this framework complicates the analysis in two important ways. First, the uncertainty itself -- which may stem from lack of prior knowledge about weed pressure, weed control efficiency, crop yield, and/or product and input prices -- needs to be modeled. Due to the complex interactions among random variables in even this simple framework, stochastic simulation is usually used to model uncertainty. The performance of each strategy under consideration is simulated in each of a large number of "states of nature" with each state of nature defined by a random draw from the multivariate distribution of random variables. The resulting set of profit levels defines an empirical distribution of profit for the strategy. This approach is used in an early bioeconomic model by King et al. (1986) and in the more recent WFARM model (Swinton and King, 1994a).

The second complication caused by introducing uncertainty into the analysis is the need to represent risk preferences.[1] Risk preferences are usually represented by a Von Neumann- Morgenstern utility function, $U(\pi)$, that can be nonlinear in profits, and the preferred strategy is that which maximizes expected utility, $E[U(\pi)]$. Individuals who are "risk neutral" have a linear utility function and will, in the framework presented here, prefer the strategy that maximizes expected profit. Individuals who are "risk averse" have a concave utility function and will be willing to sacrifice some expected profit in order to reduce the probability of large losses. Finally, individuals who are "risk loving" have a convex utility function and will be willing to sacrifice some expected profit in order to increase the probability of large gains.

Empirical work in a wide range of settings has shown that risk preferences differ considerably among individuals. This means preferred weed management strategies may also differ considerably among individuals. In general, more risk averse individuals will be willing to spend more on weed control if more expensive weed control actions work well under nearly all conditions. Often these individuals will be reluctant to shift from strategies that rely on prophylactic weed control treatments to threshold-based strategies that are more responsive to situation-specific conditions and are more likely to leave some weeds uncontrolled.

C. Adding Multiyear Effects to the Static Framework

Up to this point, the framework for analysis of weed control strategies ignores the effects of current year decisions on the potential for weed control problems in subsequent years. When these effects are considered, the performance measure shifts from profit in a single growing season to the net present value of profits for a sequence of growing seasons -- i.e.,

$$NPV=\sum_{t=0}^{T} \pi_t/(1+r)^t \tag{2}$$

where NPV is the net present value, t is a growing season index, T is the length of the planning horizon, π_t is profit in growing season t, and r is a discount factor.

For a single growing season, the decision problem for a risk neutral decision maker becomes one of choosing the weed control strategy for the current growing season that maximizes the sum of two factors: (1) current year profit, and (2) the net present value of future profits that can be realized given the choices of the current year strategy. More formally, the objective is to choose a feasible vector of control actions CA_t that maximizes:

[1]See Anderson, Dillon, and Hardaker (1977), p. 65-108, for an excellent introduction to the measurement and representation of risk preferences.

$$NPV_t=\pi_t(CA_t)+NPV^*_{t+1}(CA_t)/(1+r) \tag{3}$$

where $\pi_t(CA_t)$ is the current year profit realized when the set of control actions CA_t is chosen and $NPV^*_{t+1}(CA_t)$ is the maximum net present value of profits that can be realized in future years if CA_t is chosen in the current year.

Problems of this kind can be solved by dynamic programming, as in Taylor and Burt (1984). Alternatively, when the number of growing seasons considered is small, such problems can be solved by direct enumeration of results under all possible sequences of feasible weed control actions, as in Swinton and King (1994a,b). The dynamic programming approach is generally preferred when the number of weed species is limited to one or two and when current period actions have effects well beyond the next period. The direct enumeration approach is preferred when the need to track the dynamics of many weed species increases the dimensionality of the problem to an unmanageable level.

As demonstrated by Swinton and King (1994b), adding dynamics to the analytical framework leads to more intensive treatments in the current period -- i.e., a lower treatment threshold in the case where there is a single weed species and a single treatment. This does not mean, however, that recommendations based on a dynamic analytical framework will be more intensive, on average, over the long run. Results reported by Swinton (1991, p. 142-144) for a six-year simulation of a 480 acre corn-soybean farm indicate that average herbicide loads are only slightly higher under a dynamic decision rule using a two-year planning horizon, and differences are often not statistically significant. It appears that more intensive current year control measures tend to reduce potential future weed problems so that herbicide use over an extended planning horizon does not increase.

D. Considering Multiple Performance Measures

Up to this point, the framework for analysis of weed control decisions presented here has included only the monetary costs and benefits realized by a single decision maker. Externalities -- costs or benefits realized by others as a result of the decision maker's actions -- and nonmonetary measures of performance -- such as technical measures of environmental impacts -- have not been included. As concerns about the environmental consequences of weed control increase, interest in these aspects of performance is also increasing.

Perhaps the simplest way to factor externalities into the decision process of individual managers is to impose direct monetary costs (or rewards) that are proportionate to the size of the externality. This can be done, for example, by adding a tax to the price of herbicide products that have negative externalities or by adding subsidies that lower the cost of using measures (e.g., smother crops) that have positive externalities[2].

At the individual field level, Swinton et al. (1995) have identified the size of taxes required to eliminate the use of atrazine and of all triazine herbicides under typical conditions for Minnesota and Colorado. At the sector level, Bouzaher et al. (1992) and NAPIAP (1992) have investigated equilibrium effects of policies that restrict the use of certain herbicide products.

Wherever a monetary value is placed on nonmarket effects, establishing an appropriate value can be both difficult and controversial. One solution to this problem is to present both monetary and environmental performance measures for strategies under consideration. For example, numerous studies have presented herbicide load measures along with measures of profit. More recently, Widham (1993) developed an environmental index for use in the WEEDCAM model, and Oriade (1995) presented measures of profit and an environmental index in his analysis of site-specific weed control strategies. The environmental index used in Oriade's study reflects environmental effects more accurately than a simple measure like herbicide load.

Finally, trade-offs between potentially conflicting objectives such as maximization of profit and minimization of environmental effects can be represented more explicitly in a multiattribute utility function.[3] Unfortunately however, multi-attribute utility formations are difficult to elicit for individuals, and trade-offs among objectives are likely to differ considerably across individuals. We know of no studies that have used explicit multiattribute utility in evaluating weed management alternatives.

III. The Value of Information for Weed Management Decisions

Knowledge and information are playing an increasingly important role in crop production management decisions. A common finding from analyses of weed management decisions is that information-based, flexible strategies that tailor weed control actions to current, site-specific conditions outperform strategies based on general recommendations.

"Information" is a term that is widely used but often poorly understood. Davis and Olson (1985, p. 200) define information as:

> ... data that has been processed into a form that is meaningful to the recipient and is of real or perceived value in current or prospective actions or decisions.

[2]Externalities can also be "internalized" through mechanisms such as tradeable use permits and bans on the use of certain weed control treatments.

[3]See Anderson, Dillon, and Hardaker (1977, p. 76-88) for a good introduction to multiattribute utility analysis. Keeney and Raiffa (1976) provide a more complete, more rigorous overview on the topic.

This definition makes a distinction between data and information. Data is the "raw material" for information, but it does not become information until it is placed in the context of particular decisions. The definition also emphasizes that data becomes information only when it has value to the recipient. That value may be derived from the sense of reassurance that comes when current beliefs are confirmed, but the more important basis for the value of information is a change in behavior.

For farmers, the information derived from weed population data and knowledge of weed population thresholds has greatest value when it leads to crop management decisions that are different from and more profitable than those that would have been made in the absence of that information. The value of information is increased by improvements in accuracy and timeliness, since these affect the quality of recommendations. It is also affected by the degree of flexibility in management responses, since tailoring actions to current conditions requires a range of effective alternatives.

As Byerlee and Anderson (1982) note, the value of information can be measured for a particular message or for an information service. An example of a single message is the following statement: "this field has 7 foxtail and 2 redroot pigweed seedlings per square meter." This message has value if it causes the decision maker to do something different from what would have been done without having received it. We are more likely to be interested in the value of an information service that provides messages such as weed seedling densities and situation-specific recommendations based on those messages. A crop consulting service that makes weed control recommendations for individual fields and a site-specific management service that makes weed control recommendations for management units within a single field are examples of such services. One important feature of an information service is that the particular messages it will provide cannot be known at the time the service is acquired. Therefore, the actual value of an information service cannot be known in advance. All that can be measured is the expected value of the service over the range of possible messages.

A. A Method for Measuring the Value of an Information Service

In the following example, we use a payoff matrix to illustrate why weed population information has value and how that value can be measured. A Minnesota corn (*Zea mays* L.) producer is trying to decide whether to use a standard prophylactic postemergence weed control treatment or to base postemergence weed control decisions on weed seedling counts. The field in question was planted on May 5 and has been rotary hoed twice. Five possible "states of nature" describe the range of weed population conditions that could hold just prior to the application of postemergence weed control treatments. The states of nature are considered to be equally likely. Therefore, each has a 20% probability of occurring. Each state of nature is defined by early season weed population levels for three weed species: green foxtail [*Setaria glauca* (L) Beauv.], common lambsquarters (*Chenopodium album* L.), and redroot pigweed (*Amaranthus retroflexus* L.)(Table 1).

Table 1. Weed populations and crop yields in five states of nature under three postemergence weed control treatments

State of nature	Early season weed counts (no./m²)	End of season weed counts (no./m²) and crop yield (kg/ha): No control	Dicamba	Cyanazine
1				
Green foxtail	23	26	23	5
Lambsquarters	4	4	1	1
Pigweed	12	13	2	7
Corn yield		5200	6400	6400
2				
Green foxtail	7	8	7	1
Lambsquarters	0	0	0	0
Pigweed	2	2	0	1
Corn yield		6700	6800	6800
3				
Green foxtail	36	40	36	8
Lambsquarters	3	3	0	0
Pigweed	7	8	1	4
Corn yield		5500	6200	6500
4				
Green foxtail	55	61	56	12
Lambsquarters	5	5	1	1
Pigweed	14	15	3	8
Corn yield		4000	5800	6200
5				
Green foxtail	7	8	7	1
Lambsquarters	2	2	0	0
Pigweed	9	10	2	5
Corn yield		6000	6700	6500

Three possible postemergence weed control treatments are being considered: no control, dicamba(3,6-dichloro-2-methoxybenzoic acid), and cyanazine {2-[[4-chloro-6-(ethylamino)-1,3,5-triazin-2-yl]amino]-2-methylpropanenitrile}. Each will be supplemented by a single cultivation. End of season weed counts and crop yields associated with each treatment are also shown for each state of nature in Table 1. Note that dicamba controls lambsquarters and pigweed effectively but has little effect on green foxtail. Cyanazine provides good control for green foxtail and lambsquarters but is less effective in controlling pigweed. Therefore, the overall effectiveness of these treatments depends on the mixture of weeds present.

Table 2. Net returns in five states of nature under four weed control strategies

	Net return ($/ha)			
State of nature	No control	Dicamba	Cyanazine	Flexible strategy
1	94.10	176.30	167.90	176.30
2	220.00	211.80	207.60	220.00
3	116.10	163.50	181.50	181.50
4	-16.10	122.20	149.90	149.90
5	163.60	203.40	184.30	203.40
Mean	115.50	175.40	178.20	186.20

Suppose the price of corn is $0.089/kg ($2.25/bu) and the cost of all inputs except those used for postemergence weed control is $370.70/ha. Materials and application costs for the dicamba and cyanazine weed control treatments are $18.40/ha and $25.20/ha, respectively. There are, of course, no materials or application costs associated with the no control alternative. Under these assumptions, net returns per hectare are defined by the following expression:

$$(0.089 * Y) - 370.70 - WCC \tag{4}$$

where Y is crop yield and WCC is weed control cost.

Net returns per hectare for the three postemergence weed control strategies are shown in Table 2. The cyanazine treatment has the highest average net return per hectare. In three of the five states of nature, however, one or both of the other two treatments performs better than cyanazine. Given data on early season weed densities, knowledge of the relative efficacy of the alternative weed control treatments, and an understanding of potential yield reductions due to weed pressure, weed control actions could be tailored to actual field conditions. In the absence of that information, cyanazine would be the best fixed weed control strategy.

With perfect knowledge of prevailing weed population conditions, the farmer would use dicamba in States of Nature 1 and 5, cyanazine in States of Nature 3 and 4, and no control in State of Nature 2. As shown in Table 2, the average net return for this flexible strategy is $186.20/ha, $8.00/ha higher than the average net return for the cyanazine treatment. This $8.00/ha difference in net returns is a measure of the value of the information that allows the farmer to tailor weed control actions to specific field conditions.

In this case, information is derived from early season weed seedling counts and from knowledge of how alternative treatments perform under a wide range of conditions. Weed seedlings cannot be counted prior to the application of soil applied herbicides or the use of early season cultivation tools such as a rotary hoe. Weed seed counts from soil samples can be used, however, to make the weed population forecasts required to tailor early season weed control decisions to field conditions.

In this example, weed seedling counts are assumed to be the basis for perfect predictions of which postemergence weed control treatment will be best. In reality, of course, the performance of biological systems is difficult to forecast with perfect accuracy. As the predictive power of procedures used to process weed population data into weed control recommendations declines, the value of weed population information will also decline. Therefore, the value of weed population information is sensitive to knowledge of weed population dynamics and weed-crop competition, and the value of that information can be enhanced by research that improves understanding of these biological processes.

Measures of the value of information are also sensitive to biological conditions at the field level. Had the analysis in Tables 1 and 2 been done for a field with strong weed pressure and especially high levels of green foxtail, the cyanazine treatment (which is highly effective against green foxtail) might have been best in nearly all possible states of nature. Under these conditions, information derived from weed seedling counts would be of little value because the recommended action would almost always be the same. Similarly, if weed pressures were low and postemergence treatments were rarely needed, the value of information derived from weed seedling counts would be low.

Changes in product price or weed control costs can also affect the value of weed population information. For example, an increase in the price of corn coupled with an increase in weed control costs could make the selection of an appropriate weed control action more important, thereby increasing the value of information used for weed management decisions.

Finally, it is important to remember that there are costs as well as benefits associated with the use of weed population information. The costs can be divided into two broad categories: the costs of collecting weed seed or weed seedling counts for a particular field and the costs of processing field level data into specific weed control recommendations.

The cost of collecting weed population data depends on sampling intensity and measurement methods. As the number of samples collected per field increases, the accuracy of weed seed or weed seedling counts should increase, but at a decreasing rate. Labor costs for data collection, on the other hand, would increase at a fairly constant rate. Therefore, there will be some point at which the increased cost of collecting another sample will outweigh the increased benefit from greater accuracy. Measurement methods also affect the cost of weed population data. A range of methods has been used to count weed seeds in soil samples. These differ in both cost and accuracy. Similarly, weed seedling densities can be assessed by direct counts or by less precise and less costly visual scoring methods. At present, our knowledge of the cost/accuracy tradeoffs for both sampling intensity and alternative measurement methods is limited. This is an area where further research is needed.

B. A Review of Estimated Values for Weed Management Information

A growing body of literature reports measurements of the value of weed management recommendations based on bioeconomic models. In many studies, value estimates have been based on results from computer simulation experiments that compare performance of model recommendations to performance under some baseline strategy. There have also been several field studies designed to assess the value of model-based recommendations. Findings from selected studies on the value of weed management information are summarized in Table 3.

Studies by King et al. (1986), Lybecker et al. (1991a), and Lybecker et al. (1993) report results based on successive versions of the WEEDCAM model. The number of weed control treatments considered in the model was expanded considerably after the first simulation study, and it is noteworthy that value of information estimates increased considerably. The fact that the value of information estimate for the on-farm trials reported in Lybecker et al. (1993) is lower than that for the research plot trials reported in Lybecker et al. (1991a) may be indicative of the fact that farmers do make some adjustments to field conditions that are not made in a standard fixed weed control strategy.

Studies by Buhler et al. (1992), Forcella et al. (1993), and Swinton and King (1994b) report results based on the WEEDSIM model. It is noteworthy that the estimated value of information differs considerably in the two field studies (Buhler et al., 1992 and Forcella et al., 1993). This seems to be due to a dramatic difference in weed pressure at the two field sites. Under the high weed pressure observed by Buhler et al. (1992), model recommendations differ little from standard recommendations and so do not yield an increase in expected net return. Under the low weed pressure observed by Forcella et al. (1993), model recommendations target only the weeds present and attain comparable control at a lower cost.

The studies by Weersink et al. (1992) and by Pannell (1994) are of particular interest because they assess the value of information that can be used to adjust herbicide rates. This is an important issue from both an economic and an environmental standpoint.

Finally, the studies by Wiles et al. (1992) and by Oriade (1995) assess the value of information that can be used for site-specific management. Both studies identify moderate gains from site-specific management.

The value of information estimates in all of these studies are measures of potential willingness to pay for information. They do not include any assessment of the costs of providing that information. More empirical work is needed to assess these costs, since they will have a significant effect on the adoption of information-based strategies for weed management.

IV. Issues for Future Research

Improving the structure, accuracy, and flexibility of bioeconomic models is one important area for future research related to the economics of weed control and the value of weed management information. Bioeconomic models have helped structure multidisciplinary research, and they have been shown to be the basis for effective management recommendations. Nevertheless, existing models do sometimes yield inaccurate predictions and misleading recommendations. The

Table 3. Summary findings from selected studies on the value of weed management information

Reference	Study type	Crop	Value estimate	Comments
King et al. (1986)	Simulation	Irrigated corn	$1.08 to $3.24/ha	Value measurements were made for high and low weed densities. Treatment alternatives were limited to one soil-applied and one postemergence alternative.
Lybecker et al. (1991a)	Field study	Irrigated corn	$76.67/ha	Value estimate is the average annual difference between net returns for bioeconomic model recommendations and a fixed strategy. The study was conducted over three years on research plots.
Buhler et al. (1992)	Field study	Corn	No significant increase in net margin	Value estimate is relative to a standard herbicide treatment. This study was conducted on research plots with high weed pressure.
Weersink et al. (1992)	Simulation	Soybean	$2.29 to $4.35/ha	Value estimate is for information that can be used to adjust herbicide rates. The baseline for comparison was an if-then threshold strategy that applied the label rate.
Wiles et al. (1992)	Simulation	Soybean	From zero to $21.00/ha	Value estimates are the incremental value of site-specific management at the whole field level based on bioeconomic model recommendations. The simulation experiment design included varying levels of weed pressure and patchiness based on actual field measurements.
Forcella et al. (1993)	Field study	Corn	$32.40/ha	Value estimate is relative to a standard herbicide treatment. This study was conducted on research plots with relatively low weed pressure.

--table continued next page

Table 3. continued--

Reference	Study type	Crop	Value estimate	Comments
Lybecker et al. (1993)	Field study	Irrigated corn	$28.64/ha	Value estimate is average difference between WEEDCAM specified and farmer treatments. Fifty sites during years 1989-1992. Includes both soil-applied and postemergence treatments.
Pannell (1994)	Simulation	Wheat	$0.76 to $11.00/ha	The simulation experiment design for this study included varying levels of weed pressure and risk aversion. The value of information on potential yield was also considered.
Swinton and King (1994b)	Simulation	Corn and soybean	$7.09 to $64.10/ha	Value estimates are based on a whole farm analysis. The simulation experiment design included low and high initial weed pressure and varying levels of risk aversion.
Oriade (1995)	Simulation	Corn	From zero to $18.87/ha	Value estimates are the incremental value of site-specific management relative to management at the whole field level based on bioeconomic model recommendations. The simulation experiment design included varying levels of weed pressure and patchiness.
Oriade (1995)	Simulation	Soybean	$4.74 to $28.84/ha	Value estimates are the incremental value of site-specific management relative to management at the whole field level based on bioeconomic model recommendations. The simulation experiment design included varying levels of weed pressure and patchiness.

following are three specific areas where additional research is likely to yield significant improvements:

* Improved model designs based on new theoretical developments and more accurate model parameterization based on innovative field research are needed to improve predictions of weed population dynamics, weed control efficacy, and the effects of weed interference.

* Interactions between weed management practices and other crop production practices, most notably tillage, need to be included in bioeconomic models.

* More meaningful measures of the environmental impacts of weed control strategies need to be developed and added to the output produced by bioeconomic models.

The full value of bioeconomic models can be realized only if they are used in a field setting, yet there are significant challenges associated with field implementation. Many of these challenges center around the need to understand how formal models can be incorporated into the work processes of crop consultants. Bioeconomic model recommendations are based on a large set of carefully specified biological and economic parameters and on detailed quantitative data about the field situation. On the other hand, crop consultant recommendations are generally based on qualitative visual data, expert judgment developed through training and experience, and "mental" bioeconomic models. The following are three specific areas where research can help clarify how these two approaches can be integrated into a single, more effective delivery system:

* Field tests are needed to compare net returns for strategies based on bioeconomic models to those associated with crop consultant recommendations.

* The cost of collecting data and providing weed management information needs to be measured, both for formal bioeconomic models and for crop consultants.

* Techniques need to be developed for using field data for sequential updating of bioeconomic model parameters. This will facilitate parameterization of bioeconomic models to match local conditions and has the potential to be an important source of data to supplement that collected in controlled experiments.

Ultimately, work in these areas will transform bioeconomic models from stand-alone tools that generate recommendations without much, if any, human interaction into components of true decision support systems, that help consultants use data and models to make more effective recommendations.[4]

Finally, federal, state, and watershed level policies are likely to have increasingly important impacts on weed management decisions. Well-designed policies should provide incentives that are clearly understood and effective in achieving desired results and they should be based on consideration of both private and social costs and benefits. Bioeconomic models can be an important tool for research related to both of these key ingredients for effective policy design.

* Bioeconomic models can be used to predict preferred weed management strategies under alternative policy mechanisms, such as taxes on herbicides, tradeable herbicide use permits, herbicide bans, and subsidies for the provision of weed management information.

* Bioeconomic models can be used to assess the private and social costs and benefits associated with weed management strategies that would be preferred under alternative policies.

Additional research in each of these three areas -- improving bioeconomic models, field implementation of decision support systems for weed management, and the design of environmental policies related to weed management -- can contribute significantly to the effectiveness of farm-level management decisions. It can also help address many of the broader societal concerns about the environmental impacts of some weed control practices. As stated at the beginning of this paper, this research will be successful only if it is truly problem-oriented and is based on collaborative efforts involving scientists from weed science, economics, and other disciplines.

References

Anderson, J.R., J.L. Dillon, and B. Hardaker. 1977. *Agricultural Decision Analysis*. Iowa State University Press, Ames, IA.

Bouzaher, A., D. Archer, R. Cabe, A. Carriquiry, and J.F. Shogren. 1992. Effects of environmental policy on trade-offs in agri-chemical management. *J. Envir. Mgmt.* 36:69-80.

Buhler, D.D., R.P. King, J.L. Gunsolus, S.M. Swinton, and F. Forcella. 1992. Field evaluation of a bioeconomic weed management model for corn. *Proc. of N. Central Weed Sci. Soc.* 47:35.

[4]Keen and Scott Morton (1978) and Sprague and Carlson (1982) provide good introductions to issues related to decision support system design and development.

Byerlee, D. and J.R. Anderson. 1982. Risk, utility and the value of information in farmer decision making. *Amer. J. of Arg. Econ.* 50:231-246.

Davis, G.B. and M.H. Olson. 1985. *Management Information Systems: Conceptual Foundations, Structure, and Development*, 2nd ed., McGraw Hill Book Company, New York.

Forcella, F., D.D. Buhler, S.M. Swinton, R.P. King, J.L. Gunsolus, and B.D. Maxwell. 1993. Field evaluation of a bioeconomic weed management model for the corn belt, U.S.A. *Quantitative Approaches in Weed and Herbicide Research and Their Practical Application.* 8th European Weed Research Society Symposium, Braunschweig, Germany, p. 755-760.

Keen, P.G.W. and M.S. Scott Morton. 1978. *Decision Support Systems: An Organizational Perspective.* Addison-Wesley Publishing Company, Reading, MA.

Keeney, R.L. and H. Raiffa. 1976. *Decisions with Multiple Objectives: Preferences and Value Tradeoffs*. John Wiley & Sons, New York.

King, R.P., D.W. Lybecker, E.E. Schweizer, and R.L. Zimdahl. 1986. Bioeconomic modeling to simulate weed control strategies for weed control strategies in continuous corn (*Zea mays*). *Weed Sci.* 34:972-979.

Lybecker, D.W., E.E. Schweizer, and R.P. King. 1991a. Weed management decisions in corn based on bioeconomic modeling. *Weed Sci.* 39:124-129.

Lybecker, D.W., E.E. Schweizer, and P. Westra. 1991b. Weed management decisions in corn based on bioeconomic modeling. Abstract in *West. J. Agr. Econ.* 16:456.

Lybecker, D.W., E.E. Schweizer, and P. Westra. 1993. Computer decision aid for managing weeds in irrigated corn. *Proceedings from Conference on Agricultural Research to Protect Water Quality* 2:295-297. Soil and Water Conservation Society, Minneapolis, MN.

National Agricultural Pesticide Impact Assessment Program (NAPIAP). 1992. The effects of restricting or banning atrazine use to reduce surface water contamination in the Upper Mississippi Basin: A summary. U.S. Department of Agriculture.

Oriade, C.A. 1995. *A Bioeconomic Analysis of Site-Specific Management and Delayed Planting Strategies for Weed Control.* Unpublished Ph.D. thesis. Department of Applied Economics, University of Minnesota, St. Paul, MN.

Pannell, D.J. 1994. The value of information in herbicide decision making for weed control in Australian wheat crops. *J. of Agr. and Res. Econ.* 19:366-381.

Sprague, R.J., Jr. and E.D. Carlson. 1982. *Building Effective Decision Support Systems*. Englewood Cliffs, NJ.

Swinton, S.M. 1991. *A Bioeconomic Model for Weed Management in Corn and Soybean.* Unpublished Ph.D. thesis. Department of Agricultural and Applied Economics, University of Minnesota, St. Paul, MN.

Swinton, S.M. and R.P. King. 1994a. A bioeconomic model for weed management in corn and soybean. *Agr. Systems* 44:313-335.

Swinton, S.M. and R.P. King. 1994b. The value of pest information in a dynamic setting: the case of weed control. *Amer. J. Agr. Econ.* 76:36-46.

Swinton, S.M., D.W. Lybecker, and R.P. King. 1995. The effect of local triazine restriction policies on recommended weed management in corn. *Rev. of Agr. Econ.* 17:257-273.

Taylor, C.R. and O.R. Burt. 1984. Near-optimal management strategies for controlling wild oats in spring wheat. *Amer. J. Agr. Econ.* 66:50-60.

Weersink, A., W. Deen, and S. Weaver. 1992. Evaluation of alternative decision rules for postemergent herbicide treatments in soybean. *J. Prod. Agr.* 5:298-303.

Widham, S. 1993. *Development of a Herbicide Environmental Index for Use in a Corn-Weed Management Decision Aid Model.* Unpublished M.S. thesis. Department of Agricultural and Resource Economics, Colorado State University, Fort Collins, CO.

Wiles, L.J., G.G. Wilkerson, and H.J. Gold. 1992. Value of information about weed distributions for improving post-emergence control decisions. *Crop Protection* 11:547-554.

Wiles, L.J., R.P. King, D.W. Lybecker, E.E. Schweizer, and S.M. Swinton. 1994. *User's Guide: General Weed Management Model, Version 1.0.* Staff Paper P94-23, Department of Agricultural and Applied Economics, University of Minnesota.

Wilkerson, G.G., S.A. Modena, and H.D. Coble. 1991. HERB: decision model for postemergence weed control in soybean. *Agron. J.* 83:413-417.

Producer Attitudes and Weed Management

Micheal D.K. Owen

I. Introduction

Weeds represent the most economically important pest complex reducing world food and fiber production (Bridges, 1994). Weeds have adapted to all agronomic systems and impact all aspects of crop production. Further, weeds negatively impact livestock production, personal health, and the aesthetics of homes, parks, and commercial areas. Currently, herbicides are the primary means of weed management in industrial countries and their use is rapidly increasing in developing countries. Herbicide use in Iowa represents an estimated $500,000,000 industry and more than 95% of the row crop area receives at least one annual herbicide application (Hartzler and Wintersteen, 1991). The emphasis on herbicides for the management of weeds has limited the use of alternative strategies. Further, public sector funding has not been sufficient to generate information on weed biology and weed ecology. This, in part, has resulted in management systems that do not represent an integrated approach to weed control.

Weed management has been greatly affected by the rapid adoption of conservation tillage systems. When tillage is reduced, weed populations change, mechanical management options are reduced, and producers place a higher reliance on herbicides for weed control. Changes in tillage practices also reflect socioeconomic issues: there are fewer producers and the average farm size is increasing. Conservation tillage systems allow producers to improve time utilization and are perceived to be more economically efficient.

There is an increasing amount of land that is tenant-farmed and competition for this land is often intense. The ability of tenant producers is judged on the effectiveness of their weed control systems. Late-season weeds in fields are thought to represent poor agronomic ability. Herbicides represent a time-efficient strategy to control weeds, and producers assume herbicides to be more effective than other strategies. Consequently, producers often will apply herbicides when other strategies or no action would be the most appropriate economically and environmentally.

There are problems associated with the current weed management systems including degradation of the environment, development of herbicide-resistant weed populations, and questionable sustainability of current agriculture. Importantly, only 2% of the population in the United States is directly involved with agriculture and thus the majority of the people do not understand nor appreciate the impact of agriculture on their daily existence. Often, agricultural policy has been based on concepts and concerns without scientific substantiation. Similarly, producer attitudes about weeds, weed management, and herbicide use have been influenced by factors other than those based on science. It is important to recognize why the current systems have developed, how they are used, and what can be changed to improve them. However, producer attitudes must be considered and agronomic issues that are not based on science recognized as potentially important.

II. Producer Attitudes and Tillage

There have been numerous scientific studies investigating the effects of tillage on weed population dynamics, species succession, management strategies, and soil erosion. However, much of this research has been conducted on relatively small experimental areas and may not include typical producer practices or techniques. Currently, conservation tillage is rapidly increasing in importance and producers are implementing weed management strategies that do not necessarily reflect the biological and ecological changes in weed populations identified by research. Often, producer practices are developed as a result of on-farm demonstrations by other producers who, by popular acclaim, have been identified as experts. Regardless, herbicides are universally recognized as the primary strategy to manage weeds when tillage is reduced or eliminated. Importantly, producers identify problems in weed management as a major impediment to the successful adoption of conservation tillage practices.

Weed species composition and population dynamics are dramatically influenced by tillage and herbicide use. Wrucke and Arnold (1985) demonstrated that annual grass pressure increased significantly under no tillage systems, compared to reduced and conventional tillage systems. However, annual broadleaf weed populations did not change significantly in response to tillage. Froud-Williams et al. (1981) reported that the increased reliance upon herbicides in reduced tillage systems and the resultant changes in the physical soil environment caused modifications in the existing weed flora.

Yenish et al. (1992) noted different seed placement patterns in response to different tillage systems. Pareja et al. (1985) noted similar observations with regard to weed seed distribution and characterized the structural units in which weed seeds were located in response to tillage. Pareja and Staniforth (1985) demonstrated that the different sized soil structural units imparted significant effects on the germination of weed seeds. Typically, as tillage was reduced, the average size of the soil structural unit increased and weed seeds were less likely to germinate when located in large soil structural units.

As tillage is reduced, plant residue on the soil surface increases. This has a significant impact on the distribution of herbicides (Banks and Robinson, 1986), persistence (Banks and Robinson, 1982), and application technique (Wicks, 1985). Typically, weed control is less consistent in high residue tillage systems. Producers recognize this and substantiate these observations by suggesting that problems with weed management are the greatest concern and reason for not adopting a reduced or no tillage production system.

Chancellor (1979) observed that as agricultural practice changes, the composition of the weed flora will change. Derksen et al.(1993) suggested that changes in the weed communities were the limiting factor for the adoption of conservation tillage systems. Buhler et al. (1994) reported that hemp dogbane *(Apocynum cannabinum* L.), a perennial dicot weed, invaded no-tillage fields previously free of this species within three years of the initiation of a tillage experiment.

Generally, producers do not understand the impact that changes in tillage have on weed population dynamics. Nor do they recognize the interaction of plant residue on herbicide efficacy. Producers do, however, appear to understand the importance of soil conservation. Today, it is popular to produce soybeans (*Glycine max* L.) in narrow rows without tillage. However, the relative success to failure ratio of weed management in no-tillage systems is not good. Producers, without a clear understanding that the ecology of fields has changed dramatically in response to the elimination of tillage, attempt to control weeds with the same strategies that were successful in conventional tillage systems. They do not consider alternative strategies for weed management and expect herbicides to manage all weed problems. Given the extremely high level of weed control that is demanded by producers and the variable performance of herbicides in high residue production systems, multiple herbicide applications are likely required and the economics of the no-tillage systems are not favorable. Importantly, it is difficult to achieve a weed-free field in no-tillage. Producers unfortunately perceive that a weed-free environment is essential for profitable crop production. Thus, there is a relatively high level of dissatisfaction with no tillage and producers are more likely to abandon the system. If producers had a better understanding of the biological and ecological changes in no tillage production systems, more effective and economically efficient weed management systems could be implemented and a higher ratio of success to failure achieved.

III. Producer Attitudes and Weed Management Strategies

Producers recognize the importance of weed management in crop production. At one time, primary and secondary tillage played a major role in weed management programs. Typically, producers utilized an integrated approach for weed control that included herbicidal, cultural, and mechanical strategies. However, current trends suggest that alternative strategies such as rotary hoeing and cultivation are no longer employed effectively in weed management systems. Herbicides are now considered the main, if not sole strategy, for weed control in much of the United States.

A. Herbicidal Strategies

The rapid adoption of high residue crop production systems has increased the importance of herbicides in weed management systems. Without primary tillage, herbicides are deemed essential by producers to provide the crop with a competitive advantage over developing weed populations. Herbicide applications have shifted during the latter part of the 1980s from a high percentage of the land receiving preplant incorporated herbicides to the use of early preplant, postemergence, and soil-applied/postemergence sequential applications. Currently, a majority of the land receives more than one herbicide application per season and is treated with more than one active ingredient.

Triazine herbicide use has increased in the corn belt despite use restrictions on the labels developed to minimize the potential movement of these herbicides into ground and surface water. More than 81% of the corn (*Zea mays* L.) in Iowa received triazine herbicides, and atrazine use increased 24% between 1985 to 1990 when over 61% of the corn was treated with atrazine (Hartzler and Wintersteen, 1991).

The development of herbicide classes that inhibit the acetolactate synthase (ALS) enzyme has resulted in a major change in herbicide use practices in the last 10 years. Imidazolinone and sulfonylurea herbicides have been introduced and have rapidly gained acceptance by producers because of the high level of efficacy that they demonstrate, low use rates, the diversity of the weeds that are sensitive to these herbicides, and the application flexibility that these herbicides have. Currently, an estimated 85% of the soybean in Iowa and 64% of the soybean in the United States receive applications of ALS inhibiting herbicides (Wrubel and Gressel, 1994). Many of the experimental herbicides currently under development and potentially to be labeled within the next 10 years are ALS inhibitors. ALS inhibitor herbicides have been developed by most of the agricultural chemical companies and are represented by 15 different chemical classes (Powles and Holtum, 1994). Given the current use of ALS inhibitor herbicides, the toxicological and environmental characteristics these herbicides demonstrate, and the increasing restrictions on other herbicide classes, the ALS inhibitor herbicides represent the most important group of agricultural chemicals for the foreseeable future.

However, the widespread use of the ALS inhibitor herbicides has resulted in a number of problems. Notably, resistance to these herbicides has been documented in a number of locations and in several agronomically important weed species. Common waterhemp (*Amaranthus rudis* Sauer), Palmer amaranth (*Amaranthus palmeri* S. Wats.), and common cocklebur (*Xanthium strumarium* L.) populations have been identified as resistant to ALS inhibitor herbicides (Horak and Peterson, 1995; Schmitzer et al., 1993). Common waterhemp is the most prevalent pigweed species in Iowa, and numerous locations have been identified that have ALS resistant populations (Robert G. Hartzler, Iowa State Univ., personal communication). These populations have demonstrated consistent cross resistance to ALS inhibitor herbicides other than the one that initially was the selective agent (John R.R. Hinz, Iowa State Univ., personal communication). Producers recognize that weeds resistant to ALS inhibiting herbicides are increasing in frequency, but they have not, to date, changed their use practices.

Recent initiatives in the government and private sector stress the need to reduce herbicide use. Concerns about the contamination of groundwater and surface water by agricultural chemicals, particularly herbicides, have resulted in increasing label restrictions. Consistent and widespread detections of herbicides, albeit typically below the concentrations suggested by the Environmental Protection Agency (EPA) to be a health concern, have resulted in educational programs designed to allow the reduction of herbicide use. Hartzler (1993) presented producers with a number of strategies that will allow both the reduction of herbicide inputs and a high level of weed management. In Europe, various governments have mandated pesticide use reductions (Belinder et al., 1994; Lawson, 1994). Three of ten Canada provinces have policies to reduce pesticide use (Hamill et al., 1994). However, there has been considerable debate as to how the desired reductions in herbicide use should occur.

Efforts to reduce herbicide use in Sweden have not resulted in a reduced land area treated with herbicides (Belinder et al., 1994). Rather, reductions were accomplished by adopting newer herbicide technology in which the active ingredients are used at extremely low doses, compared to older herbicides. Similar results were reported in the United Kingdom (Lawson, 1994).

In Iowa, while the percentage of land treated with atrazine has increased, the amount of product used per unit area has declined significantly (Hartzler and Wintersteen, 1991). Banding herbicides, utilizing reduced rates of herbicides in conjunction with mechanical strategies, and spot applications are suggested strategies that would allow a significant reduction in herbicide use (Hartzler, 1993). However, producers have not adopted these practices due to the increased risk of poor weed control, management skills required, and lack of time needed to accomplish these strategies. Producers perceive the savings on herbicides to be outweighed by the risk of weed control failures. Generally, producers rely entirely on herbicides for weed management, and problems in weed control that develop are dealt with by applying another herbicide.

B. Mechanical Strategies

Mechanical strategies have historically been important components of weed management programs. Thill et al. (1991) stressed the need for mechanical weed control practices in an integrated weed management program, thus reducing the environmental impact of agriculture. Similarly, Buhler et al. (1995) suggested that cultivations and rotary hoeing, in combination with reduced herbicide rates would maintain effective weed control in conservation tillage systems and minimize the environmental impacts of crop production. They indicated that often multiple mechanical weed control strategies were needed, in addition to low herbicide rates, to ensure effective control of weeds. It is important to note that this research was conducted in fields with low to moderate weed populations. Further, the timing of the mechanical weed control practices was critical to achieve effective results without increasing soil erosion.

Owen (1994) suggested that the use of mechanical weed control strategies would minimize the development of herbicide resistant weeds. Lovely et al. (1958) demonstrated that the rotary hoe could significantly reduce weed populations in soybeans. The herbicide industry also endorses mechanical strategies as an important means of managing herbicide resistant weeds (Anonymous, 1994). Gunsolus (1990) summarized mechanical weed management research conducted over the last 35 years and indicates that highly refined management skills are essential for successful weed control with mechanical techniques. However, he reports that significant crop stand reductions may occur from the mechanical practices.

Hartzler et al. (1993) reported that weed populations were similar in 59% of the fields where mechanical weed management was compared with herbicidal treatments. Importantly, corn yields were reduced in only 22% of the fields where weeds were mechanically managed compared to herbicides.

Despite the fact that mechanical weed management techniques are economically effective, environmentally appropriate, and agronomically acceptable, producers have not readily embraced these strategies. A survey conducted in 1991 (Hartzler and Wintersteen, 1991) indicated that 75 and 21% of Iowa corn were cultivated and rotary hoed, respectively. More recently, Owen and Pecinovsky (1993) reported a similar percentage cultivated, but rotary hoeing had declined to 9% of the producers surveyed. Producer acceptance of no tillage systems, narrow row soybean production, and larger farming operations has resulted in a lesser emphasis on mechanical weed control. Producer attitudes indicate that in no tillage systems, no mechanical soil disturbance can be allowed. The perception is that rotary hoes will not effectively control weeds in high residue systems. However, equipment has been designed to operate in no tillage systems and likely has a demonstrable economic advantage over respray herbicide treatments.

One factor that has influenced producer attitudes toward mechanical weed control is the competitiveness within the agricultural chemical industry. Essentially all row crops in the corn belt receive herbicide treatments. Thus, to improve market share, companies must acquire land that was treated with competitive products. To facilitate this, companies have guaranteed the performance of the herbicides and producers have no obligation nor reason to supplement herbicidal weed control with mechanical practices. This has allowed producers to effectively mediate the risk of weed control failures without investing time and without making any management decisions. Thus, producers have minimized the use of cultivators and rotary hoes and opt instead for rescue herbicide applications.

C. Cultural Strategies

Cultural weed management is an integral component of crop production. Whether or not producers are consciously aware, they include cultural weed management in all crop production systems. Holt (1994) suggested the need for increased research on cultural weed management strategies. Coble (1994) proposed that important areas for future weed science research include the study of crop/weed interactions and alternative weed control methodology. Liebman and Dyck (1993) stress the need for an ecological approach to weed management emphasizing nonchemical strategies for weed control. Unfortunately, producers have not taken full advantage of cultural weed management opportunities in weed control programs primarily because they do not understand the principles.

Lueschen and Anderson (1980) evaluated seven cultural practices and reported that velvetleaf (*Abutilon theophrasti* Medicus) seed population declined to 10% of the original seedbank after intensive tillage for 4 years. However, where tillage was not done, a much higher percentage of the velvetleaf seedbank remained. Where alfalfa (*Medicago sativa* L.) was grown continuously, 56% of the velvetleaf seedbank remained, and under chemical fallow without soil disturbance, 37% of the velvetleaf seedbank was found after 4 years. Similar trends were reported for wild mustard [*Brassica kaber* (DC.) L.C. Wheeler] (Warnes and Andersen, 1984).

Crop rotation has historically been an important cultural strategy for weed control. A highly diversified crop rotation, utilizing summer annual crops, spring grains, and perennial forages, did not allow the establishment of the complementary weed community adapted for each crop (Liebman and Janke, 1990). The simpler the rotation strategy or the more similar the growth habit of rotational crops, the greater the likelihood for adapted weeds to become established. MacHoughton (1973) conducted a six year crop rotation study and found that the three-crop rotation resulted in fewer giant foxtail (*Setaria faberi* Herrm.) and velvetleaf compared to continuous soybeans. However, with the current production system typically found in the corn belt, crop rotation opportunities are severely limited. Typically, continuous corn or corn-soybean rotations are the only systems used by producers. Given that corn and soybeans are summer annual crops planted early in the spring and harvested in the fall, a complementary community of aggressive summer annual weeds has proliferated.

Differences in the competitive abilities of crop varieties and hybrids represents an opportunity to improve cultural weed control. Wyse (1994) recommended the development of crops with more competitive growth characteristics. He points out, however, that traditional plant breeding has not produced crops developed specifically to be competitive with weeds. Jordan (1993) also indicated the importance of crop competition with weeds as a nonchemical weed management strategy. However, insufficient information concerning the mechanisms of crop interference currently exists. Thus, a thorough evaluation of the biological costs and benefits of crop interference should be accomplished prior to the initiation of any breeding programs.

Developments in biotechnology have resulted in crops that allow the merging of cultural and herbicidal strategies. Owen (1994) suggested that crops genetically modified to have herbicide resistance represented an excellent opportunity to improve weed control through cultural and chemical strategies. Goldburg et al. (1990) expressed concern that herbicide resistant crops are a serious threat to agriculture and will result in considerable environmental and economic problems. Seed and herbicide industries, however, have accelerated the development of herbicide resistant crops and have emphasized the environmental and economic benefits of this technology (Beale, 1994; Krueger, 1994; Wells, 1994; Hughes, 1994; Pieters, 1994).

Thus far, imazethapyr-resistant corn hybrids are the only herbicide resistant crop that has achieved acceptance with producers. Generally, producers have not embraced this technology as an appropriate cultural strategy to manage weeds. However, when glyphosate-resistant soybean varieties become widely available, it is anticipated that there will be a major adoption of this technology. Whether or not this represents an economic and environmental opportunity in cultural weed management remains to be seen.

Crop planting date is another cultural factor that can influence weed management. Research has demonstrated that many weeds have periodic germination events (Stoller and Wax, 1973). By delaying planting, and if tillage or herbicides are used, a high percentage of the weed population can be eliminated prior to planting and thus managed. Iowa producers who manage weeds without herbicides were surveyed and found to delay planting one week, on the average, when compared with producers who use herbicides (Pecinovsky and Owen, 1994).

In Minnesota, producers plant corn in late April and soybeans in early May (Gunsolus, 1990). Iowa producers typically plant one to two weeks earlier than in Minnesota. The reasons for planting earlier are that corn yield potential quickly declines with delayed planting, and unfavorable weather typically minimizes available planting days during the early spring. While soybean yields are less responsive to delayed planting, potentially unfavorable weather encourages producers to plant crops as early as possible. The result is that planting date, while an effective strategy to manage weeds culturally, is not widely used as a weed control technique.

Crop row spacing is also a production factor that producers can utilize as a cultural weed management technique. Producers are rapidly adopting narrow row spacings for soybean production because, in part, of the weed control advantages that this planting pattern demonstrates when compared to conventional row spacings (D. Keith Whigham, Iowa State Univ., personal communication). However, the success of narrow row spacing for weed control requires the elimination of early germinating weeds, thus allowing the soybeans to grow long enough in a weed-free environment and ultimately forming a canopy over the inter-row spaces.

If the crop does not form a canopy soon enough, or if weeds are not effectively controlled at planting, there is a significant risk of yield losses and harvestibility problems. Producers therefore may want to compromise on row spacings and use an intermediate width that allows the option of mechanical weed control (Gunsolus, 1990). Interestingly, Iowa producers who managed weeds with fewer herbicides had an average row spacing of 96.5 cm while producers who relied upon herbicides had 76.2 cm rows (Owen and Pecinovsky, 1993). The wider row spacing provided less opportunity for crop interference with weeds but did facilitate late-season mechanical control strategies.

Other opportunities for cultural weed management include smother plants, living mulches, and cover crops (Wyse, 1994). These strategies have not been widely adopted by producers and may have risks associated with their adoption that are unrelated to weed control. For example, cover crops such as rye *(Secale cereale* L.) may serve as a refuge for insect pests or may impart moisture stress upon the crop (Robert G. Hartzler, Iowa State Univ., personal communication).

With changes in tillage practices, farm size, and herbicide development, cultural and mechanical weed management strategies have not been effectively utilized. The emphasis on herbicides as the primary strategy for weed control continues despite the development of problems from this technology. However, utilization of multiple weed management strategies is important and should lessen the likelihood of herbicide resistant weed populations and environmental concerns associated with herbicides.

IV. Producer Attitudes and Weed Thresholds/Models

Numerous researchers have proclaimed the need for effective weed thresholds and predictive models (Coble, 1994; Holt, 1994; Wyse, 1994). Recently, considerable resources have been dedicated to research in weed biology and ecology with specific emphasis on developing predictive models. A North Central Regional Project (NC-202) has developed regional research protocols to investigate various seedbank phenomena, thus supplying data for a predictive model (James J. Kells, Mich. State. Univ., personal communication). This project involves a number of weed scientists working cooperatively on the research objectives and has resulted in the expenditure of considerable scientific efforts.

Coble and Mortensen (1992) described various weed threshold concepts and suggested that short-term thresholds should be considered differently than multiple-season thresholds, given the risks associated with the respective weed populations, resultant seed production, and predictive abilities. Wilkerson et al. (1991) reported on a predictive model for postemergence weed control in soybeans. The model was based on considerable weed interference research and utilized estimated yield values, soybean selling price, and cost of treatment. The model was useful for single season thresholds but did not account for the future cost of escaped weeds. This model was adapted to the Midwest and has been available to producers (Mortensen et al., 1993).

There are several significant problems affecting the development of useful weed threshold values and the eventual inclusion in predictive models. Notably, the weed threshold value must account for the ability of the weed to interfere with crop production. Typically, this value has been expressed as reduction of crop yield. However, the competitive ability of weeds varies in response to environmental conditions (Staniforth, 1958). Further, most research has reported on the competitive ability of a single species at varying populations with different periods of interference (Knake and Slife, 1962). For a predictive model to be effective, it must account for multiple species and should accommodate the economic cost of seed production from weed population levels below the economic threshold.

King et al. (1986) described a bioeconomic model in which weed management strategies in continuous corn were based on weed densities, seed numbers, and other agronomic factors. This model served as the basis for the NC-202 project. However, the long-term implications of model decisions on weed seedbank and future control options were lacking. Swinton et al. (1994) used multiple data sets to model crop yield losses attributable to mixed weed infestations. They found that the calculated competition coefficients were stable over years but varied for locations. Concerns were expressed about the accuracy of information describing various biological processes that affect weed interference and they recommended that future research focus on weed seedbank dynamics.

Lindquist et al. (1995) studied demographic processes affecting velvetleaf population dynamics in soybeans and found that crop competition reduced velvetleaf survival and seed production up to 82%. However, the study site was infested with *Verticillium* spp., which significantly reduced initial velvetleaf populations. Importantly, velvetleaf seedling recruitment was consistent during the two year study.

Hartzler (1995) investigated the long-term implications of velvetleaf population dynamics based on predicted economic thresholds and found that the seed rain from one year resulted in a 24-fold increase in velvetleaf population. Velvetleaf population increased the second year after seeding but declined the following two years. Regardless, the populations represented increased costs of management for future years. Also, the reported differences between locations illustrates the problems of predictive models to be useful over a wide geographic area.

Cousens (1987) suggested that concerns about the precision of predictive models indicated that subjective observations of weed populations were appropriate for effective weed management. While the efforts in modeling and research on weed interference, seedbank dynamics, seedling recruitment, and other variables have provided an improved understanding about weed management decisions, producer decisions must account for other critically important considerations. Given the dramatic increase of Midwest cropland farmed under a rental agreement, producers often attempt to control weeds beyond an economically appropriate level. Weed escapes are perceived to be evidence of poor agronomic abilities for producers and jeopardize future rental contracts. The lack of information concerning weed seedbank processes reinforces producer concerns about dramatic increases in weed populations due to weed escapes. Importantly, conservation tillage systems likely increase dramatic weed population fluctuations. This lack of knowledge about weed population dynamics and the current agricultural systems has lessened the willingness of producers to accept predictive models and weed thresholds.

V. Producer Attitudes and Sustainability

Concerns about the environmental and economic sustainability of current agricultural production systems have been important topics of discussion in public forums, news media, and scientific meetings. Generally, agriculture is not seen to be a responsible industry with regard to environmental, sociological, or economic issues. Specifically, weed management practices have been associated with herbicide drift, groundwater and surface water contamination, and eroding environmental quality. Zimdahl (1994) points out that weed science has been a specialized discipline emphasizing herbicide technology while minimizing the study of the weeds. While weed science has been extraordinarily successful in solving producer problems with weeds, the technology has become too focused and the resultant biological and economic problems are now increasingly apparent. As a result, there has been considerable effort in developing integrated, sustainable weed management systems. However, one must question whether or not these new strategies will receive endorsement from producers. Is the current agricultural production system truly amiss?

Many scholars have questioned the current agricultural system. Titi (1992) stressed the need for European agriculture to be responsive to challenges in economics, politics, and the environment. Wyse (1994) suggested that current production systems, with accompanying herbicide technology, are not capable of adapting and thus resolving associated problems of soil erosion, water contamination, and concerns for the quality of rural life. Weeds define crop production practices and while weed science has solved producer's weed problems, the solutions have been based primarily on herbicides. Thus, weeds and the resultant herbicide technologies are considered to be the major problem limiting the development of a sustainable agriculture. To resolve these issues, Wyse recommended expanding research efforts in weed and crop ecology, dramatically changing current production systems, and including novel crops. The key is to provide a viable, environmentally sensitive economy for rural America that includes agricultural systems based on fewer herbicide inputs and conservation tillage practices that minimize soil erosion.

Thill et al. (1991) reviewed weed management practices with regard to the use of integrated approaches and suggested that considerable progress has been made in the development of integrated weed management concepts, and producers have adopted these practices. However, it was stressed that there was considerable need for further advances. For example, herbicides have dominated the scientific literature and accounted for 80% of papers published in *Weed Science* through 1977. However, a significant upward trend of weed biology papers was noted in recent history. It was clear, however, that the authors felt it critical to develop improved integrated weed management systems that stress sustainability.

Clearly there are questions whether or not current agricultural systems are sustainable. What defines a sustainable system and do they result in differences with regard to weed management issues? Research conducted by Owen and Pecinovsky (1993) quantified differences in weed management strategies, weed populations, and perceptions of weed management for conventional and sustainable producers. Producers defined as conventional attributed 83% of weed control to herbicides and 16% to tillage practices while sustainable growers reported that 44% of weed control was due to herbicide use and tillage was responsible for 55% of the weed control in their systems. Sustainable producers utilized banded herbicide applications more than conventional producers and they did not use preplant incorporated applications as often as conventional producers. Sustainable producers (65%) reported that herbicide performance was poor or average while all conventional producers reported average or better herbicide performance. Sustainable producers utilized more mechanical weed control techniques and made multiple rotary hoeing and cultivations while conventional producers typically did not cultivate nor use a rotary hoe.

Significant weed population differences between sustainable and conventional producers were noted. Sustainable fields had higher and more diverse weed populations than conventional fields; however, no differences in velvetleaf and yellow foxtail [*Setaria glauca* (L.) Beauv.] were observed. Soil cores demonstrated that sustainable fields had 2.5 times more giant foxtail and 4 times more pigweed (*Amaranthus* spp.) seeds in the soil seedbank when compared with conventional fields (Owen and Pecinovsky, 1993). While yields were generally lower for the sustainable producers, production costs resulted in profitability similar to conventional systems. The long-term implications of weed population differences, however, suggest that economically effective weed management may be a serious problem in the future for the sustainable producers. Importantly, average farm size was 116 ha for sustainable producers compared to 284 ha for conventional producers.

Sustainable weed management, as described by Owen and Pecinovsky (1993), required considerably more management skills and time inputs compared to conventional systems. The basis of sustainable weed management was mechanical strategies that supplemented cultural practices. Given the scale of farm operations, it is highly unlikely that most producers can or would adopt sustainable practices that minimize herbicidal weed control. However, there are opportunities to improve the efficiency of current weed management, placing emphasis on environmental and economic considerations. With improved technology, it is anticipated that producers would move quickly toward more sustainable weed management. However, herbicides will continue to be an important component of weed management for the foreseeable future.

VI. Producer Attitudes and the Pesticide Industry

The agricultural chemical industry has directed weed management attitudes over the last three decades. Currently, a majority of row crop acres receive herbicide applications. When a new herbicide is introduced, market share must be captured from another herbicide. Thus, herbicide sales have become highly competitive and marketing campaigns extremely important in influencing producer decisions. An objective review of current herbicides suggests that products are more similar than different agronomically and economically. Thus, the industry must try to differentiate their proprietary products through advertising. This has resulted in the proliferation of herbicide commercials, gimmicks, and advertisements that create the impression that herbicides will control all weeds under all conditions without exception.

In promotional brochures, fields treated with specific products are depicted as weed-free for the entire growing season. These campaigns have effectively convinced producers that herbicides are formulated to provide complete weed control regardless of environmental conditions, weed seedbank populations, or tillage regimes. These advertisements have also convinced producers that weed control is synonymous with weed-free, and producers have adopted a zero-tolerance level of weed escapes.

Another consideration is that due to the competitiveness of the herbicide market, the similarities of products, and producer expectations, the herbicide industry has chosen to guarantee herbicide performance. Essentially, producers purchase a herbicide anticipating a weed-free field without any obligation to supplement the herbicide with mechanical control strategies. Herbicide manufacturers have resorted to this marketing strategy to maintain and increase market share but must also respond to the producer impressions of herbicide performance and the likely unobtainable weed-free condition that their marketing programs have created. Herbicide performance guarantees have dramatically affected herbicide market share and most companies will guarantee herbicide performance via some marketing program.

Thus, because of the herbicide performance guarantees, producers are demanding herbicide resprays at an alarming rate. This represents a significant misuse of herbicide technology and is an environmental and economic problem of major proportions. The author estimates that as much as 25 to 30% of Iowa corn was retreated in 1994 because of herbicide guarantee programs. There is no way of determining whether these resprays were agronomically warranted. However, the timing of these applications typically was later than what would normally be recommended for effective weed control. Further, crop size likely increased the potential of herbicide injury for these late applications.

Producers see these guarantee programs as an excellent way to minimize the risk of weed management. Intuitively, producers recognize that various agronomic and environmental factors influence herbicide performance. Further, they understand the need for supplemental weed management, whether by mechanical strategies or cultural manipulation of the system. Unfortunately, producers likely do not have the time to accomplish mechanical weed control in a timely fashion. Producers thus recognize that weed control will not be as good or consistent as assumed necessary for highest yields. However, herbicide manufacturers who guarantee herbicide performance are seen by producers as effectively managing this risk of poor weed control. Consequently, producers are less likely to adopt alternative weed management strategies and will continue to rely solely on herbicides for weed control.

VII. Summary and Conclusions

Producers deem weed control as the critical part of crop production. Further, producers recognize the importance of reducing soil erosion and are adopting conservation tillage systems. The resulting changes in weed biology in conservation tillage systems are perceived to represent a major impediment to the successful adoption of these environmentally appropriate crop production systems. The general public perceives agricultural pesticides, specifically herbicides, as significant environmental contaminants and are demanding reductions in use. Thus, producers are expected to rely less on herbicides while adopting tillage systems that potentially increase weed populations and reduce control options. Weed science has not supplied the research information that improves the potential for producers to utilize herbicides in conjunction with alternative weed management strategies. However, given the current research emphasis on weed biology and ecology, alternative weed management strategies may become available in the near future.

Producers also face challenges with weed management decisions that are based more on the perceptions of weed populations rather than the actual economic impact. Given the current average farm size, it is difficult to use alternative weed management strategies effectively. Weed escapes, while not competitive with crop yield, which result from alternative strategies are perceived to be economically important and are used to judge the agronomic skills of producers. Given the importance of rental farmland in the current systems, it is likely important for producers to control all weed escapes if they want to continue farming rental property.

Finally, the herbicide industry has contributed to the current perception of weed management. Herbicides are guaranteed to provide a weed-free field regardless of agronomic or environmental conditions, and fields have been retreated as requested. As a result, producers are less likely to adopt alternative weed management strategies, and herbicide use will continue at the same high level. Given that many of these herbicide treatments may be unwarranted biologically and that the level of weed control perceived by producers necessary for crop production is not realistic, guaranteed herbicide performance represents a major problem for agriculture. The public and private sectors should work toward more economically efficient and environmentally responsible weed management systems by minimizing herbicide misuse and increasing the adoption of alternative weed management practices.

References

Anonymous. 1994. Guidelines for managing resistance to ACCase-inhibiting herbicides. ACCase Inhibitor Resistance Working Group. 8 p.

Banks, P.A. and E.L. Robinson. 1982. The influence of straw mulch on the soil reception and persistence of metribuzin. *Weed Sci.* 30:164-168.

Banks, P.A. and E.L. Robinson. 1986. Soil reception and activity of acetochlor, alachlor, and metolachlor as affected by wheat (*Triticum aestivum*) straw and irrigation. *Weed Sci.* 34:607-611.

Beale, M.W. 1994. Development of sethoxydim tolerance in corn. *Proc. North Cent. Weed Sci. Soc.* 49:169.

Belinder, R.R., G. Gummeson, and C. Karlsson. 1994. Percentage-driven government mandates for pesticide reduction: the Swedish model. *Weed Technol.* 8:350-359.

Bridges, D.C. 1994. Impact of weeds on human endeavors. *Weed Technol.* 8:392-395.

Buhler, D.D., J.D. Doll, R.T. Proost, and M.R. Visocky. 1995. Integrating mechanical weeding with reduced herbicide use in conservation tillage corn production systems. *Agron. J.* 87:507-512.

Buhler, D.D., D.E. Stoltenberg, R.L. Becker, and J.L. Gunsolus. 1994. Perennial weed populations after 14 years of variable tillage and cropping practices. *Weed Sci.* 42:205-209.

Chancellor, R.J. 1979. The long-term effects of herbicides on weed populations. *Ann. Appl. Biol.* 91:141-144.

Coble, H.D. 1994. Future directions and needs for weed science research. *Weed Technol.* 8:410-412.

Coble, H.D. and D.A. Mortensen. 1992. The threshold concept and its application to weed science. *Weed Technol.* 6:191-195.

Cousens, R. 1987. Theory and reality of weed control thresholds. *Pl. Prot. Quart.* 2:13-20.

Derksen, D.A., G.P. Lafond, A.G. Thomas, H.A. Loeppky, and C.J. Swanton. 1993. Impact of agronomic practices on weed communities: tillage systems. *Weed Sci.* 41:409-417.

Froud-Williams, R.J., R.J. Chancellor, and D.S.H. Drennen. 1981. Potential changes in weed flora associated with reduced-cultivation systems in cereal production in temperate regions. *Weed Res.* 21:99-109.

Goldburg, R., J. Rissler, H. Shand, and C. Hassebrook. 1990. Biotechnology's bitter harvest: Herbicide-tolerant crops and the threat to sustainable agriculture. Biotechnology Working Group Report. 73 p.

Gunsolus, J.L. 1990. Mechanical and cultural weed control in corn and soybeans. *Am. J. Alter. Agric.* 5:114-119.

Hamill, A.S., G.A. Surgeoner, and W.P. Roberts. 1994. Herbicide reductions in North America: In Canada, an opportunity for motivation and growth in weed management. *Weed Technol.* 8:366-371.

Hartzler, R.G. 1993. Reducing herbicide use in field crops. Iowa State University Extension Report IPM-36.

Hartzler, R.G. 1995. Velvetleaf (*Abutilon theophrasti*) population dynamics following a single year's seed rain. *Weed Technol.* (accepted for publication).

Hartzler, R.G. and W. Wintersteen. 1991. A survey of pesticides used in Iowa crop production in 1990. Iowa State University Extension Report Pm-1441.

Hartzler, R.G., B.D. Van Kooten, D.E. Stoltenberg, E.M. Hall, and R.S. Fawcett. 1993. On-farm evaluation of mechanical and chemical weed management practices in corn *(Zea mays*). *Weed Technol.* 7:1001-1004.

Holt, J.S. 1994. Impact of weed control on weeds: New problems and research needs. *Weed Technol.* 8:400-402.

Horak, M.J. and D.E. Peterson. 1995. Biotypes of Palmer amaranth (*Amaranthus palmeri*) and common waterhemp (*Amaranthus rudis*) are resistant to imazethapyr and thifensulfuron. *Weed Technol.* 9:192-195.

Hughes, M.R. 1994. Update on DuPont herbicide tolerant crop programs. *Proc. North Cent. Weed Sci. Soc.* 49:171.

Jordan, N. 1993. Prospects for weed control through crop interference. *Ecol. Applic.* 3:84-91.

King, R.P., D.W. Lybecker, E.E. Schweizer, and R.L. Zimdahl. 1986. Bioeconomic modeling to simulate weed control strategies for continuous corn (*Zea mays*). *Weed Sci.* 34:972-979.

Knake, E.L. and F.W. Slife. 1962. Competition of *Setaria faberi* with corn and soybeans. *Weeds* 10:26-29.

Krueger, R.W. 1994. Weed control in genetically modified crops: American Cyanamid. *Proc. North Cent. Weed Sci. Soc.* 49:169-170.

Lawson, H.M. 1994. Changes in pesticide usage in the United Kingdom: Policies, results and long-term implications. *Weed Technol.* 8:360-365.

Liebman, M. and E. Dyck. 1993. Weed management: A need to develop ecological approaches. *Ecol. Applic.* 3:39-41.

Liebman, M. and R.R. Janke. 1990. Sustainable weed management practices. In: C. Francis, C.B. Flora, and L.D. King (eds.), *Sustainable Agriculture in Temperate Zones*. John Wiley and Sons, Inc., New York. p. 111-143.

Lindquist, J.L., B.D. Maxwell, D.D. Buhler, and J.L. Gunsolus. 1995. Velvetleaf (*Abutilon theophrasti*) recruitment, survival, seed production, and interference in soybean (*Glycine max*). *Weed Sci.* 43:226-232.

Lovely, W.G., C.R. Weber, and D.W. Staniforth. 1958. Effectiveness of the rotary hoe for weed control in soybeans. *Agron. J.* 50:621-625.

Lueschen, W.E. and R.N. Andersen. 1980. Longevity of velvetleaf (*Abutilon theophrasti*) seed in soil under agricultural practices. *Weed Sci.* 28:341-346.

MacHoughton, J. 1973. Ecological changes in weed populations as a result of crop rotations and herbicides. Ph.D. dissertation, University of Illinois at Champaign-Urbana.

Mortensen, D.A., A.R. Martin, F.W. Roeth, T.E. Harvill, R.W. Klein, G.A. Wicks, D.L. Holshouser, and J.W. McNamara. 1993. NebraskaHERB: Computer-aided postemergence weed management. Coop. Exten. Div., Univ. Nebraska, Lincoln.

Owen, M.D.K. 1994. Impact of crop tolerance specific herbicides on weed management systems: Corn and soybeans. *Proc. North Cent. Weed Sci. Soc.* 49:167-168.

Owen, M.D.K. and K.T. Pecinovsky. 1993. Effect of alternative management strategies on weed populations. *Proc. North Cent. Weed Sci. Soc.* 48:97-98.

Pareja, M.R. and D.W. Staniforth. 1985. Seed-soil microsite characteristics in relation to weed seed germination. *Weed Sci.* 33:190-195.

Pareja, M.R., D.W. Staniforth, and G.P. Pareja. 1985. Distribution of weed seed among soil structural units. *Weed Sci.* 33:182-189.

Pecinovsky, K.T. and M.D.K. Owen. 1994. Efecto de estrategias alternas de manejo en poblaciones de malezas. *Proc. V Congreso Internacional de Manejo Integrado de Plagas.* 5:164.

Pieters, E. 1994. Use of glufosinate for weed control in genetically transformed plants. *Proc. North Cent. Weed Sci. Soc.* 49:171.

Powles, S.B. and J.A.M. Holtum. 1994. Herbicide resistance in plants. *Biology and Biochemistry*. CRC Press, Inc., Boca Raton, FL. 353 pp.

Schmitzer, P.R., R.J. Eilers, and C. Cseke. 1993. Lack of cross-resistance of imazaquin-resistant *Xanthium strumarium* acetolactate synthase to flumetsulam and chlorimuron. *Plant Physiol.* 103:281-283.

Staniforth, D.W. 1958. Soybean-foxtail competition under varying soil moisture conditions. *Agron. J.* 50:13-15.

Stoller, E.W. and L.M. Wax. 1973. Periodicity of germination and emergence of some annual weeds. *Weed Sci.* 21:574-580.

Swinton, S.M., D.D. Buhler, F. Forcella, J.L. Gunsolus, and R.P. King. 1994. Estimation of crop yield loss due to interference by multiple weed species. *Weed Sci.* 42:103-109.

Thill, D.C., J.M. Lish, R.H. Callihan, and E.J. Bechinski. 1991. Integrated weed management - a component of integrated pest management: A critical review. *Weed Technol.* 5:648-656.

Titi, A.E. 1992. Integrated farming: an ecological farming approach in European agriculture. *Outlook on Agriculture* 21:33-39.

Warnes, D.D. and R.N. Andersen. 1984. Decline in wild mustard (*Brassica kaber*) seeds in soil under various cultural and chemical practices. *Weed Sci.* 32:214-217.

Wells, B.H. 1994. Glyphosate tolerant crop update. *Proc. North Cent. Weed Sci. Soc.* 49:170-171.

Wicks, G.A. 1985. Early application of herbicides for no-till sorghum (*Sorghum bicolor*) in wheat (*Triticum aestivum*) stubble. *Weed Sci.* 33:713-716.

Wilkerson, G.G., S.A. Modena, and H.D. Coble. 1991. HERB: Decision model for postemergence weed control in soybean. *Agron. J.* 83:413-417.

Wrubel, R.P. and J. Gressel. 1994. Are herbicide mixtures useful for delaying the rapid evolution of resistance? A case study. *Weed Technol.* 8:635-648.

Wrucke, M.A. and W.E. Arnold. 1985. Weed species distribution as influenced by tillage and herbicides. *Weed Sci.* 33:853-856.

Wyse, D.L. 1994. New technologies and approaches for weed management in sustainable agriculture systems. *Weed Technol.* 8:403-407.

Yenish, J.P., J.D. Doll, and D.D. Buhler. 1992. Effects of tillage on vertical distribution and viability of weed seed in soil. *Weed Sci.* 40:429-433.

Zimdahl, R.L. 1994. Who are you and where are you going? *Weed Technol.* 8:388-391.

Integrating the Management of Weeds and Impacts on the Environment: High Tech Research or Education Solutions

Roger L. Becker

I. Introduction

Agriculture is more closely scrutinized by the general public today than ever. Environmental concerns associated with pesticide use in crop production are at the forefront of our social and political agendas. Integration of weed management principles, techniques, and technologies into mainstream crop production systems is necessary to address these environmental concerns. No change in an individual production practice can, in and of itself, remedy these concerns. Only a linkage of management practices through whole farm planning will allow sustainable production of food and fiber, while maintaining the integrity of environmental resources. As academicians living in the information age, we tend to seek solutions for complex crop production problems that are information intensive and highly technical. In essence, agriculture has evolved to high information and technology input to achieve low input farming. Low input farming in this context is low input of labor and time, not of chemicals or capital. This paper will explore a research driven, information intensive decision aid approach to integrating weed management and environmental concerns. An appropriate alternative may be increased educational efforts towards adoption of known, improved management practices within a systems thinking process of whole farm planning, and refocusing research towards those educational needs.

II. Background

At face value, addressing environmental concerns associated with weed management seems simplistic. Simply reduce the amount of herbicide use, and the amount of herbicide detected in our natural resources will also be reduced. However, when trying to integrate this principle into current agricultural production practices, the solutions quickly become less straightforward. Thill et al. (1991) reviewed the integration of weed management into the broader array of pest management. Their work outlines a solid foundation from which to integrate weed science into other pest disciplines, but to effectively integrate weed management and environmental impacts, the scope must be broadened even further. Whole farm planning is considered one means to do so (Bridge, 1993; El Titi, 1992).

Whole farm planning in a pure sense is not achievable through a decision aid approach. The reason is the complexity of each discipline that decision aids characterize, and the confounding nature of the inherent variability of each when numerous disciplines are combined for whole farm planning. Whole farm planning is a process, not decision aid tools or specific research outcomes. Whole farm planning is a systems thinking approach that integrates various management needs into a decision making process for which computer software can enhance, but not replace, the critical thinking skills of the farmer. Many farmers already perform a whole farm plan in a sense when developing management strategies. For them, a more formal approach to whole farm planning with educational materials, knowledgeable people, and decision aid tools will improve their management even further. For other farmers, just the process of whole farm planning is a paradigm shift in their approach to farming, and available knowledge, experiences, and research tools will vastly improve their management skills. Most farmers would embrace a process that improves their economic position, and that of the rural community and environment in which they live. That should be the goal of the process of whole farm planning.

Herbicides are currently a vital tool in crop production. Farmers have reduced the risk of crop losses from weed competition through the use of herbicides. However, weed competition still results in an estimated $1.8 billion economic loss in corn (*Zea mays* L.) and soybean (*Glycine max* L. Merr.) annually, using current best management practices that include herbicides (Bridges, 1992). Weeds would cause an estimated $8.9 billion economic loss in these two crops if current best management practices, but no herbicides, were used. In the U.S., herbicides account for an estimated 85% of total pesticide use on the major agronomic crops. Herbicide use on corn and soybean accounted for 82% of the estimated 394 million pounds of herbicides used on major agronomic crops in 1989 (USDA/ERS, 1990).

Herbicides have been detected in water resources from normal agricultural practices, other nonpoint sources (Goodrich et al., 1991), and transport through atmospheric deposition (Nations and Hallberg, 1992; Buser, 1990). Increased human cancer risk and damage to aquatic ecosystems are a concern for herbicides in water resources (Humenik et al., 1987). The movement of herbicides with water can be reduced, but cannot be prevented (Baker and Michelson, 1994). To maintain public trust and support for U.S. agriculture, weed management in crop production must be integrated with management to protect the environment to reduce herbicide contamination of our water resources.

Current agriculture production includes large- or small-scale, high- or low-input farm operations. Herbicides have become an integral part of most row-crop production systems. Total replacement of herbicides with other weed management technologies is not possible without major restructuring of labor allocation or development of new technologies in agriculture. Alternatives to herbicides may have impacts on the environment as well. Research to address these issues is underway. During the transition while effective alternatives to herbicides are being developed, improved management of herbicides is necessary. To meet these future challenges, Coble (1994) described the need for changes in public funding, for regional efforts, for private sector involvement, and for development of information delivery systems in weed science research. This paper will expand his recommendations to include the role of education, farmers, and extension to address the environmental concerns arising from the use of herbicides.

III. Implementation Experiences

A few studies have linked environmental concerns such as water quality with weed management. Franklin et al. (1994) approached the problem from the perspective of the farmer. They used available herbicide transport and monitoring data linked to weed management decisions to explore the use of information intensive technologies to minimize the leaching of herbicides to groundwater. They selected herbicides for use in cropping systems based on a relative leaching potential index system (Hornsby, 1992; Hornsby and Augustijn-Beckers, 1991). They compared this relatively simple potential index rating approach to the use of model simulations (MYGRT™ [1]). Local data were used in MYGRT™ to predict the degree and amount of herbicide movement to specific soil depths. Relative leaching potential index ratings did not correlate well with model predictions of herbicide movement to select herbicides for use. In addition, the authors found little relevancy of recommendations based on a risk index rating system to determine the toxicological impact of a herbicide if it did contaminate groundwater, compared with using stated federal Maximum Contamination Levels (MCLs) and the predicted concentration of a herbicide that moved to groundwater based on MYGRT. The overall impact on farmers was that weed control using recommended alternative herbicides based

[1] Electric Power Research Institute's Electric Power Software Center, 930 Hi Line Dr., Dallas, TX 75207.

Table 1. Runoff potential of corn herbicides and their comparative risk to human health from surface water

Pesticide	Relative runoff potential[a]	Risk index[b]
Butylate (Sutan)	0.98	4,200
Atrazine (Aatrex)	1.00	160,000
Cyanazine (Bladex)	1.23	140,000
Simazine (Princep)	1.21	220,000
Alachlor (Lasso)	0.79	210,000
Metolachlor (Dual)	1.17	86,000
Metribuzin (Sencor/Lexone)	0.64	7,200
Propachlor (Ramrod)	0.64	7,200
Nicosulfuron (Accent)	1.37	--[c]
Primisulfuron (Beacon)	1.13	--

[a] Relative to atrazine = 1.0.
[b] Based on concentration in runoff and MCL or equivalent. Higher values incur more risk.
[c] No measure of risk to human health from drinking water was available.
(Adapted from Nelson and Jones, 1994.)

on information intensive, site-specific management cost significantly more than the herbicide program replaced. The authors point out that considering all the uncertainties regarding actual benefits to the environment or humans, farmers would be justified in asking what benefits to society would occur if they assume the burden of higher expenses and management input. The authors suggested that they did not have a credible answer, nor does anyone else at this time.

The EPA develops risk assessment strategies for use in policy and planning that can be used in broad geographic areas, and for watershed management. One such effort was a model simulation for corn production on a Kossuth County, Iowa farm (Nelson and Jones, 1994). This was a strategic planning approach, not the farmer-specific approach taken by Franklin et al. (1994). Herbicide inputs were indexed for risk based on exposure through drinking water as determined by drinking water health guidelines and predicted presence in water resources. There was a striking difference in the estimated risk associated with the use of various herbicides (Table 1). By design, this approach is useful for broad-based policy planning, not site-specific farm level weed management. Broad restrictions on a geographic or watershed basis through a strategic planning approach may unduly penalize farmers who may not contribute significant contamination to the water resource, yet farm within restricted use boundaries, if site-specific issues are not also addressed (Kellogg et al., 1994).

To avoid costly regulation through a voluntary approach to effectively address water quality issues, individual farmers must be able to make real-time, site-specific herbicide management decisions. The reference to real-time site-specific decisions in this paper is choosing the actual management practice used at the time of implementation, as compared with strategic planning that offers management options that may or may not be implemented due to uncontrollable or changing factors. The number of real-time site-specific options are limited and the outcome of a wrong decision more damaging compared with strategic planning, as were the experiences of Franklin et al. (1994). Considering the complexity of the decisions to be made, computer software decision aids seem a logical delivery method to integrate environmental and weed management technologies within a whole farm planning context.

IV. Whole Farm Planning Decision Aids to Integrate Site-Specific Weed Management with Environmental Concerns

To gain adoption of management of weeds and environmental concerns, foreseeable barriers and challenges must be overcome. The potential strengths and limitations of the use of decision aids to integrate information intensive, site-specific management of weeds with the management of environmental concerns in a voluntary approach will be addressed in five topic areas: a) weed ecology and biology decision aids, b) related databases to integrate weed management decision aids, c) toxicology, d) pesticide transport and behavior, and e) whole farm planning potential in the near future.

A. Weed Ecology and Biology Decision Aids

Two major areas of research in weed science are needed to integrate the management of weed and environmental concerns. Characterizing the pest through weed biology and ecology will improve the accuracy of, and reduce herbicide use. Characterizing herbicide behavior in the environment will do so as well, but for different reasons. Some weed scientists have focused on weed biology and ecology; others on characterizing herbicide behavior in the environment. Both research arenas are necessary. However, since many disciplines in addition to weed science are well equipped and are actively researching herbicide transport and behavior, public sector weed scientists could contribute the most with a relatively scarce level of staffing if focused on weed biology and ecology. Other disciplines will characterize herbicide behavior, but precious few will do the weed biology and ecology necessary to address this multi-faceted environmental issue if weed scientists do not.

Weed ecology and biology information can minimize the environmental impacts of herbicide use with improved targeting of application through information intensive techniques such as weed scouting, defining weed emergence patterns, and defining weed seed bank dynamics. Figure 1 shows typical components of current weed management decision aids. Current weed management decision aids are narrowly focused compared to whole farm planning. Despite this narrow focus, implementing a weed management decision aid program in and of itself is difficult. Getting resources and personnel to maintain a decision aid program once up and

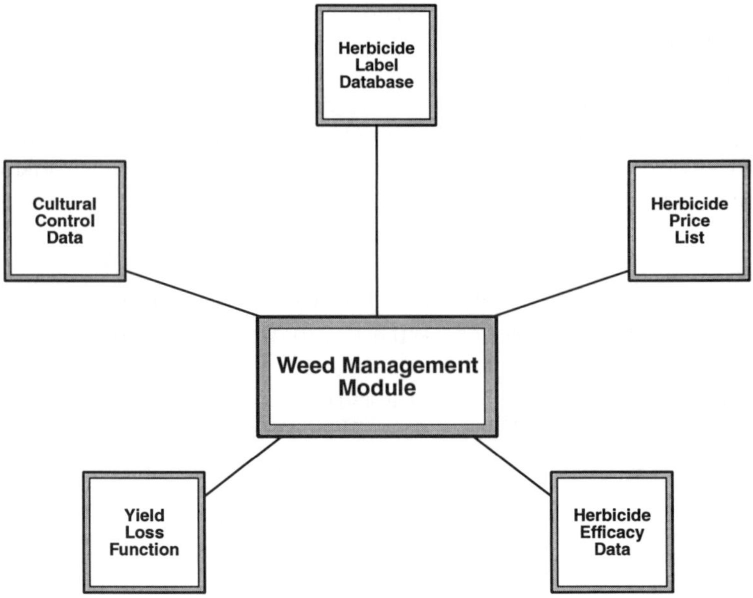

Figure 1. Components of current weed management decision aid efforts.

running is even more challenging. One begins to glimpse the daunting task ahead to broaden the scope to develop a whole farm planning decision aid.

Several herbicide decision aid software programs are in use or under development (Table 2). Programs currently used in agricultural production that address yield loss and use weed thresholds are HERB from North Carolina State University (Wilkerson et al., 1991; Wilkerson et al., 1995) with adaptations of HERB used by other states, and NebraskaHERB from the University of Nebraska (Mortensen et al., 1993) with ancestral linkages to HERB. Both programs have a track record in the marketplace and are currently expanding to include soil applied herbicide use and to incorporate environmental assessment modules to complement current efficacy and profit-based herbicide use recommendations. Another program with implementation potential is the General Weed Management model (GWM) (Wiles et al., 1994). GWM has evolutionary ties to WEEDSIM (Swinton and King, 1994) and WEEDCAM (Lybecker et al., 1994).

Weed thresholds can be used successfully in the short term to reduce herbicide use (Coble, 1985; Coble and Mortensen, 1992; King et al., 1986; Mortensen and Coble, 1995; Swinton and King, 1994; and Swinton et al., 1994). However, few field studies have explored the long-term implications of tolerance of low-level weed infestations based on the use of thresholds in a yearly economic yield loss assessment. This could result in a seed bank buildup and eventual increases in weed populations (Lindquist et al., 1995).

Prophylactic use of herbicides in cereal crops at reduced rates were compared with threshold driven spray, no spray decisions in a study in in England (Lawson,

Table 2. Examples of weed management software

CORNHERB/SOYHERB	Michigan State Univ., East Lansing
GWM	USDA/ARS, Colorado State Univ., Fort Collins
HERB/PreHERB	North Carolina State Univ., Raleigh
HERBICIDE SELECTOR	Ohio State Univ., Columbus
NebraskaHERB/WeedSOFT	Univ. of Nebraska, Lincoln
SELOMA	Univ. Bari, Metaponto, Italy
WEED	North Carolina State Univ., Raleigh
WEEDCAM	USDA/ARS, Colorado State Univ., Fort Collins
WEEDSIM	Univ. of Minnesota, St. Paul

1994). After four years, the yearly prophylactic application of herbicide at reduced rates out-performed the use of thresholds. Prophylactic herbicide use returned higher profits and had lower weed seed bank populations after four years of study. In addition, over 80% of the locations studied with the threshold management approach did not reduce herbicide use. Due to the low economic tolerance of weeds in cereal grain cropping systems, a spray, no spray assessment generally resulted in the decision to spray. However, the use of thresholds to assess herbicide application is critical. As threshold technologies improve, their use will become more common. If the decision is to use a herbicide, the use of reduced rates should also be explored.

Despite advances in weed modeling and use of thresholds, extension weed scientists typically feel more inclusive programs are needed. Programs listed in Table 2 offer attributes that would be useful to construct a comprehensive, regional weed decision aid, building on the HERB and NebraskaHERB efforts (Becker et al., 1995). The components required for comprehensive, regional weed management decision aid strategic planning tools are shown in Figure 2. This decision aid includes all of the components in Figure 1. Additional critical components are anti-quality factors caused by weeds such as eastern black nightshade (*Solanum ptycanthum*) in soybean, or poisonous or high fiber weeds in forage crops. Inclusion of cultural and alternative weed control technologies, herbicide resistant crops, and herbicide resistant weed management modules is needed. In addition, environmental and toxicology modules that address human toxicology, leaching, surface runoff, atmospheric deposition, and nontarget organism toxicity are needed. Weed identification, an applied component often overlooked, must be included.

Perhaps the most important module lacking in current weed management software is that of management risks assumed by the producer. Even if we had complete, accurate information on weed biology, thresholds, seed banks, alternative weed control strategies, and environmental impacts associated with herbicide use, weed management decisions by producers are strongly influenced by the need to manage economic risk and labor constraints. These management risks include variables that are uncontrollable in biological systems such as climate and weather, the number of field working days available to perform weed management operations, the time and labor available in a farming unit, cost, and perhaps most

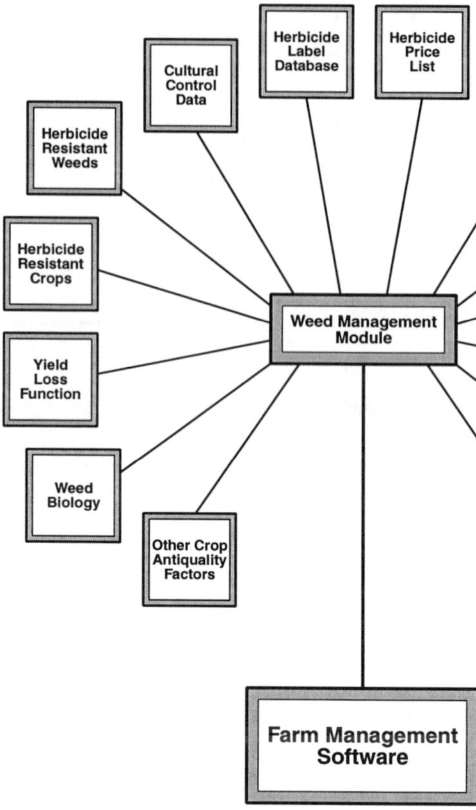

Figure 2. Weed management decision aid requirements for comprehensive planning.

importantly, the consistency or dependability of a management practice over time. The use of HERB to determine herbicide use in soybeans provided reasonable prediction for an economic return approximately 60% of the time (Monks et al., 1995). Management of risks and consistency were discussed by Becker and Staniforth (1981), Gunsolus (1990), and Gunsolus et al. (1994) relevant to herbicide and tillage alternatives for weed control. Well-conducted science can enhance, but not replace, the management skills nor eliminate the economic constraints farmers integrate into successful weed management programs.

Current farming systems typically operate within a narrow profit margin. Inconsistencies of threshold use and of biological alternatives such as smother or cover crops pose management challenges for widespread adoption by farmers. As these technologies build a performance history, and therefore dependability across varied environments and cropping systems, they will gain acceptance by farmers. However, considerable time will pass before these technologies will greatly impact the current use of herbicides.

Efforts in weed ecology and weed biology must be linked with private efforts to achieve whole farm planning targets to improve the speed and breadth of knowledge acquisition. The private sector typically lacks the infrastructure, personnel, and profit guarantees to generate regional weed ecology and biology databases. Businesses that do will approach universities individually to develop their proprietary databases. Weed scientists should combine efforts and provide an information clearing house to which various industry groups could bring their issues and funding for development of public databases on weed ecology and biology. Otherwise, we will get piecemeal grants for "information screening" from industry much like current herbicide screening grants of the present.

This trend has already begun. Private sector software that will evolve without a cooperative effort will be marketed as the best weed management module because it contains brand X yield loss functions of university Y, for example. Farmers and crop consultants will then come to extension and applied research weed scientists for information to determine which product to purchase. This will hinder team efforts as weed scientists struggle to meet these demands and those of their specific weed ecology and biology grant requirements. This will slow advances in weed ecology and biology as the research needs are too complex and demanding for the limited number of public sector weed scientists to address individually. With the limited time and resources available to the few weed scientists available, we should proactively be working towards regional databases positioned to be used by the private sector. Otherwise, weed science resources will be depleted trying to validate each other's information marketed by company X in our local environments in a reactive mode, further detracting from our mission of generating databases robust enough to be used in the real world.

If public weed scientists succeed in proactively developing a weed biology and ecology information clearing house with robust regional databases, the private sector will be more comfortable in taking risks to move to information product development and marketing. The private sector can distinguish their whole farm planning software in the marketplace through benefits of their program based on their business strengths such as accounting and tax modules, yet support regional public databases on weed ecology and biology. Combined funding from public and private sources could flow through a Weed Science Society of America Weed Biology and Ecology Information Clearing House. This clearing house could attract funds for regional projects and expedite the generation of information to drive weed management decision aids. Public sector weed scientists should not sell ourselves short on the potential to influence the direction of private sector development and implementation of research-based information systems. Weed scientists can influence how private and public resources will be focused. The choice to do so is ours alone.

B. Related Databases to Integrate Weed Management Decision Aids

To gain adoption through effective integration of weed management and environmental concerns into cropping systems, a broad array of supportive databases must be developed. A user-friendly, computerized national pesticide label database in a matrix format is lacking, and is a critical limitation to development of site-specific environmental and pest management programs. Currently, weed scientists struggle every year to update herbicide label information for extension bulletins and applied computer software. Many other agencies, companies, and individuals compile this information annually for other efforts. Tremendous resources and time are essentially wasted because of the lack of a national database that could be accessed by all. Special needs such as Section 24 (C), Section 2 (ee) and Section 18 labels must be included to make the database comprehensive. Preliminary discussions have occurred between the EPA and the National Crop Protection Association (NCPA) towards this goal. Historically, internal resources and computer hardware are lacking within the EPA to develop this program. Collaboration between herbicide manufacturers and the EPA must be expedited to resolve this important issue. Information intensive management of herbicides, whether for site-specific, precision, or prescription farming, is not possible until this impediment is resolved.

Table 3 lists several pesticide label databases. NPIRS and PEST-BANK™ provide pesticide label reference information in a text format. The Pesticides Properties Database (PPD) (Herner, 1992) is probably the most inclusive database containing information on pesticide half-lives and other chemical properties. Yet, for regional or farm-specific weed management decision aids, regionalized herbicide property databases must be developed. The influence of regional differences in climate, soil type, microbial activity, and soil pH on herbicide properties must be incorporated before site-specific decision aids can be implemented.

Some private sector databases have the potential to link with public sector weed management programs. One such database that formats data in a matrix rather than a text-based format is the Crop Data Management Systems[2] (CDMS). Computerized information compiled and stored in a text format does not allow access and transfer of information within a computer program to derive progressive outcomes building on previous decision choices. Databases must be designed in a matrix format for true decision-making capabilities. CDMS is designed for efficient computer access of pesticide data that is used in other computer program subroutines and has a history of use in the market place.

Climate and soil databases are the most accurate and inclusive of databases discussed in this paper. Working outside these disciplines, I considered the soil database used for the Soil Pesticide Interaction Screening Procedure (SPISP) and the Soils-5 database (Table 4) as ready for use in site-specific environmental decision making. However, people working within this field feel that these databases, like databases in other disciplines, have inadequacies that require further development before use to make site-specific, information intensive decisions. Climate databases have long histories in some locales and relatively complete data sets. Yet, integrating climate databases into other computer programs to add probability or frequency boundaries to herbicide transport, weed biology, or

[2] Crop Data Management Systems, Inc., Hart Building, 423 Fourth Street, Marysville, CA 95901.

Table 3. Examples of public and private sector pesticide label and properties databases

CDMS	Crop Data Management Systems, Inc., Marysville, CA
EPA Environmental Fate One-Liner	USEPA, Washington, D.C.
NPIRS	Nat. Pesticide Information Retrieval System, Purdue Univ., West Lafayette, IN
PEST-BANK™	Silver Platter Information, Inc., Newton Lower Falls, MA
PPD	Pesticide Properties Database, USDA - NRCS/ARS/CES, Washington, D.C.

Table 4. Examples of soil, climate, and hydrology databases

SOILS-5	EarthInfo, Inc., Boulder, CO
SPISP	Soil/Pesticide Interaction Screening Procedure, USDA/NRCS, Washington, D.C.
ClimatDATA	National Climatological Data Center, NOAA/NCDC, Asheville, NC
HydroDATA	U.S. Geological Survey/USDI, MS804 National Center, Reston, VA

management risks related to available field working days will require considerable developmental work.

C. Toxicology

Toxicity of herbicides must be considered in the selection process for site-specific integration of the management of weeds and environmental concerns. Toxicology is heavily influenced by public policy and public perceptions, in addition to scientific findings. Some background on this is necessary before discussing site-specific information needs. Uncertainties in estimating risks from pesticide exposure to humans shape public policy and public perceptions of regulatory needs (Barrett et al., 1993). The challenges and difficulties in determining and applying pesticide exposure human risk estimates to policy is discussed by Beall et al. (1991). From a public policy standpoint, should the goal for protecting water quality be to reduce detectable levels of offending pesticides? Many feel that the detection of pesticides in water resources is not tolerable at any level. It is unrealistic to expect that pesticide use on millions of crop acres will not result in detection in ground and surface water. If intentional nonpoint source introductions of pesticides to the landscape for agricultural production is to continue, chronic toxicity risk assessments must be improved and communicated to the public by epidemiologists and toxicologists, not weed scientists.

Regulatory efforts are underway to remove some of the most acutely toxic pesticides from the market. Regulatory efforts are also underway to reduce the presence of the most commonly detected pesticides in the environment, some of which may have relatively low acute toxicity but pose a high environmental risk due to their widespread presence. The toxicological impacts may be debatable in some cases. A current example of a regulatory approach to reduce the environmental risks of an offending herbicide is to restrict or prohibit the use of atrazine in geographic regions or near vulnerable landscape features. Removing high risk herbicides without consideration of the risks of the weed management replacement alternatives may be substituting one environmental risk for another. Replacing the use of herbicide with an increase in primary or secondary soil tillage may jeopardize the primary goal of surface water protection, a reduction of soil sediment loading. Without viable alternatives to herbicides, removing high risk herbicides in geologically sensitive areas would mean an increase in the use of another herbicide that will likely be detected at some level in water resources. Once analytical techniques improve for new herbicide chemistry, these too will likely be detected at some level. What the toxicological implications of the presence of these replacement herbicides at low levels means will be debatable as well.

Federal and state policies and regulations most visible at the farm level are focused on pesticide use reduction to reduce risk. Simply reducing the amount of herbicides used without targeting individual products based on toxicity to nontarget organisms or to people may not result in actual risk reductions, despite achievement of pesticide use reduction goals. Site-specific management of herbicide use that also assesses toxicity would more effectively reduce risk compared with a general use reduction policy. For site-specific farm management, specific toxicity information must be available. Toxicology databases exist (Table 5), but with limitations to site-specific implementation similar to those discussed previously for databases in other disciplines. Toxicology databases often are for the active ingredient, however, which is only part of the toxicology information needed. To implement site-specific management, exposure to the applicator or nontarget organisms must be based on formulated product, as the most limiting toxicity concern may differ from that when based on active ingredient. The use of glyphosate as the formulated product Roundup™ compared with Rodeo™ is a case in point. The surfactant, not the active ingredient glyphosate, used in Roundup™ is potentially toxic to shellfish in aquatic use as well as a potential eye irritant to the applicator. Rodeo™ is marketed for aquatic use, and surfactants labeled for aquatic use are added by the applicator and have different toxicity as reflected in different signal words. The toxicology decision points are not active-ingredient-based in this example.

The Extension Toxicological Network (EXTOXNET) (Hotchkiss et al., 1989) provides a general reference on pesticide toxicity to humans and nontarget organisms in a text-based format that is available on the Internet. The Integrated Risk Information System (IRIS)[3] is a database used primarily by state and federal health officials. It provides information on systemic and carcinogenic pesticide

[3] USEPA, Washington, D.C. 20460.

Table 5. Examples of pesticide monitoring and toxicology databases

AQUIRE	Aquatic Toxicity Information Retrieval Database, USEPA, Duluth, MN, and USDA/NRCS, Duluth, MN
DATATOX™	Spectrum Research, Inc., Duluth, MN
EXTOXNET	Extension Toxicological Network, Cornell Univ., Ithaca, NY
IRIS	Integrated Risk Information System, USEPA, Washington, D.C.
PESTFATE™	Pesticides: Fate and Persistence, Spectrum Research Inc., Duluth, MN
PMI	Pesticide Monitoring Inventory, USEPA OPP, Washington, D.C.
TERRATOX	USEPA, Washington, D.C.

toxicity. It contains reference doses and chemical potency slopes that have undergone thorough and standardized scientific review and are listed in the *Federal Register*. Values in IRIS are used to derive federal and state health exposure limits for drinking water. Many pesticides are not currently listed in this information resource, however. The Aquatic Information Retrieval Toxicity Database (AQUIRE)[4] is a database evolving within the EPA that offers pesticide toxicology data on nontarget organisms. This database currently includes only fresh and salt water organisms, but the concept and format provide a model for more inclusive nontarget organism databases. For site-specific decision making, the EPA, pesticide manufacturers, and the U.S. Fish and Wildlife Service should pursue development of an inclusive, national computerized pesticide toxicology database and consolidate existing databases analogous to development of a national pesticide label database discussed previously.

The EPA has successfully used toxicological risk indices for policy and regulatory planning on a broad scale (Nelson and Jones, 1994). The Environmental Impact Quotient (EIQ) (Kovach et al., 1992) is an index used for broad, strategic planning. Risk indices can be misleading, however, if applied to site-specific whole farm planning. An example would be the use of pyrethroid insecticides that have a relatively high risk index, due in part to high toxicity to fish. If a pyrethroid insecticide was applied in a habitat with extremely low surface runoff potential, the more appropriate risk question may be the risk to the pesticide applicator. Compared with alternative organophosphate insecticides, pyrethroid insecticides have moderate to low chronic toxicity risks, but relatively low acute toxicity to the applicator. These subtle differences are not readily apparent in indices, which limits the usefulness of indices for site-specific farm management. Teague et al. (1995) studied economic and water quality trade-offs comparing pesticide strategies in alternative production systems. They compared three indices: a) a modification of the EIQ index based on the characteristics of the chemical, b) the chemical environmental index, and c) a chemical concentration index. Both b) and c) have

[4] S. Bradbury and C. Russom, USEPA, Duluth, MN, and R. Schroeder, USDA/NRCS, Duluth, MN.

the addition of an estimate of the concentration of a pesticide in runoff or percolation to weight the toxicity-based EIQ index. The addition of the concentration of pesticide in runoff or percolation offered the greatest reduction in environmental risks without significant income losses.

Human imposition of food production upon the landscape will have some negative impact on the environment. With current pesticide detection capabilities, zero tolerance is not an acceptable framework from which to manage or regulate pesticide use if widespread use continues. Also, toxicity-derived accepted levels of pesticides are continually changing as more information is developed. Therefore, reducing detections of pesticides in a water resource towards a toxicity-based "acceptable" level of contamination, or replacing an "unacceptable" pesticide with an "acceptable" pesticide based on toxicity-derived MCLs may not truly reduce risk. This approach does provide a formal process from which to promulgate regulatory policy, however, and no regulatory alternative other than a general use reduction policy is currently available. In the near term, while databases are being developed that will enable decision aids to make meaningful site-specific decisions based on toxicity, general pesticide use reduction is a useful policy or regulatory goal.

D. Pesticide Transport and Behavior

Considerable research is directed towards modeling the behavior of pesticides in the environment (Table 6). A symposium, The Role of Modeling in Regulatory Affairs, sponsored by the Weed Science Society of America (Wauchope and Duffy, 1992) reviews pesticide behavior models and many issues discussed in this paper. Zubkoff (1992) discussed the regulatory approach of the USEPA and use of models, and was the most inclusive paper presenting an overview of databases and models available for determination of environmental fate of pesticides.

Internally, the EPA and the USDA tend to use pesticide behavior models that evolved through their respective systems. This is understandable, for when my colleagues and I attempted to select a research model to assess surface runoff and leaching to use for real-time decision-making at the farm site-specific level, we quickly became gridlocked with too many viable opportunities, each with its own strengths and weaknesses. Researchers who develop and work with models are quick to point out that no model currently exists that can accurately predict pesticide fate in site-specific applications due to limitations common in modeling efforts, a lack of data to drive capable models or a lack of capable models. A lack of coordinated data formats for existing data, and a lack of a coordinated database development are significant underlying problems.

One program to assess the potential for pesticide movement off-site that has moved beyond research applications is the effort of the Natural Resource Conservation Service (NRCS) with SPISP (Goss, 1992). SPISP provides relative leaching and surface runoff potential ratings for various pesticides based on the Ground Water Loading Effects of Agricultural Management Systems model (GLEAMS) (Knisel et al., 1989) using the Pesticide Properties Database (PPD).

Table 6. Examples of surface runoff and/or leaching models

CMLS	Chemical Movement in Layered Soils (Nofzinger and Hornsby, 1985)
CREAMS	Chemical, Runoff, and Erosion from Agricultural Management Systems (Knisel, 1980)
DRASTIC	Depth to water, recharge, aquifer media, soil media, topography, impact of the unsaturated aqueous zone between the soil media and water table, and conductivity of the aquifer (Aller et al., 1985)
GLEAMS	Groundwater Loading Effects of Agricultural Management Systems (Leonard et al., 1987)
HSPF	Hydrological Simulation Program - FORTRAN (Bicknell et al., 1993)
LEACHMP	Leaching Estimation and Chemistry Model - Pesticides (Wagenet and Hutson, 1989)
NAPRA	National Agricultural Pesticide Risk Assessment (Getter et al., 1993)
NPURGS	National Pesticide/Soils Database and User Decision Support System for Risk Assessment, Ground and Surface Water Contamination, USDA/NRCS, Washington, D.C.
PATRIOT	Pesticide Assessment Tool for Rating Investigations of Transport, AQUA TERRA Consultants, Quray, CO and USEPA, Athens, GA
PRZM	Pesticide Root Zone Model (Mullins et al., 1993)
RZWQM	Root Zone Water Quality Model (Ahuja et al., 1993)
SPISP	Soil/Pesticide Interaction Screening Procedure (Wauchope et al., 1990)
STREAM	Stream Transport and Agricultural Runoff of Pesticides Exposure Assessment (Donigian and Mulkey, 1992)

NRCS soil information is linked to this program combining an assessment of the impact of pesticide selection and the soil to which it is applied to determine the potential for off-site movement. This approach has been used successfully for education efforts in Minnesota (Becker et al., 1990); however, considerable work is needed to regionalize the NRCS effort before it can be used for real-time site-specific decision making.

E. Whole Farm Planning Potential in the Near Future

The concept of site-specific or farmer level management has been discussed several times in this paper. Ultimately to protect water resources, in lieu of broad, geographically based herbicide use restrictions, individual farmers collectively must make specific management choices on their individual fields. In light of the

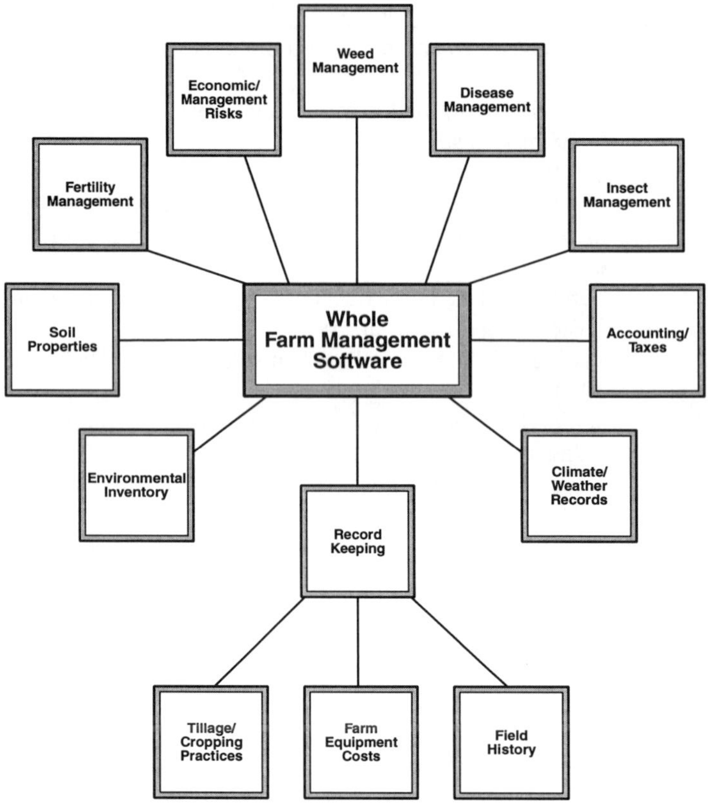

Figure 3. Whole farm planning decision aid for cropping systems.

discussion presented so far, an information intensive decision aid approach to real-time, site-specific management may not be feasible. Whole farm planning decision aid needs for a crop-based farming operation are shown in Figure 3. Note that weed management issues discussed in Figure 2 comprise just a small portion of the management challenges facing individual farmers, all of which are critical to the financial success of the operation. Inclusion of livestock would further increase management complexity and risk for the manager. Considering the research and development needed to overcome the inadequacies of modeling efforts to implement an integrated whole farm planning decision aid, it may be presumptuous to expect a highly technical, information intensive decision aid program will be available in the near future. Remember, whole farm planning is a process, not a decision aid tool.

We are on a collision course with the patience of the public to address environmental issues and our ability to deliver the information intensive technologies necessary for a successful voluntary approach to address contamination of the environment from herbicide use. If the patience of the public diminishes, a regulatory approach with the requirement of prescription farming to address these concerns may be implemented. Precedence exists for this in California. Requirements of prescription farming would expedite the development of decision aids and the expansion of the crop consultant business. It also would add to production costs, which may or may not return a profit to producers or benefits to the environment because of the information gaps discussed.

To expeditiously develop systems to meet these challenges voluntarily, a different focus is required of public education and research. The current research model is to have several different research teams working on separate projects with common research goals to build robustness and constructive criticism within the discipline. That serves basic research well, but is it the appropriate system to address the applied problem solving task at hand? Historically, it has been a more achievable and fundable research goal to generate a research model than to generate the complex and elusive biological information needed to implement and maintain that model. We may not have ample public resources to support duplicative applied research in some areas and underfund critical research needs in other areas. But how should efforts be refocused and consolidated?

This brings us back to the issue of how do farmers make decisions? The overriding concern for the farmer may not be detailed research information, but rather the overall management risks involved in the farming operation. With the degree of uncertainty of economic risk management, what level of accuracy do we need in individual modeling efforts to implement site-specific management? The issues of trade-offs between model accuracy and capabilities to address the environmental impacts have been reviewed (DeCoursey, 1992; Donigian and Carsel, 1992; Wauchope, 1992). The more accurate a model becomes, the more information it demands, and the more likely it cannot be used for site-specific, integrated farm decision making since the necessary data are not available, and will not be available in the foreseeable future. A more limiting concern may be that even if the necessary data could be gathered, the inherent variability in biological systems may relegate the use of models to only strategic planning, not real-time site-specific decision making. Such may be the case for economic yield loss functions for weed competition or for pesticide movement. In nonirrigated systems, the variability in rainfall alone may prevent the use of models for real-time, site-specific decision making.

One solution is to refocus research towards farmer needs and build research programs from this grass roots foundation back through the system to answer those needs. Such is the rhetoric in research funding circles, but it is rarely practiced. Farmer focus groups should be the starting point to define the problem, and consequently the research questions. Truly applied efforts would be most successful by linking farmer-driven research efforts across universities and agencies towards a common goal. Various teams could focus on contributions they could make to the overall goal and thereby achieve a higher quality and quantity of data in a shorter amount of time.

Success will require a rethinking of how funding is allocated to land grant universities and agencies for outreach and implementation of research findings. Competitive grants philosophy by its very design stifles team efforts. Team efforts must truly be supported and funded at higher levels than individual projects. This means a departure from the practice of numerous allocations of inadequate resources to instead adequately fund a few, broadly integrated team projects. This also requires cooperation at the federal agency level, such as the USDA and the USEPA agreeing on, and directing resources towards a joint program that will be utilized at the farm level to address environmental issues. An attempt at a comprehensive whole farm planning model is underway within the USDA Agricultural Research Service (ARS) (Ahuja et al., 1994). In light of the scope of the task at hand, I would suggest that even this group, the most comprehensive and integrated at present, is still too narrowly defined and not integrated broadly enough among universities, agencies, extension, and farmers.

Federal and state agency and university personnel have a presence in the local counties to build cooperative efforts with farmers. The NRCS and Soil and Water Conservation Districts (SWCDs) will likely be the ones charged with enforcing regulations, or promoting incentive programs to address environmental issues at the farm level. The USDA Cooperative Extension Service (CES) will be charged with education efforts. These should be the lead public agencies to build inclusive infrastructure at the county level and facilitate the development of the process of whole farm planning. These groups can then communicate research tools needed within this process to integrated national and regional research teams to develop appropriate research-based technologies.

Implementation of the whole farm planning process would more appropriately be termed product development if in the private sector. Public sector implementation to gain adoption, and to support efforts once adopted is and will continue to be the most challenging of any impediments to whole farm planning. Education and decision-making programs fall into the gray area where the public sector starts to retreat from support of implementation programs that are viewed as projects that should be in the private sector, yet private sector incentives are tenuous at best.

Private and public sector programs that have potential to evolve into comprehensive whole farm planning are listed in Table 7, although individually this may or may not be their goal. These programs are notable because they bring a diverse array of farm management issues together through various delivery technologies. They start at the farmer or dealer/consultant level addressing management needs. They generally are integrating management needs across disciplines. Some have a business infrastructure to develop, market, and support their programs. Decision aids are ideal products for agribusiness in the information age. One growth area for agribusiness in the future is information intensive, service oriented products, rather than traditional manufactured products. These are logical entry points to build cooperative programs within the public and private sector to develop joint visions, goals, funding, and individual roles to expedite the development of whole farm planning in the information age.

Table 7. Examples of private and public sector efforts with potential for whole farm planning

AgriMAPP	AgriCAD, DeKalb, IL
AgriSource	Cenex/Land O'Lakes, St. Paul, MN
AgrData Resources, Inc.	Milwaukee, WI
CDMS	Crop Data Management Systems, Inc., Marysville, CA
INFIELDER	Monsanto Co., St. Louis, MO
MAX	Farming for Maximum Efficiency, Ciba-Geigy Corp., Agricultural Division, Greensboro, NC; Successful Farming Meredith Corp., Des Moines, IA; CTIC (Conservation Technology Information Center), West Layayette, IN
PEET	Pesticide Economic and Environmental Tradeoffs, Univ. of Florida, Gainesville
Planetor	Center for Farm Financial Management, Univ. of Minnesota, St. Paul

Risk indices, economics, soils information, and pest management are linked in some decision aids such as Planetor[5] and PEET (Hoag et al., 1994). These appear to be the most progressive, broad-based efforts to date that integrate weed management and environmental concerns and that are being used in the field. These could provide the base program to which improved modules in the topic areas discussed are added once limitations are overcome, with the overall goal of integration into a whole-farm planning decision aid as depicted in Figure 3.

V. Research or Education Solutions

This brings us to a closing paradox, what to do to integrate the management of weeds and the environment. Demands are being placed on resources for research and education, and both are worthy of increased support. For the near term, say the next five years, more progress may be made towards fulfilling society's demands for action by increasing efforts and funding in education and implementation instead of research on information intensive technologies to address the environmental problems at hand. An example is the Future 2002 Program proposed to achieve herbicide use reduction in Ontario, Canada (Hamill et al., 1994). Atypically, funding for outreach and implementation of pest management would be increased to $4.4 million, at least respectable compared with the proposed $5.6 million investment in research. Targeted funding for research typically exceeds that for

[5] Planetor, A Whole Farm Planning System for SMART (Sustaining and Managing Agricultural Resources for Tomorrow). Center for Farm Financial Management, Minnesota Extension Service, University of Minnesota, St. Paul, MN 55108.

education by a ratio of 5:1 to 10:1. In the Ontario plan, in addition to addressing research needs, considerable resources are directed towards adding personnel to implement programs and gain adoption of applied management principles by farmers, a traditional extension role.

This assumes that many water quality situations could be improved through innovative programs to work with individual farmers at the watershed level, implementing research and extension information that already exists. Widespread adoption by farmers of weed management practices that protect water resources from contamination by herbicides has not yet occurred. This is in part due to a lack of readily available management information for site-specific situations. Information intensive, research based decision aids to do this have been slow in coming and are better suited to strategic planning rather than real-time decision making. Currently, the best decision aid to integrate management of weeds and the environment for real-time decision making is a knowledgeable person.

Immediate, visible progress through voluntary implementation of existing research knowledge would improve our image with nonagricultural sectors, and improve the quality of water resources. Best management practices described in Baker and Michelson (1994) are examples of research-based management information that could gain wider adoption through education programs. This requires people such as farmers, consultants, and educators to integrate weed management and environmental concerns on specific farms, not new research programs.

Programs in place such as the Anoka Sand Plain Demonstration Project in Minnesota, one of eight in the original President's Water Quality Initiative Program, are using this approach. A similar success is the Nutrient and Pest Management Project in Wisconsin that provides fertility and pest management education and consultation to farmers, targeting water quality issues. These are examples of investing in personnel to implement research based information into specific farming systems to voluntarily address water quality issues. Numerous educational materials have been developed for Wisconsin, one of which is Nutrient and Pesticide Best Management Practices for Wisconsin Farms (Schmidt and Sturgul, 1989). Becker et al. (1990) developed educational materials that were also included in Minnesota's USDA/NRCS Field Office Technical Guide to address environmental concerns. Training sessions were conducted across the state as a joint effort with Minnesota Extension, NRCS, and SWCD personnel. A review of the impact of herbicides on water resources and best management practices to prevent contamination that details policy, research, and educational needs has recently been completed for Minnesota (Yenish et al., 1995). The New York City - Watershed Agriculture Council has set a goal of securing whole farm plans from 95% (500) of farms in its watershed by 1997. Economic mitigation and education are important components of New York's program.

This is not to say that research efforts should be gutted. I am suggesting that to expedite implementation of research-based, management programs to address environmental issues, resources currently directed toward research could be refocused to accomplish key applied research needs economically in a shorter time frame. Building research programs for strategic planning at the whole farm level must be achieved long before the use of decision aids to make real-time site-specific management decisions can be fully implemented. Considering narrow profit margin constraints and consistency of success required for adoption of management options in corn and soybean crop production, site-specific management will require a level of accuracy and consistency that is not possible in the near future. In the near term, the addition of educational modules to whole farm strategic planning computer decision aids may be the most valuable contribution of researchers to farmers. Building in educational subroutines into strategic planning software tools with the learning skills of the customer in mind will educate farmers and consultants as to why decisions were made, and will build their capacity to assess management challenges that arise. Otherwise, the rote output of decision aid recommendations will inevitably fall short when uncontrollable variables or lack of data limit decision aid success.

VI. Conclusions

To effectively integrate the management of weeds and environmental concerns, research and education efforts in the public sector must be integrated through an agreed-upon shared mission, goals set to achieve that mission, and funds redirected towards broad-based team efforts to achieve that mission. Regional efforts by weed scientists to develop weed ecology and biology databases must continue, be expanded, and be funded. Public sector databases must be developed for pesticide label use and toxicity to allow their inclusion in decision-making tools. Regionalized pesticide transport models must be developed that utilize local soil and climate databases. The pesticide properties databases must be regionalized for pesticide transport models to be useful in local site-specific management. Whole-farm planning is a nebulous task. It is a process, not a research outcome, and research should develop tools to support that process. Whole-farm planning is already being practiced by all successful farmers. The focus of research and education efforts should be to enhance this systems-thinking approach. Computer decision aid programs are a vital subcomponent within this systems-thinking approach. The human element cannot be overlooked in such a broad-based, integrated decision-making process. While research is underway to provide strategic whole-farm planning tools, immediate public and producer concerns can be met in part by increased education efforts through implementation of the process of whole-farm planning. Only then can we truly integrate the management of weeds and environmental concerns into a whole-farm planning system that addresses the questions society is demanding of us.

References

Ahuja, L.R., D.G. DeCoursey, B.B. Barnes, and K.W. Rojas. 1993. Characterization of macropore transport studied with the ARS root zone water quality model. *Trans. ASAE* 36:369-380.

Ahuja, L.R., P.N.S. Bartling, D. Edmunds, J. Ascough II, M. Shaffer, J. Hanson, G. McMaster, L. Wiles, and D. Woodson. 1994. Development of a decision support system for farmers and ranchers in the great plains, 1994 Report. USDA/ARS Northern Plains Area, National Resources Research Center, Fort Collins, CO 80526.

Aller, L., B. Truman, J.H. Lehr, and R.J. Petty. 1985. DRASTIC: A standardized system for evaluating groundwater pollution potential using hydrogeologic settings. U.S. Environmental Prot. Agency, EPA/600/2-85/018. Washington, D.C.

Baker, J.L. and S.K. Michelson. 1994. Application Technology and Best Management Practices for Minimizing Herbicide Runoff. *Weed Technol.* 8:862-869.

Barrett, M.R., W.M. Williams, and D. Wells. 1993. Use of ground water monitoring data for pesticide regulation. *Weed Technol.* 7:238-247.

Beall, G.A., C.M. Bruhn, A.L. Craigmill, and C.K. Winter. 1991. Pesticides and your food: How safe is "safe"? *Calif. Agric.* 45:4-11.

Becker, R.L. and D.W. Staniforth. 1981. Cost-benefit assessments of herbicide use. *Proc. N. Cent. Weed Contr. Conf.* 36:68-71.

Becker, R.L., D. Herzfeld, K.R. Ostlie, and E.J. Stamm-Kotovich. 1990. Pesticides: Surface runoff, leaching, and exposure concerns. Minnesota Extension Service, University of Minnesota, St. Paul, MN. AG-BU-3911. 32 p.

Becker, R.L., D.A. Mortensen, A.R. Martin, T.E. Harvill, G.G. Wilkerson, H.D. Coble, and K.A. Renner. 1995. The Need for a Multi-state Decision Aid for Weed Management in Corn and Soybean. *Proc. Weed Sci. Soc. Amer.* 35:96.

Bicknell, B.R., J.C. Imhoff, J.L. Kittle Jr., A.S. Donigian Jr., and R.C. Johanson. 1993. Hydrological simulation program - FORTRAN. User's manual for release 1.0. USEPA, Athens, GA.

Bridge, G. 1993. Is whole farm planning the answer? *J. Soil Water Conserv.* 48:295-298.

Bridges, D.C. (ed.). 1992. Crop losses due to weeds in the United States. 1992. Weed Sci. Soc. of Am., Champaign, IL. 398 p.

Buser, H. 1990. Atrazine and other *s*-triazine herbicides in lakes and rain in Switzerland. *Environ. Sci. Technol.* 24:1049-1058.

Coble, H.D. 1985. The development and implementation of economic thresholds for soybeans. p. 295-307. In: R.E. Frisbie and P.L. Adkisson (eds.), *Integrated Pest Management on Major Agricultural Systems*, Texas A & M University.

Coble, H.D. 1994. Future directions and needs for weed science research. *Weed Technol.* 8:410-412.

Coble, H.D. and D.A. Mortensen. 1992. The threshold concept and its application to weed science. *Weed Technol.* 6:191-195.

DeCoursey, D.G. 1992. Developing models with more detail: Do more algorithms give more truth? *Weed Technol.* 6:709-715.

Donigian, A.S. and R.F. Carsel. 1992. Developing computer simulation models for estimating risks of pesticide use: research vs. user needs. *Weed Technol.* 6:677-682.

Donigian, A.S., Jr. and L.A. Mulkey. 1992. STREAM - An exposure assessment methodology for agricultural pesticide runoff. Chapter 18. In: J.L. Schnook (ed.), *Fate of Pesticides and Chemicals in the Environment*, John Wiley & Sons, Inc., New York, NY.

El Titi, A. 1992. Integrated farming: an ecological farming approach in European agriculture. *Outlook Agric.* 21:33-39.

Franklin, R.E., V.L. Quisenberry, B.J. Gossett, E.C. Murdock. 1994. Selection of herbicide alternatives based on probable leaching to ground water. *Weed Technol.* 8:6-16.

Getter, W.F., S. Plotkin, J.K. Bagdon, and E.S. Hesketh. 1993. National agricultural pesticide risk assessment. Proc. USA/CIS joint conference on environmental hydrology and hydrogeology, Industrial and agricultural impacts on the hydrologic environment. Y. Eckstein and A. Zaporozec (eds.), Water Environment Federation, Alexandria, VA. Vol. 2:179-191.

Goodrich, J.A., B.W. Lykins, Jr., and R.M. Clark. 1991. Drinking water from agriculturally contaminated groundwater. *J. Environ. Qual.* 20:707-717.

Goss, D.W. 1992. Screening procedure for soils and pesticides for potential water quality impacts. *Weed Technol.* 6:701-708.

Gunsolus, J.L. 1990. Mechanical and cultural weed control in corn and soybean. *Amer. J. Alt. Agric.* 5:114-119.

Gunsolus, J.L., D.D. Buhler, W.E. Lueschen, and R.M. Schmitt. 1994. Integrated weed management techniques to reduce herbicide inputs and economic risk. *Amer. Soc. Agron. Abstr.* 1994:80.

Hamill, A.S., F.A. Surgeoner, and W.P. Roberts. 1994. Herbicide reduction in North America: in Canada, an opportunity for motivation and growth in weed management. *Weed Technol.* 8:366-371.

Herner, A.E. 1992. The USDA-ARS pesticide properties database: A consensus data set for modelers. *Weed Technol.* 6:749-752.

Hoag, D.L., A.G. Hornsby, and D.L. Nofziger. 1994. PEET: Pesticide Economic and Environmental Tradeoffs. Proc. Int. Conf. Computers in Agric., Feb. 6-9, Orlando, FL. Florida Cooperative Extension Service, Univ. of Florida, Gainesville, FL.

Hornsby, A.G. 1992. Site-specific pesticide recommendations: The final step in environmental impact prevention. *Weed Technol.* 6:736-742.

Hornsby, A.G. and P.W. Augustijn-Beckers. 1991. Handbook, Managing pesticides for crop production and water quality. Florida Coop. Ext. Serv., Inst. Food Agric. Sci., Univ. Florida, Gainesville, FL.

Hotchkiss, B.E., J.W. Gillett, M.A. Kamrin, J.W. Witt, and A. Craigmill. 1989. EXTOXNET, Extension Toxicology Network. Cornell Univ., Ithaca, NY.

Humenik, F.J., M.D. Smolen, and S.A. Dressing. 1987. Pollution from nonpoint sources. *Environ. Sci. Technol.* 21:737-742.

Kellogg, L.R., M.S. Maizel, and D.W. Goss. 1994. The potential for leaching of agrichemicals used in crop production: A national perspective. *J. Soil and Water Cons*. 49:294-298.

King, R.P., D.W. Lybecker, E.E. Schweizer, and R.L. Zimdahl. 1986. Bioeconomic modeling to simulate weed control strategies for continuous corn (*Zea mays*). *Weed Sci.* 34:972-979.

Knisel, W.G. (ed.). 1980. CREAMS: A field-scale model for Chemical, Runoff, and Erosion from Agricultural Management Systems. U.S. Dept. Agric., Sci. Educ. Adm., Conserv. Rep. No. 26.

Knisel, W.G., R.A. Leonard, and F.M. Davis. 1989. GLEAMS User Manual. Southeast Watershed Res. Lab., USDA/ARS, Tifton, GA. 25 p.

Kovach, J., C. Petzoldt, J. Degni, and J. Tette. 1992. A method to measure the environmental impact of pesticides. New York's Food and Life Sciences Bulletin, Geneva, NY, 8 p.

Lawson, H.M. 1994. Changes in pesticide usage in the United Kingdom: Policies, results, and long-term implications. *Weed Technol.* 8:360-365.

Leonard, R.A., W.G. Knisel, and D.A. Still. 1987. GLEAMS: Groundwater Loading Effects of Agricultural Management Systems. *Trans. ASAE* 30:1403-1418.

Lindquist, J.L., B.D. Maxwell, D.D. Buhler, and J.L. Gunsolus. 1995. Velvetleaf (*Abutilon theophrasti*) recruitment, survival, seed production, and interference in soybean (*Glycine max*). *Weed Sci.* 43:226-232.

Lybecker, D.W., E.E. Schweizer, and P. Westra. 1994. WEEDCAM Manual. Transfer weed management expert system technology for reduced corn herbicide use to farmers, extension agents and crop consultants. July 1994. Colorado State Univ., USDA-ARS, Fort Collins, CO.

Monks, C.D., D.C. Bridges, J.W. Woodruff, T.R. Murphy, and D.J. Berry. 1995. Expert system evaluation and implementation for soybean (*Glycine max*) weed management. *Weed Technol.* 9:535-540.

Mortensen, D.A. and H.D. Coble. 1995. Economic thresholds for weeds: management implications. In: L.G. Higley and L.P. Pedigo (eds.), *Economic Thresholds for Integrated Pest Management*. University of Nebraska Press (in press).

Mortensen, D.A., A.R. Martin, R.W. Roeth, T.E. Harvill, R.W. Klien, G.A. Wicks, R.G. Wilson, D.L. Holshouser, and J.W. NcNamara. 1993. NebraskaHERB User's Manual, CP-11, Ver. 3.0., December 1993, Dept. of Agronomy, Univ. of Nebraska, Lincoln, NE.

Mullins, J.A., R.F. Carsel, J.G. Scarbrough, and A.M. Ivery. 1993. PRZM-2, A model for predicting pesticide fate in the crop root zone and unsaturated soil zones: User's manual for release 2.0. EPA/600/R-93/046. USEPA, Athens, GA.

Nations, B.K. and G.R. Hallberg. 1992. Pesticides in Iowa precipitation. *J. Environ. Qual.* 21:486-492.

Nelson, H. and R.D. Jones. 1994. Potential regulatory problems associated with atrazine, cyanazine, and alachlor in surface water source drinking water. *Weed Technol.* 8:852-861.

Nofzinger, D.L. and A.G. Hornsby. 1985. Chemical movement in soils: IBM PC User's Guide. FL Coop. Ext. Serv. Circ. No. 654. Univ. of Florida, Gainesville.

Schmidt, K.P. and S.J. Sturgul (tech. eds.). 1989. Nutrient and Pesticide Best Management Practices for Wisconsin Farms. WDATCP Tech. Bullet. ARM-1., Univ. of Wisconsin and Wisconsin Dept. Agric. Trade and Consumer Prot., Madison, WI. 174 p.

Swinton, S.M. and R.P. King. 1994. A bioeconomic model for weed management in corn and soybean. *Agric. Syst.* 44:313-335.

Swinton, S.M., D.D. Buhler, F. Forcella, J.L. Gunsolus, and R.P. King. 1994. Estimation of crop yield loss due to interference by multiple weed species. *Weed Sci.* 42:103-109.

Teague, M.L., H.P. Mapp, and D.J. Bernardo. 1995. Risk indices for economic and water quality tradeoffs: An application to Great Plains agriculture. *J. Prod. Agric.* 8:405-415.

Thill, D.C., J.M. Lish, R.H. Callihan, and E.J. Bechinski. 1991. Integrated weed management - a component of integrated pest management: a critical review. *Weed Technol.* 5:648-656.

USDA/ERS. AR-17. 1990. Agricultural resources - inputs situation and outlook report. February.

Wagenet, R.J. and J.L. Hutson. 1989. LEACHM: A model for simulating the leaching and chemistry of solutes in the plant root zone. Water Resources Institute, New York, NY. *Contunuum* Vol. 2, June.

Wauchope, R.D. 1992. Environmental risk assessment of pesticides: Improving simulation model credibility. *Weed Technol.* 6:753-759.

Wauchope, R.D. and M.J. Duffy. 1992. Role of modeling in regulatory affairs, Introduction to the symposium. *Weed Technol.* 6:670-672.

Wauchope, R.D., A.G. Hornsby, D.W. Goss, and J.P. Burt. 1990. The SCS/ARS/CES pesticide properties database: I, A set of parameter values for first-tier comparative water pollution risk analysis. In: Pesticides in the next decade: the challenges ahead, Virginia Water Resources Research Center, Blacksburg, VA.

Wiles, L.J., R.P. King, D.W. Lybecker, E.E. Schweizer, and S.M. Swinton. 1994. General Weed Management Model, Ver. 1.0. Staff Paper Series, P94-23. December 1994. Dept. of Applied Econ., Univ. of Minnesota, St. Paul, MN.

Wilkerson, G.G., H.D. Coble, and S.A. Modena. 1995. HERB User's Manual, Ver. 5.0, April 1995, Dept. of Crop Sci., North Carolina State Univ., Raleigh, NC.

Wilkerson, G.G., S.A. Modena, and H.D. Coble. 1991. HERB: Decision model for postemergence weed control in soybean. *Agron. J.* 83:413-417.

Yenish, J.P., R.L. Becker, and J. Spetzman. 1995. Pesticides in Minnesota water resources and best management practices for prevention of contamination. Report to the Minnesota Department of Agriculture, St. Paul, MN (in press).

Zubkoff, P.L. 1992. The use of runoff and surface water transport and fate models in the pesticide registration process. *Weed Technol.* 6:743-748.

Tillage Systems and Crop Residue Management Impacts on Soil Physical Properties: Implications for Weed Management

Birl Lowery and David E. Stoltenberg

I. Introduction

In the past several decades, there has been considerable research and public debate on tillage systems, crop residue management, and pesticide use. One of the main reasons for this flurry of activity is concern for the environment. Retention of crop residue on the soil surface has been shown by numerous researchers to provide positive environmental impacts because of the resulting reduction of soil erosion.

Crop residue management is controlled to a large extent by the type of primary tillage used. There are any number of conservation tillage systems (CT) that result in a considerable amount of residue remaining on the soil surface, such as ridge-tillage, to little remaining with a twisted-shank chisel plow followed by multiple disking and disk systems. Conservation tillage has been defined as a form of non-inversion tillage that retains protective amounts of crop residue on the soil surface (SCSA, 1982). In addition to CT, it is possible to produce crops with no-tillage whereby soil disturbance is minimized and the previous crop residue is retained on the soil surface. However, no-tillage (considered a form of CT) systems require a higher degree of management. When a large percentage of the previous crop residue is left on the soil surface, the micro-environment is altered when compared to tillage systems that bury crop residue. The main properties directly impacted are the soil moisture and thermal regimes. In general, crop residue has a greater albedo (ability to reflect sunlight) and its thermal conductivity is much less than that of mineral soils. The combination of these changes in surface boundary conditions causes reduced soil temperature and evaporation of water from the soil surface. Surface residue acts as a buffer to raindrop impact, thereby reducing soil erosion and water runoff. Therefore, soil water content and profile water recharge are increased when plant residue remains on the soil surface. Surface roughness is another means of reducing runoff by increasing surface storage. Some tillage systems, such as chisel and ridge-tillage, promote surface roughness and retain residue on the surface resulting in multiple means of increasing soil water recharge.

In addition to the impact of crop residue on soil physical properties, soil disturbance by tillage independently produces an array of different physical properties, and this combined with residue effects, results in a very complicated system. The effects of a given tillage system on soil physical properties are strongly influenced by the length of time a system has been in place. There is a critical adjustment or equilibration period for a tillage system during which time the system dynamics become established. This establishment period, although not well defined, must be completed before a conservation tillage system can be operated under a steady-state management plan. Infiltration of water is one of the physical properties that has been found to be strongly related to the length of time a tillage system has been in place. Other properties that have been associated with the age of a tillage system are soil density, strength, and aggregate stability, although the adjustment period and resulting soil properties seem to vary as a function of soil type and climatic conditions.

Over the past several decades, scientists have gained a better understanding of the impact of different tillage systems, and to some extent climatic conditions, on soil physical properties. However, there is not a good understanding of all the mechanisms and interactions among tillage systems, soil physical properties, and weed ecology. In this chapter, we have attempted to present a rudimentary assessment of different tillage/residue systems and their influence on soil micro-climate and weed management/ecology. Tillage impact on the soil physical environment has been divided into several sections for the purpose of presentation. However, we recognize that the impact of all aspects of the micro-environment on weed seed bank dynamics and weed growth are interrelated. Soil temperature and moisture regimes, strength, compaction, and aggregation, along with the chemical environment, are all interrelated with respect to seed demography and plant growth. However, complex interactions of these parameters exceed the investigative ability of scientists in a single discipline. Since scientists tend to not stray much beyond their disciplinary bounds, these subjects are presented separately as this is how they were investigated. In future research, there is a pressing need for an integrated research approach on the impact of various micro-environmental conditions on weed ecology and management.

II. Soil Thermal Regime

A. Temperature

As the percentage of the soil surface covered by plant residue increases, there is a decrease in soil temperature and heat flux (van Wijk et al., 1959; Phillips et al., 1980; Johnson and Lowery, 1985; Al-Darby and Lowery, 1987; Azooz et al., 1995). In some areas, such as the midwestern USA, with inherent cold soil conditions at planting in the spring, cold soil temperatures associated with crop residue from reduced tillage systems result in a negative impact on plant performance. Reductions in soil temperature under different tillage systems can be substantial. Johnson and Lowery (1985) reported 5.9, 2.3, and 1.8°C cooler soil temperatures (differences in maximum temperatures) in the seed zone (5-cm depth) of no-tillage (NT), chisel plow (CH), and ridge-tillage (RT), respectively, when compared with moldboard plow (MB) treatment for corn (*Zea mays* L.) production. The surface corn residue cover that resulted in these temperature differences were 36, 25, and 59% for CH, RT, and NT, respectively.

B. Thermal Properties

The amount of crop residue that remains on the soil surface is an important factor impacting the thermal regime. However, this is not the only cause of differences in soil temperature between tillage systems that leave residue on the soil surface and moldboard plowing, which buries residue and surface-deposited weed seeds. Thermal properties of the plow layer of different tillage systems may differ (Wierenga et al., 1982). Bulk density and water content of the plow layer can be amended by tillage, which in turn influences soil thermal conductivity and heat capacity, and therefore, thermal diffusivity (Allmaras et al., 1977; Hay et al., 1978; Johnson and Lowery, 1985). Volumetric heat capacity (C_V) (the amount of heat required to increase the temperature of a unit volume of soil one degree centigrade) varies among tillage systems. The C_V for NT has been found to be greater than that for MB and other CT systems. The thermal conductivity (a measure of how much heat will be conducted through the soil under a given condition) is often greater in CT, especially NT, than MB (Johnson and Lowery, 1985). Thus, under conditions with the same amount of heat input, more heat may penetrate to greater depths in

NT than MB. This means that there is a potential that the surface horizon will be cooler in NT than MB regardless of residue differences.

C. Plant Performance

Poor crop performance with CT has been attributed to the reduced soil temperature (Burrows and Larson, 1962; Mock and Erbach, 1977; Siemens and Oschwald, 1978; Al-Darby and Lowery, 1986, 1987). There is reduced corn seed germination and early growth with CT compared to MB (Schneider and Gupta, 1985; Al-Darby and Lowery, 1987). Al-Darby and Lowery (1987) reported corn germination rate ranking as MB > RT = CH > NT. This reduced soil temperature has also contributed to reduced early plant growth. Walker (1969) and Barlow et al. (1977) showed that a decrease in soil temperature as small as 1°C would reduce corn growth rate. Assuming a similar impact of temperature on weed seed germination, we would expect a delay in germination and early growth with CT.

D. Weed Seed Germination and Weed Emergence

Although there has been considerable research on the impact of reduced soil temperature on crop germination and growth, there have been few studies relating the impact of CT and associated reduced soil temperature in the soil micro-environment on weed seed germination and weed emergence. In addition to the gaseous environment (oxygen, carbon dioxide, volatile germination inhibitors), light, allelochemicals, seed-related factors (innate dormancy, impermeable seed coat), and depth of seed burial, weed seed germination and dormancy are functions of many soil physical properties including temperature and water regimes (Egley and Duke, 1985; Egley, 1986; Taylorson, 1987; Wilson, 1988). Even though weed seed germination may occur under suitable soil temperature conditions, other physical characteristics of the soil, in addition to those previously noted, such as density, strength, and aggregate stability may prevent or reduce the rate of weed emergence.

Plant growth is closely related to growing degree days (GDD). The minimum soil temperature required for corn germination is 10°C (Cummins and Parks, 1961). The mean seed zone temperature or GDD has been suggested as an index for corn emergence predictions (Gupta, 1985; Schneider and Gupta, 1985; Swan et al., 1987). Schneider and Gupta (1985) found that an average of 40 GDD is necessary for corn emergence. This was consistent for several CT systems over a range of different residue covers.

Similar uses of GDD have been investigated for weed seed germination and weed emergence, but GDD has been a less accurate predictor of weed than crop emergence (Stoller and Wax, 1973; Boydston, 1990; Forcella, 1993). This is attributable to greater variability of weed seed genetic, physiological, and physical characteristics in weed seed banks than in crop cultivars. Unlike crops that have been selected for uniform germination and emergence, weed seed banks contain a complex population of weed seed, usually of several species. Weed seed is found at various depths in the soil profile and may be germinable or nongerminable (viable but dormant). In addition, seed banks in cultivated soils consist of weed seed with a complex age classes. Germinability as well as requirements for germination may vary among these age classes of seed (Simpson et al., 1989). Thus, weed seed banks are influenced by agronomic and environmental factors that result in a complex system in which germination, emergence, and other aspects of weed seed fate (dormancy, predation, mortality) can vary greatly and be difficult to characterize.

Given the aforementioned conditions, it is easy to see why total seasonal emergence of weeds from the seed bank is not predicted accurately by GDD. Forcella (1992) found that annual emergence of yellow foxtail [*Setaria glauca* (L.) Beauv.], green foxtail [*S. viridis* (L.) Beauv.], redroot pigweed (*Amaranthus retroflexus* L.), and common lambsquarters (*Chenopodium album* L.) varied between years, but emergence was related in a parabolic manner to GDD (base 10°C air temperature) in April. At several locations in the Corn Belt, annual emergence of redroot pigweed and common lambsquarters was related inversely to GDD (base 10°C air temperature) and rainfall in April; this response was attributed to induction of secondary dormancy in nondormant seeds by the lack of oxygen and/or high temperatures (Forcella et al., 1992).

The variable relationship between GDD and time of weed emergence has been reported by other scientists. Stoller and Wax (1973) found that cumulative heat units in the soil above 10°C were not correlated with initiation of emergence for several weed species. Emergence time of longspine sandbur [*Cenchrus longispinus* (Hack.) Fern] was not predicted by GDD (Boydston, 1990). Relative to GDD, soil temperature regime appears to predict more accurately the time and rate of weed emergence. Boydston (1990) found that average soil temperature at 2.5 cm was a more accurate predictor of early emergence of puncturevine (*Tribulus terrestris* L.) than GDD. Similarly, a daily germination rate constant based on soil temperature regime and water potential predicted accurately the emergence of velvetleaf (*Abutilon theophrasti* Medicus) (Forcella, 1993) and common lambsquarters (Harvey and Forcella, 1993) (Figure 1).

Even though several models (Bridges et al., 1989; Benech Arnold et al., 1990; Alm et al., 1993; Forcella, 1993; Harvey and Forcella, 1993) have been developed to predict weed seed germination and weed emergence based on soil temperature, most of these models were not developed and have not been validated under CT. Given the amount of data available and models (Schneider and Gupta, 1985; Shaffer and Larson, 1987; Gupta et al., 1991) of soil temperature under different tillage systems, it may be possible to combine the temperature and weed emergence models and predict accurately the emergence of weeds under different CT.

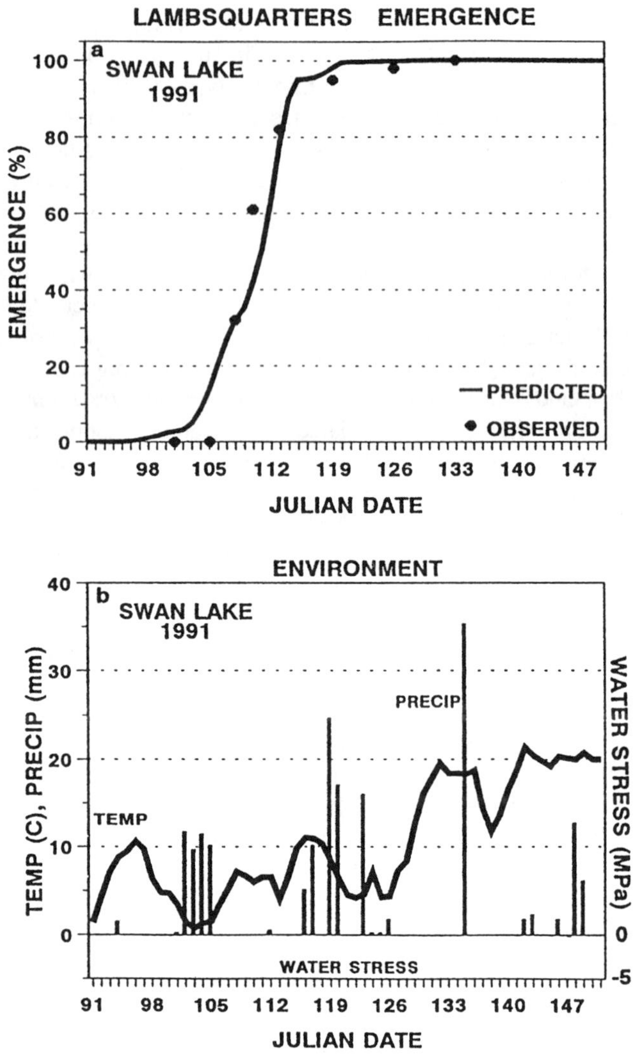

Figure 1. Simulated and observed emergence of common lambsquarters (a) and recorded soil temperature, precipitation, and estimated soil water potential (stress) (b) at the Swan Lake research site near Morris, MN, in April and May, 1991. Correspondence for the simulation = 0.98 ($R^2 = 0.98$, $P < 0.001$). (From Harvey and Forcella, 1993.)

III. Soil Moisture Regime

A. Water Content in Conservation Tillage Systems

Leaving crop residue on the soil surface results in a decrease in soil erosion (Wischmeier, 1973; Laflen et al., 1978). This reduction in erosion is attributed to interception of raindrops by the residue, but there is also an increase in infiltration of rain water. The impact of different tillage systems on infiltration and runoff varies considerably. While some researchers have found decreased runoff from conservation tillage (Romkens et al., 1973; Siemens and Oschwald, 1978; Johnson and Moldenhauer, 1979; Andraski et al., 1985), others have reported increase in runoff (Free and Bay, 1969; Onstad, 1972; Johnson et al., 1979; Langdale et al., 1979; McGregor and Greer, 1982; Burwell and Kramer, 1983; Angle et al., 1984). Regardless of the runoff condition, it is generally accepted that CT results in greater soil moisture during the growing season because of reduced evaporation (Moody et al., 1963; Triplett et al., 1968; Jones et al., 1968, 1969; Blevins et al., 1971; Johnson et al., 1984).

B. Water Content Impact on Weed Seed Germination and Weed Growth

Water retention increases in the surface horizon with NT because of increased bulk density (Johnson et al., 1984). Evaporation of water from the soil surface is decreased because of surface residue. If the impact of greater soil moisture on weed seed fate (germination, predation, and mortality), weed emergence, and weed growth is known, weed management could be improved. It should be noted that with the increase in soil water content more heat is needed to warm soil since the specific heat capacity of water is much greater than that of soil. Little is known about the growth and development of weed species in CT and the associated greater soil water content relative to MB. DeFelice et al. (1988) found that velvetleaf growth rate, dry weight, leaf area, height, and capsule production were similar between MB and NT; however, soil water content or age of these tillage systems were not reported.

Seeds of many weed species have specific requirements for temperature, moisture, and aeration for germination (Egley and Duke, 1985; Egley, 1986; Taylorson, 1987; Wilson, 1988). Under these conditions, germination occurs after a seed imbibes sufficient water. The rate at which water is taken up by the seed is a function of the seed-soil contact and the hydraulic conductivity on the soil. At water content less than field moisture capacity, a fine seedbed can provide conditions for greater water retention and hydraulic conductivity than a coarse seedbed (Johnson et al., 1984; Pareja and Staniforth, 1985). A fine seedbed is often considered essential for good seed-soil contact and germination of small seeds.

Seed germination can be a direct function of soil water content. Even though increased soil temperature is mainly responsible for the germination of many weed species in the spring (Forcella, 1993; Harvey and Forcella, 1993), later emergence of weeds during the growing season is related to rainfall (Stoller and Wax, 1973;

Roberts, 1984). Water stress results in decreased seed germination and plant emergence (Hagin, 1952; Miller and Mazurak, 1958; Roberts and Ricketts, 1979; Pareja and Staniforth, 1985; Weaver et al., 1988). Pareja and Staniforth (1985) found that germination of crop and weed seed in contact with soil macropore spaces was mainly a function of moisture availability. Corn, soybean [*Glycine max* (L.) Merr.], velvetleaf, and giant foxtail (*Setaria faberi* Herrm.) seeded among large soil aggregates (coarse seedbed) showed decreased seedling emergence with less frequent irrigation or decreasing water content. Unaggregated soil (fine seedbed) provided better conditions for germination for these species at low irrigation frequencies. In contrast, when seeded inside large soil aggregates, seedling emergence of these species decreased with increased soil water content (Figure 2). Decreased emergence was attributed to poor aeration inside fully moist soil aggregates.

On a larger scale, soil moisture level has a major influence on the size and nature of the seed bank (Cavers and Benoit, 1989). The relative abundance of individual species is dependent on soil type; seeds of some species germinate more rapidly in well-drained, coarse-textured soil than in medium- or poorly-drained, fine-textured soils. Greater fluctuations in soil moisture in coarse-textured soils causes more hydration-dehydration cycles in weed seeds and can increase germination of certain grass and broadleaf weeds species (Cavers and Benoit, 1989).

IV. Soil Density and Aggregate Size

A. Compaction and Conservation Tillage

There is some degree of confusion over the impact of reduced tillage on soil strength. Many researchers have reported increases in soil bulk density and cone penetration resistance with some of the popular CT (such as NT), while others have shown a decrease (Gantzer and Blake, 1978; Hill and Cruse, 1985; Hilfiker and Lowery, 1988; Unger and Kaspar, 1994). We should note that greater bulk density means there is a decrease in the total porosity, and in fine-textured soils there is often a reduction in the size of pores. Therefore, as compaction increases, soil water content will be greater under unsaturated conditions.

Soil compaction alters the thermal properties of soil. The C_V increases with compaction and is greater in CT compared to MB (Johnson and Lowery, 1985). Soil thermal conductivity is a function of both water content and bulk density; its values become larger as the water content and/or bulk density increase.

B. Soil Aggregates and Compaction Impact on Plant Growth

Plant emergence can be significantly altered by soil compaction. Corn emergence can be delayed as much as 8 days by compaction from heavy farm equipment (Lowery and Schuler, 1991). Increasing soil bulk density or aggregate size delayed emergence rate and reduced emergence of hard red spring wheat (*Triticum*

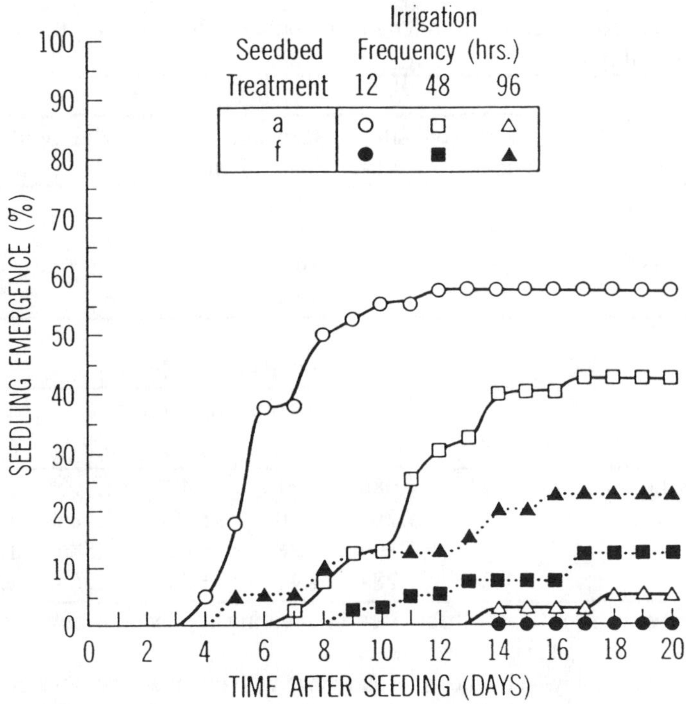

Figure 2. Cumulative seedling emergence of velvetleaf seeded among aggregates 2.49 to 1.25 cm (Treatment a) and inside 3-g aggregates (Treatment f) under three irrigation frequencies in the growth chamber. (From Pareja and Staniforth, 1985.)

aestivum L.) (Nasr and Selles, 1995). The rate of emergence and total seedling population were affected by the interaction of soil density and aggregate size. Thus, we anticipate that weed species with seeds of similar size and morphology as wheat will be similarly affected by compaction.

Total porosity of a soil is inversely related to its bulk density; thus, as the bulk density increases, the porosity decreases. The bulk density in CT is often greater than that of MB (Hilfiker and Lowery, 1988). Thus, soil compaction associated with CT means a reduction in the total soil porosity of these systems, which in turn suggests that the volume of soil available for use by plant roots is smaller with CT. Hilfiker and Lowery (1988) reported decreases in total corn-seedling root length for CT compared with MB on some soil types and increases on other soils (Table 1). Reports of a direct relationship between CT and root growth of weeds were not found, thus we must infer that a similar root-soil relationship occurs for common weed species.

The growth and reproductive tissue development of some weeds grown in pots with restricted soil volumes have been found to be adversely affected (McConnaughay and Bazzaz, 1991). Growth of common lambsquarters (moderately deep taproot), giant foxtail (diffuse, shallow root system), ivyleaf

Table 1. Total corn-seedling root length and bulk density for four tillage systems from three soil types

	Root length (m)					
Tillage system	Griswold silt loam (5 week)		Kewaunee silt loam (7 week)		Plainfield loamy sand (6 week)	
Moldboard plow	19.82[a]		11.80		6.68	
Chisel plow	25.79		17.99		7.62	
Ridge-tillage	11.75		19.63		5.27	
No-tillage	29.21		18.42		4.23	
	Bulk density ($Mg\ m^{-3}$)					
	Depth ranges (cm)					
	0-8	8-16	0-8	8-16	0-8	8-16
Moldboard plow	1.22	1.08b[b]	1.31	1.42	1.39b	1.44
Chisel plow	1.13	1.26a	1.30	1.52	1.48ab	1.48
Ridge-tillage	1.21	1.30a	1.38	1.45	1.58a	1.59
No-tillage	1.26	1.28a	1.41	1.50	1.56a	1.59

[a]There were no significant differences in seedling root length between tillage treatments at the 0.10 significance level.
[b]Column values followed by the same and those with no letters are not significantly different at the 0.10 probability level, according to the protected LSD test.
(From Hilfiker and Lowery, 1988.)

morningglory [*Ipomoea hederacea* (L.) Jacq.] (variable, shallow taproot), and velvetleaf (moderately deep taproot) was found to be sensitive to the volumes of soil in which their roots were grown. These species differed in patterns of response to soil physical space in both vegetative and reproductive parameters, suggesting that these weed species will be affected by compaction. The vegetative growth of all these species increased when they were grown in greater soil volumes. However, when velvetleaf was grown in larger soil volumes, McConnaughay and Bazzaz (1991) found an increase in reproductive tissue relative to vegetative tissue, whereas giant foxtail flowered earlier and produced greater reproductive tissue in smaller soil volumes. In addition, the mass of individual velvetleaf seed from plants grown in larger pot volumes was greater. This suggests that seedlings from these seeds may have potential for greater establishment and survival. The authors concluded that species that deploy roots across a larger soil volume and have more even distributions of root growth and activity throughout the season are more sensitive to root restriction via limited rooting space.

Tillage impacts the size of soil aggregates and the distribution of weed seed among and within soil aggregates. In fields that had been under MB or CT for more than 12 years, Pareja et al. (1985) found proportionally more unaggregated soil under MB than CT in the surface layer (0 to 5 cm) of soil and that the proportion of unaggregated soil decreased with depth under MB but increased under CT. However, in the surface layer of soil, MB incorporated weed seed (mostly foxtail species) more uniformly among various soil aggregate classes, whereas with CT more seed accumulated in the unaggregated fraction of the soil.

Some primary tillage systems such as CH create considerable surface roughness and result in rather large soil aggregates (Johnson et al., 1984). Weed seeds will be distributed within and between these large soil aggregates. The micro-environment within and between aggregates is different. This variable micro-environment will impose different influences among weed seed. The population of weed seed within the soil may be viewed as being composed of several subpopulations occupying different seed-soil microsites. Although the requirements for germination of seed of a species may be similar, the time at which these requirements are met in the soil may be different for each subpopulation (Pareja and Staniforth, 1985). Thus, within and between aggregates, the micro-environment will strongly influence weed seed fate since the water content is greater within aggregates but oxygen is often less than between aggregates.

V. Weed Management as Impacted by Tillage

A. Surface Residue

Mulching has been an important weed management practice in vegetable cropping systems for many years; however, it has been utilized most widely in specialized or small-scale systems (Patriquin, 1988; Swaider et al., 1992). Therefore, it is anticipated that surface residue cover with any tillage system will have an impact on reducing weed seed germination, weed emergence, and/or growth in agronomic cropping systems. Johnson et al. (1989) found that as the rate of surface corn residue increased under crop-free conditions and without herbicides, control of green foxtail, giant foxtail, and common lambsquarters increased. At the highest rate of corn residue (10.08 Mg ha^{-1}) greater than 50% control of these species was attained. They concluded that surface corn residue contributed to overall weed control even when herbicides were applied.

B. Weed Species Shifts

A substantial amount of research has been conducted on crop and weed management in CT and results have been extensively and recently reviewed (Froud-Williams, 1988; Forcella and Burnside, 1994; Forcella et al., 1994; Lindwall et al., 1994; Wicks et al., 1994). Emerging generalizations identified by Forcella et al. (1994) regarding tillage systems and weed species were 1) a shift towards annual grass, e.g., foxtails, and perennial weed species in CT, 2) relatively large-seeded broadleaf weed species, e.g., velvetleaf, tend to have higher densities in MB, and 3) small-seeded broadleaf weed species, e.g., common lambsquarters, appear more variable in their adaptations to tillage systems. In addition to these trends, species diversity has been found to decline with increased intensity of soil disturbance. After 25 years, the seed bank under grass sod had three to four times as many species as MB and the number of species in NT was generally greater than MB (Cardina et al., 1991).

Few long-term, detailed studies have been conducted on the ecology of weed species shifts in CT (Swanton et al., 1993), but one critical factor appears to be the differential ability among weed species to survive, germinate, and emerge from different depths within the soil profile (Cousens and Moss, 1990; Mester and Buhler, 1991; Mohler, 1993). Tillage impacts the vertical distribution of weed seed within the soil profile and thus, it impacts weed seed fate, including weed seed germination and weed emergence. Annual grass weed species such as foxtails have greater germination and establishment at shallower soil depths than large-seeded, broadleaf species (Mester and Buhler, 1991).

Unlike MB which buries surface-deposited weed seed due to soil inversion, many CT result in only partial inversion of the soil and weed seed burial. After 12 years of tillage, Pareja et al. (1985) found 85% of the total weed seed in the upper 5 cm of soil in CT but only 28% in MB. With greater tillage and residue burial, weed seed distribution tends to be more uniform in the soil profile. Even though MB buries surface-deposited weed seed, soil inversion returns viable, previously buried weed seed to the surface and upper portion of the soil profile. Yenish et al. (1992) found weed seeds uniformly distributed in the upper 19 cm of the soil profile in MB, but over 30 and 60% of the seeds were in the upper 1 cm of the soil in CH and NT, respectively. Similarly, at least 40% of giant and green foxtail seedlings emerged from the upper 1 cm of the soil in NT compared to about 25% in CH and 15% in MB (Buhler and Mester, 1991). Thus, in contrast to MB, NT tends to position grass weed seed more favorably in the soil profile for germination than large-seeded, broadleaf weed species.

C. Weed Seed Bank Depletion

Germination, predation by pathogens and animals, and mortality because of natural processes deplete or reduce viable weed seed banks (Simpson et al., 1989). However, only a small percentage of the large reservoir of seed in soil emerges annually and contributes to seed bank depletion (Roberts and Ricketts, 1979; Roberts and Boddrell, 1984; Forcella et al., 1992). In the Midwest, seasonal emergence of weeds from viable seed banks was low and was greater for grass (8.9%) than broadleaf (3.7%) weed species (Forcella et al., 1992). Forcella et al. (1992) found that from 50 to 90% of all seed in the seed bank were dead.

The rate of weed seed bank depletion varies among species and with the type and frequency of soil disturbance, but in general the rate of seed decline is less in nondisturbed than disturbed soil (Froud-Williams, 1988). In the absence of velvetleaf seed influx into the soil, Lueschen et al. (1993) found that after 17 years, the soil in several MB cropping systems contained 0.8 to 2.5% of the initial viable velvetleaf population, while in NT systems the number of velvetleaf seed ranged from 15 to 25% of the original seed population; in both MB and NT, nearly all velvetleaf seed was viable. Similarly, when wild mustard [*Brassica kaber* (DC.) L.C. Wheeler] plants were not allowed to produce seed during seven growing seasons, about 50% of the initial seed population in the soil remained in nondisturbed soil, whereas less than 3% remained in MB (Warnes and Andersen, 1984).

Increased weed seed germination and emergence associated with soil disturbance may be because of several factors, including improved soil aeration, exposure to light, increased loss of volatile germination inhibitors, or greater diurnal range in temperature near the soil surface (Egley, 1986; Wilson, 1988). Germination of some weed species, however, is not enhanced by soil disturbance; this may be attributable to movement of seed to the soil surface where moisture may be inadequate or exposure of photosensitive seeds to light. Movement of seeds from near the soil surface to greater depths may also inhibit germination.

The magnitude of weed seed influx into the soil during a growing season may substantially impact the long-term weed population density of a tillage system. A model developed by Mohler (1993) predicted that in the first year following input of seed into the soil, NT will have more seedlings than other CT or MB, but in later years NT will likely have fewer seedlings unless seed dormancy or seed survival near the soil surface is high. If weed seed input to the soil occurs annually, CT will have more seedlings than MB.

VI. Conclusions and Research Needs

There is a profusion of data on moldboard plow and crop phenology, weed management and ecology, and soil behavior. There is also a mounting database on the impact of CT systems on crop growth and soil chemical and physical properties, but less so for weed ecology and management. Even though our knowledge of seed bank and population dynamics of weeds has been increased in recent years, there are many aspects of the weed life cycle in CT that are poorly understood. Limited research has been conducted on the impact of soil physical properties on the soil micro-environment and weed seed fate in CT. Ecological research in natural ecosystems has generally not extended to agroecosystems (Swanton et al., 1993). Comprehensive ecological research, including physical properties of the soil system, would help characterize weed ecology in CT. Along with the development of bioeconomic weed management models (King et al., 1986; Wilkerson et al., 1991) and the recent introduction of a dynamic, multi-species, multiple-control bioeconomic weed management model (Swinton and King, 1994), edaphic factors, including soil physical properties, that influence weed seed fate (e.g., germination, predation, and mortality) and growth processes (e.g., rate of emergence and development) must be fully characterized. Increased collaboration between soil science, ecology, and weed science disciplines in the future will be necessary if a better understanding of weed ecology and management is to be attained.

The following research needs, related to interactions between tillage systems, soil physical environment, and weed management, are based upon our review of current and previous research findings. The most pressing need is for a multi-disciplinary research approach to attain a better understanding of weed ecology and develop weed management alternatives for an array of micro-environmental and tillage conditions. Specific research needs are: 1) to elucidate the impact of changes in soil physical properties on soil micro-environment and weed seed fate; 2) to obtain a better understanding of agroecosystems with respect to their physical environment; and 3) to integrate soil physical environmental parameters into bioeconomic weed management models.

References

Al-Darby, A.M. and B. Lowery. 1986. Evaluation of corn growth and production with three conservation tillage systems. *Agron. J.* 78:901-907.

Al-Darby, A.M. and B. Lowery. 1987. Seed zone soil temperature and early corn growth with three conservation tillage systems. *Soil Sci. Soc. Am. J.* 51:768-774.

Allmaras, R.R., E.A. Hallauer, W.W. Nelson, and S.D. Evans. 1977. Surface energy balance and soil thermal property modifications by tillage-induced soil structure. Minn. Agric. Exp. Stn. Tech. Bull. 306.

Alm, D.M., E.W. Stoller, and L.M. Wax. 1993. An index model for predicting seed germination and emergence rates. *Weed Technol.* 7:560-569.

Andraski, B.J., T.C. Daniel, B. Lowery, and D.H. Mueller. 1985. Runoff results from natural and simulated rainfall for four tillage systems. *Trans. ASAE* 28:1219-1225.

Angle, J.S., G. McClung, M.S. McIntosh, P.M. Thomas, and D.C. Wolf. 1984. Nutrient loss in runoff from conventional and no-till corn watersheds. *J. Environ. Qual.* 13:431-435.

Azooz, R.H., B. Lowery, and T.C. Daniel. 1995. Tillage and residue management influence on corn growth. *Soil Tillage Res.* 33:215-227.

Barlow, E.W.R., L. Boersma, and J.L. Young. 1977. Photosynthesis, transpiration, and leaf elongation in corn seedling at suboptimal soil temperatures. *Agron. J.* 69:95-100.

Benech Arnold, R.L., C.M. Ghersa, R.A. Sanchez, and P. Insausti. 1990. A mathematical model to predict *Sorghum halepense* (L.) Pres. seedling emergence in relation to soil temperature. *Weed Res.* 30:91-99.

Blevins, R.L., D. Cook, S.H. Phillips, and R.R. Phillips. 1971. Influence of no-tillage on soil moisture. *Agron. J.* 63:593-596.

Boydston, R.A. 1990. Time of emergence and seed production of longspine sandbur (*Cenchrus longispinus*) and puncturevine (*Tribulus terrestris*). *Weed Sci.* 38:16-21.

Bridges, D.C., H-I. Wu, P.J.J. Sharpe, and J.M. Chandler. 1989. Modeling distributions of crop and weed seed germination time. *Weed Sci.* 37:724-729.

Buhler, D.D. and T.C. Mester. 1991. Effect of tillage systems on the emergence depth of giant (*Setaria faberi*) and green foxtail (*Setaria viridis*). *Weed Sci.* 39:200-203.

Burrows, W.C. and W. E. Larson. 1962. Effect of amount of mulch on soil temperature and early growth of corn. *Agron. J.* 54:19-23.

Burwell, R.R. and L.A. Kramer. 1983. Long-term annual runoff and soil loss from conventional tillage of corn. *J. Soil Water Conserv.* 36:315-319.

Cardina, J., E. Regnier, and K. Harrison. 1991. Long-term tillage effects on seed banks in three Ohio soils. *Weed Sci.* 39:186-194.

Cavers, P.B. and D.L. Benoit. 1989. Seed banks in arable lands. p. 309-328. In: M.A. Leck, V.T. Parker, and R.L. Simpson (eds.), *Ecology of Soil Seed Banks*. Academic Press, San Diego, CA.

Cousens, R. and S.R. Moss. 1990. A model of the effects of cultivation on the vertical distribution of weed seeds within the soil. *Weed Res.* 30:61-70.

Cummins, D.G. and W.L. Parks. 1961. The germination of corn and wheat as affected by various fertilizer salts at different soil temperatures. *Soil Sci. Soc. Am. Proc.* 25:47-49.

DeFelice, M.S., W.W. Witt, and M. Barrett. 1988. Velvetleaf (*Abutilon theophrasti*) growth and development in conventional and no-tillage corn (*Zea mays*). *Weed Sci.* 36:609-615.

Egley, G.H. 1986. Stimulation of weed seed germination in soil. *Rev. Weed. Sci.* 2:67-89.

Egley, G.H. and S.O. Duke. 1985. Physiology of weed seed dormancy and germination. p. 27-64. In: S.O. Duke (ed.), *Weed Physiology. Volume I. Reproduction and Ecophysiology.* CRC Press, Boca Raton, FL.

Forcella, F. 1992. Prediction of weed seedling densities from buried seed reserves. *Weed Res.* 32:29-38.

Forcella, F. 1993. Seedling emergence model for velvetleaf. *Agron. J.* 85:929-933.

Forcella, F. and O.C. Burnside. 1994. Pest management-weeds. p. 157-197. In: J.L. Hatfield and D.L. Karlen (eds.), *Sustainable Agriculture Systems*. CRC Press, Boca Raton, FL.

Forcella, F., D.D. Buhler, and M.E. McGiffen. 1994. Pest management and crop residues. p. 173-189. In: J.L. Hatfield and B.A. Stewart (eds.), *Crops Residue Management*. Lewis Publishers, Boca Raton, FL.

Forcella, F., R.G. Wilson, K.A. Renner, J. Dekker, R.G. Harvey, D.A. Alm, D.D. Buhler, and J. Cardina. 1992. Weed seedbanks of the U.S. Corn Belt: Magnitude, variation, emergence, and application. *Weed Sci.* 40:636-644.

Free, G.R. and C.E. Bay. 1969. Tillage and slope effects on runoff and erosion. *Trans. ASAE* 12:209-215.

Froud-Williams, R.J. 1988. Changes in weed flora with different tillage and agronomic management systems. p. 213-236. In: M.A. Altieri and M. Liebman (eds.), *Weed Management in Agroecosystems: Ecological Approaches*. CRC Press, Boca Raton, FL.

Gantzer, C.J. and G.R. Blake. 1978. Physical characteristics of the Le Seur clay loam soil following no-till and conventional tillage. *Agron. J.* 70:853-857.

Gupta, S.C. 1985. Predicting corn planting dates for moldboard and no-tillage systems in the Corn Belt. *Agron. J.* 77:446-455.

Gupta, S.C., B. Lowery, J.F. Moncrief, and W.E. Larson. 1991. Modeling tillage effects on soil physical properties. *Soil Tillage Res.* 20:293-318.

Hagin, J. 1952. Influence of soil aggregation on plant growth. *Soil Sci.* 74:471-478.

Harvey, S.J. and F. Forcella. 1993. Vernal seedling emergence model for common lambsquarters (*Chenopodium album*). *Weed Sci.* 41:309-316.

Hay, R.K.M., J.C. Holmes, and E.A. Hunter. 1978. The effects of tillage, direct drilling and nitrogen fertilizer on soil temperature under a barley crop. *J. Soil Sci.* 29:174-183.

Hilfiker, R.E. and B. Lowery. 1988. Effect of conservation tillage systems on corn root growth. *Soil Tillage Res.* 12:269-283.

Hill, R.L. and R.M. Cruse. 1985. Tillage effects on bulk density and soil strength of two Mollisols. *Soil Sci. Soc. Am. J.* 49:1270-1273.

Johnson, M.D. and B. Lowery. 1985. Effect of three conservation tillage practices on soil temperature and thermal properties. *Soil Sci. Soc. Am. J.* 49:1547-1552.

Johnson, C.B. and W. C. Moldenhauer. 1979. Effect of chisel versus moldboard plowing on soil erosion by water. *Soil Sci. Soc. Am. J.* 43:177-179.

Johnson, M.D., B. Lowery, and T.C. Daniel. 1984. Soil moisture regimes of three conservation tillage systems. *Trans. ASAE* 27:1385-1390,1395.

Johnson, M.D., D.L. Wyse, and W.E. Lueschen. 1989. The influence of herbicide formulation on weed control in four tillage systems. *Weed Sci.* 37:239-249.

Johnson, H.P., J.L. Baker, W.D. Shrader, and J.M. Laflen. 1979. Tillage system effects on sediment and nutrients on runoff from small watersheds. *Trans. ASAE* 22:1110-1114.

Jones, J.N., Jr., J.E. Moody, and J.H. Lillard. 1969. Effects of tillage, no-tilling, and mulch on soil water and plant growth. *Agron. J.* 61:719-721.

Jones, J.N., Jr., J.E. Moody, G.W. Shear, W.W. Moschler, and J.H. Lillard. 1968. The no-tillage system for corn. *Agron. J.* 60:17-20.

King, R.P., D.W. Lybecker, E.E. Schweizer, and R.L. Zimdahl. 1986. Bioeconomic modeling to simulate weed control strategies for continuous corn (*Zea mays*). *Weed Sci.* 34:972-979.

Laflen, J.M., J.L. Baker, R.O. Hartwig, W.F. Buchele, and H.P. Johnson. 1978. Soil and water loss from conservation tillage systems. *Trans. ASAE* 21:881-885.

Langdale, G.W., A.P. Barnett, R.A. Leonard, and W.G. Fleming. 1979. Reduction of soil erosion by the no-till system in the Southern Piedmont. *Trans. ASAE* 22:82-86,92.

Lindwall, C.W., F.J. Larney, A.M. Johnston, and J.R. Moyer. 1994. Crop management in conservation tillage systems. pp. 185-209. In: P.W. Unger (ed.), *Managing Agricultural Residues*. Lewis Publishers, Boca Raton, FL.

Lowery, B. and R.T. Schuler. 1991. Temporal effects of subsoil compaction on soil strength and plant growth. *Soil Sci. Soc. Am. J.* 55:216-223.

Lueschen, W.E., R.N. Andersen, T.R. Hoverstad, and B.K. Kanne. 1993. Seventeen years of cropping systems and tillage affect velvetleaf (*Abutilon theophrasti*) seed longevity. *Weed Sci.* 41:82-86.

McConnaughay, K.D.M. and F.A. Bazzaz. 1991. Is physical space a soil resource? *Ecology* 72:94-103.

McGregor, K.C. and J.D. Greer. 1982. Erosion control with no-till and reduced till corn for silage and grain. *Trans. ASAE* 25:154-159.

Mester, T.C. and D.D. Buhler. 1991. Effects of soil temperature, seed depth, and cyanazine on giant foxtail (*Setaria faberi*) and velvetleaf (*Abutilon theophrasti*) seedling development. *Weed Sci.* 39:204-209.

Miller, S.A. and A.P. Mazurak. 1958. Relationships of particle and pore size to the growth of sunflowers (*Helianthus annuus* L.). *Soil Sci. Soc. Am. Proc.* 22:275-278.

Mock, J.J. and D.C. Erbach. 1977. Influence of conservation-tillage environments on growth and productivity of corn. *Agron. J.* 69:337-340.

Mohler, C.L. 1993. A model of the effects of tillage on emergence of weed seedlings. *Ecol. App.* 3:53-73.

Moody, J.E., J.N. Jones, Jr., and J.H. Lillard. 1963. Influence of straw mulch on soil moisture, soil temperature and the growth of corn. *Soil Sci. Soc. Am. Proc.* 27:700-703.

Nasr, H.M. and F. Selles. 1995. Seedling emergence as influenced by aggregate size, bulk density, and penetration resistance of the seedbed. *Soil Tillage Res.* 34:61-76.

Onstad, C.A. 1972. Soil and water loss as affected by tillage practices. *Trans. ASAE* 15:287-289.

Pareja, M.R. and D.W. Staniforth. 1985. Seed-soil microsite characteristics in relation to weed seed germination. *Weed Sci.* 33:190-195.

Pareja, M.R., D.W. Staniforth, and G.P. Pareja. 1985. Distribution of weed seed among soil structural units. *Weed Sci.* 33:182-189.

Patriquin, D.G. 1988. Weed control in organic farming systems. p. 303-317. In: M.A. Altieri and M. Liebman (eds.), *Weed Management in Agroecosystems: Ecological Approaches*. CRC Press, Boca Raton, FL.

Phillips, R.R., R.L. Blevins, G.W. Thomas, W.W. Fryer, and S.H. Phillips. 1980. No-tillage agriculture. *Science* 208:1108-1113.

Roberts, H.A. 1984. Crop and weed emergence patterns in relation to time of cultivation and rainfall. *Ann. Appl. Biol.* 105:263-275.

Roberts, H.A. and J.E. Boddrell. 1984. Seed survival and seasonal emergence of seedlings of some ruderal plants. *J. Appl. Ecol.* 21:617-628.

Roberts, H.A. and M.E. Ricketts. 1979. Quantitative relationships between the weed flora after cultivation and the seed population in the soil. *Weed Res.* 19:269-275.

Romkens, M.J.M., D.W. Nelson, and J.V. Mannering. 1973. Nitrogen and phosphorus composition of surface runoff as affected by tillage method. *J. Environ. Qual.* 2:292-295.

Schneider, E.C. and S.C. Gupta. 1985. Predicting corn emergence from soil temperature, matric potential, and aggregate size distribution. *Soil Sci. Soc. Am. J.* 49:415-422.

Shaffer, M.J. and W.E. Larson (ed.). 1987. *NTRM, a soil-crop simulation model for nitrogen, tillage, and crop-residue management*. USDA Conserv. Res. Rep. 34-1. Natl. Tech. Inform. Serv., Springfield, VA.

Siemens, J.C. and W.R. Oschwald. 1978. Corn-soybean tillage systems: Erosion control, effects on crop production, costs. *Trans. ASAE* 21:293-302.

Simpson, R.L., M.A. Leck, and V.T. Parker. 1989. Seed banks: General concepts and methodological issues. p. 3-8. In: M.A. Leck, V.T. Parker, and R.L. Simpson (eds.), *Ecology of Soil Seed Banks*. Academic Press, San Diego, CA.

Soil Conservation Society of America (SCSA). 1982. *Resource Conservation glossary*. Soil Conserv. Soc. Am., Ankeny, IA.

Stoller, E.W. and L.M. Wax. 1973. Periodicity of germination and emergence of some annual weeds. *Weed Sci.* 21:574-580.

Swaider, J.M., G.W. Ware, and J.P. McCollum. 1992. *Producing Vegetable Crops*. Interstate Publishers, Inc., Danville, IL.

Swan, J.B., E.C. Schneider, J.F. Moncrief, W.H. Paulson, and A.E. Peterson. 1987. Estimating corn growth, yield, and grain moisture from air growing degree days and residue cover. *Agron. J.* 79:53-60.

Swanton, C.J., D.R. Clements, and D.A. Derksen. 1993. Weed succession under conservation tillage: A hierarchical framework for research and management. *Weed Technol.* 7:286-297.

Swinton, S.M. and R.P. King. 1994. A bioeconomic model for weed management in corn and soybean. *Agric. Sys.* 44:313-335.

Taylorson, R.B. 1987. Environmental and chemical manipulation of weed seed dormancy. *Rev. Weed Sci.* 3:135-154.

Triplett, G.B., Jr., D.M. Van Doren, Jr., and B.L. Schmidt. 1968. Effect of corn stover mulch on no-tillage corn yield and water infiltration. *Agron. J.* 60:236-239.

Unger, P.W. and T.C. Kaspar. 1994. Soil compaction and root growth: A review. *Agron J.* 86:759-766.

van Wijk, W.R., W.E. Larson, and W.C. Burrows. 1959. Soil temperature and the early growth of corn from mulched and unmulched soil. *Soil Sci. Soc. Am. Proc.* 23:428-434.

Walker, J.M. 1969. One-degree increments in soil temperatures affect maize seedling behavior. *Soil Sci. Soc. Am. Proc.* 33:729-736.

Warnes, D.D. and R.N. Andersen. 1984. Decline of wild mustard (*Brassica kaber*) seeds in soil under various cultural and chemical practices. *Weed Sci.* 32:214-217.

Weaver, S.E., C.S. Tan, and P. Brain. 1988. Effect of temperature and soil moisture on time of emergence of tomatoes and four weed species. *Can. J. Plant Sci.* 68:877-886.

Wicks, G.A., O.C. Burnside, and W.L. Felton. 1994. Weed control in conservation tillage systems. p. 211-244. In: P.W. Unger (ed.), *Managing Agricultural Residues*. Lewis Publishers, Boca Raton, FL.

Wierenga, P.J., D.R. Nielsen, R. Horton, and B. Kies. 1982. *Tillage Effects on Soil Temperature and Thermal Conductivity*. Spec. Pub. 44. American Society of Agronomy, Madison, WI.

Wilkerson, G.G., S.A. Modena, and H.D. Coble. 1991. HERB: Decision model for postemergence weed control in soybean. *Agron. J.* 83:413-417.

Wilson, R.G. 1988. Biology of weed seeds in the soil. p. 25-39. In: M.A. Altieri and M. Liebman (eds.), *Weed Management in Agroecosystems: Ecological Approaches*. CRC Press, Boca Raton, FL.

Wischmeier, W.W. 1973. Conservation tillage to control water erosion. p. 133-141. In: Proc. Natl. Conf. on Conservation Tillage, Des Moines, Iowa. Soil Conserv. Soc. Am., Ankeny, IA.

Yenish, J.P., J.D. Doll, and D.D. Buhler. 1992. Effects of tillage on vertical distribution and viability of weed seed in soil. *Weed Sci.* 30:429-433.

Mechanisms of Herbicide Leaching and Volatilization and Innovative Approaches for Sampling, Prediction, and Control

T.J. Gish, A. Shirmohammadi, C.S. Helling, K.-J.S. Kung, B.J. Wienhold, and W.J. Rawls

I. Introduction

Although there is a general consensus that our nation's water resources are improving, there is still a widespread concern that more needs to be done (van Schilfgaarde, 1991). Today, many of the water quality issues that address environmental degradation can be attributed to nonpoint sources of pollution of which agriculture has been identified as a major contributor (EPA, 1992).

Once applied to a field, agricultural chemicals can enter an environment where they were not intended - by runoff into surface streams, leaching to ground water, or by volatilization and redeposition to any land or water body via rain, fog, or dry deposition. Since pesticides can impact the environment by a number of pathways, a comprehensive approach must be used in the development and evaluation of new management practices or chemical formulations that may be used by agriculture.

Currently, most research efforts are focused on the development and promotion of farming practices and systems that both improve water quality and increase chemical efficacy. Effective interaction among disciplines is critical if we are to ensure that a new agricultural practice or chemical alternative does not simply shift the pollutant from one part of the environment to another. On the other hand, cost and scale of observation (plot, field, or watershed scales) limit how many chemical loss pathways can be monitored. In addition, there is still some ambiguity associated with what processes should be monitored and how those measurements should be made.

The objective of this chapter is to present a perspective that allows the major loss pathways of a herbicide to be evaluated so that field-scale behavior can be ascertained and effective farming practices developed. Accordingly, methods for evaluating chemical leaching, characterization of field-scale chemical flux variability, and volatilization will be briefly discussed.

II. Pesticide Leaching

In the early 1960s, water and solutes were conceptualized as moving uniformly through the soil with the entire soil matrix being used in the transport process. Consequently, pesticide leaching was thought to be negligible since pesticides generally have a high affinity for the soil matrix and are susceptible to biological degradation. In the early 1970s, the notion that water and solute transport was nonuniform was confirmed. Nielsen et al. (1973) found that soil water diffusivity was highly variable, exhibiting coefficients of variation (CV) between 284 to 412%. Additionally, Libardi et al. (1980) observed that the unsaturated conductivity of a silty clay loam soil was also highly variable with CVs ranging from 235 to 447%. Consequently, a lognormal distribution is typically used to describe the spatial variability of the hydraulic conductivity (Biggar and Nielsen, 1976; Dagan and Bresler, 1979; Yeh et al., 1985; Cushman, 1987). Eventually, preferential flow processes were observed and believed to be at least partially responsible for the variability in water and solute fluxes typically observed in field soils (Richards and Steenhuis, 1988; Gish et al., 1991). Recently, preferential flow simulations in a layered soil demonstrated that the preferential flux patterns were clearly a function of funnel flow mechanisms and not the lognormally distributed hydraulic conductivity (Ju and Kung, 1993).

A. Measurement and Evaluation

Preferential flow is the process where only a small fraction of the available pore space is used to move water and solutes (Luxmoore, 1991). Conceptually, preferential flow can be grouped into three categories, the first being a function of structural voids like soil cracks, worm holes, and root channels. In the literature, flow occurring in structural voids has often been referred to as macropore flow

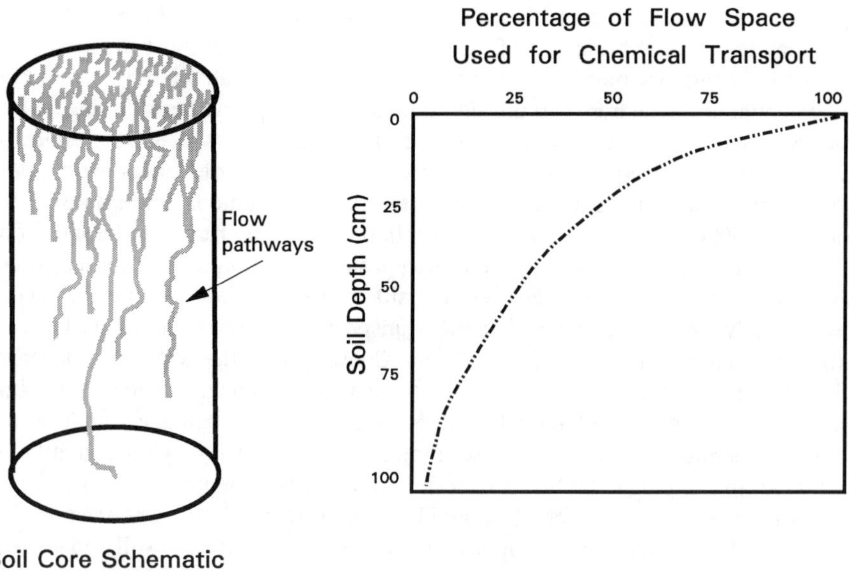

Figure 1. Preferential flow volume as a function of depth. Representative soil core schematic showing preferential flow volume relative to the volume of soil not used in the flow process.

(Bouma, 1981; Germann and Beven, 1981; Gish and Jury, 1983). The second category of preferential flow involves any process that results in the formation of an unstable wetting front. Unlike structured soils that contain structural voids, unstable wetting fronts can occur seemingly homogeneous and are frequently referred to as fingering (Glass et al., 1989; Baker and Hillel, 1990). The third type of preferential flow occurs in heterogeneous soils containing tilted soil lenses. This latter type of flow is frequently referred to as funnel flow (Kung, 1990). In many fields, all three types of preferential flow processes may be occurring simultaneously, and it may be extremely difficult or impossible to differentiate among them. In all cases, preferential flow processes are responsible for transporting water and solutes to great depths much faster than can be described by classical fluid dynamics subjected to average moisture contents and soil water potentials.

A critical characteristic of preferential flow is that the percentage of available pore space used to transport solutes decreases with depth (Figure 1). Quisenberry et al. (1991) showed that in a loamy soil, water and solutes initially moved uniformly in the surface Ap horizon. However, in the lower horizons more than 75% of the water and solutes moved through less than 25% of the soil matrix. In 0.3 by 0.3-m soil columns that had been excavated from the field, Edwards et al. (1992) observed that after 30 mm of water had been applied, <2% of the cross-sectional soil area had been used to transport 54% of the transported atrazine. In a 2-D simulation of water flow through layered soil, Ju and Kung (1993) showed that 90% of water utilized <25% of the soil matrix. Furthermore, Ju and Kung (1993) indicated that the soil volume contributing to water flow in the field might be much less than 25% since the funnel flow process becomes more drastic in 3-D. Other investigators have shown that extremely small water inputs can initiate preferential pesticide transport through field soils, with pesticides utilizing as little as 3% in the upper 1 m of soil (Gish et al., 1991; Isensee et al., 1990). Since the volume of soil used to preferentially conduct solutes decreases with depth, preferential transport is difficult to detect with some methods, and impossible to measure with others. For example, the probability of capturing a preferential flow process is low with randomly located solution samplers (Biggar and Nielsen, 1976). However, even if a continuous preferential flow pathway was sampled, soil cores by themselves cannot detect preferential transport. Soil core samples are typically homogenized prior to herbicide extraction. Consequently, the herbicide concentration determined for the sample is actually an averaged concentration for the bulk sample. However, as preferential flow becomes more important, the volume of soil used for chemical transport decreases dramatically with depth. Consequently, the actual sample-averaged herbicide concentration may be as much as 2 to 3 orders of magnitude less than the actual herbicide concentrations in the preferential flow pathway. This hypothesis is supported by a comparison between soil and solution phase atrazine residue concentrations from samples collected from a 0.3-ha no-tillage field in Maryland (Figure 2). Six solution samplers were randomly located at each of six depths in the field for a total of 36 solution samplers. Six soil cores were removed and segmented into 5-cm increments. For this illustration, the herbicide concentrations were normalized to a herbicide mass per volume of soil. Normalized solution phase concentrations were calculated as the product of the volumetric water content and the herbicide solution phase concentration. The normalized herbicide concentrations associated with the soil core data are actually a total herbicide concentration, C_T, since the soil samples contain some herbicide-solution in addition to any sorbed herbicide. Consequently, the normalized herbicide concentrations associated with the soil cores can be mathematically represented as

$$C_T = \rho_b\, C_S + \Theta\, C_1$$

where C_T is the total pesticide concentration (μg cm^{-3} soil), C_s is the soil-adsorbed atrazine concentration (μg g^{-1} soil), C_l is the liquid phase atrazine concentration (μg cm^{-3} water), ρ_b is the soil bulk density (g soil cm^{-3} soil), and θ is the soil water content (cm^3 water cm^{-3} soil), and each parameter is determined as a function of depth. Since solution samples were removed about 24 hr after the rain event, the volumetric water content was greater for these samples than for soil cores which were removed several days later. The partition coefficient between the solution and sorbed phases can be represented by the linear relationship

$$C_S = K_d\, C_1$$

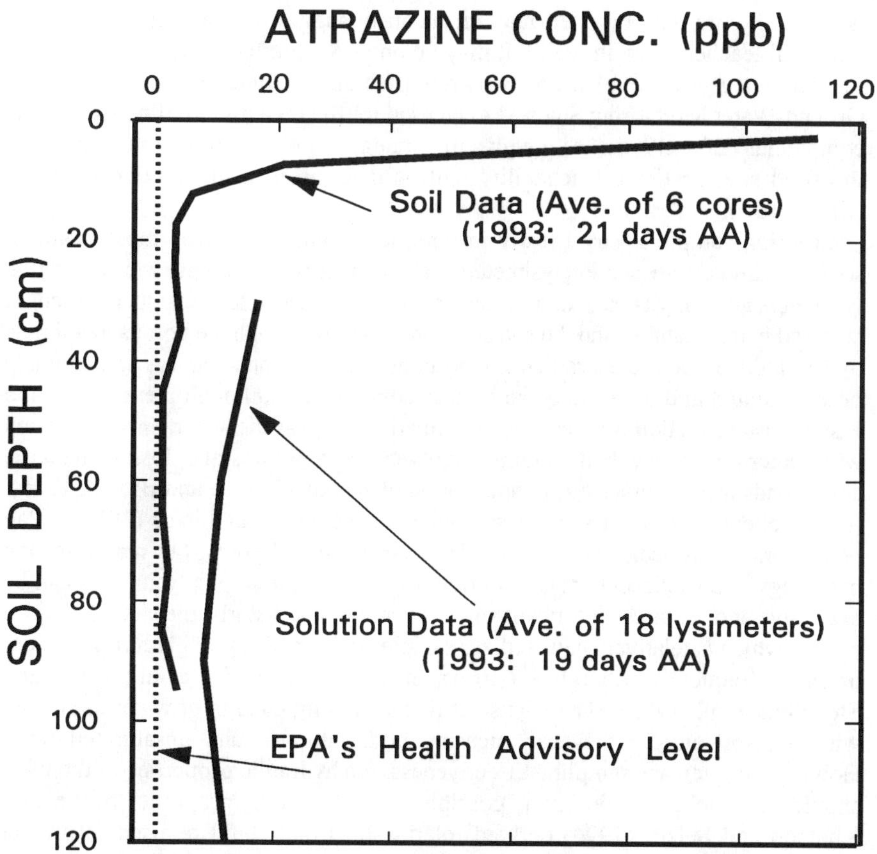

Figure 2. Comparison of atrazine concentrations from soil and solution phases. Data collected from a 0.3-ha no-tillage field in Maryland. No additional rainfall had occurred between days 19 and 21.

where K_d is the atrazine distribution coefficient (cm^3 g^{-1}). For this Maryland soil, all of the distribution coefficients were >1. Consequently, if the entire soil volume is used for pesticide transport (uniform flow), the normalized atrazine concentration associated with the soil cores should be greater than or equal to the atrazine mass associated with the solution phase. However, the normalized atrazine solution phase concentrations below 0.3 m were always greater than those measured with soil cores. These data indicate that only a fraction of the soil matrix was used for atrazine transport and that atrazine transport was water flux-dominated. Furthermore, these data suggest that soil cores do not adequately measure that fraction of the soil-water used in transporting chemicals. If a mobile-immobile region concept is utilized (van Genuchten and Wagenet, 1989), the soil core concentrations would be influenced more by the immobile soil-water fraction, while the solution lysimeter measurements would be dominated by the mobile soil-water fraction. The net result is that soil core data by itself may erroneously suggest that a chemical is not a potential leacher when in reality it may be one. Accordingly, the Environmental Protection Agency (EPA) has recently rewritten their "Guidelines for Small-Scale Ground-Water Monitoring Studies" to be used for the registration of new pesticide compounds (EPA, 1995). Although soil core data will be required to characterize the field site, pesticide leachability will be determined from solution samplers (EPA, 1995).

Solution samplers vary greatly in complexity and cost. The most common solution samplers are suction lysimeters, pan lysimeters, and sample wicks. Suction lysimeters generally consist of a porous membrane attached to a chamber. Vacuum is stored in the chamber and the soil solution is drawn through the porous membrane by the induced pressure gradient. Consequently, solution samplers can evaluate both saturated and unsaturated soil water conditions. Although inexpensive and easy to install, suction lysimeters have limitations such as the uncertainty associated with determining the soil volume involved in sampling, and loss of volatile compounds from samples due to application of suction (Barbee and Brown, 1986). An excellent review on suction samplers can be found in Litaor (1988). Pan lysimeters theoretically collect samples from saturated pores, but may alter the hydrology of the adjacent area. Additionally, installation of pan lysimeters can be expensive and requires significant soil excavation that disturbs the site. The third option, which is relatively new, utilizes sample wicks made up of glass rope, nylon, or more frequently, fiberglass (Brown et al., 1986; Holder et al., 1991; and Steenhuis et al., 1996). The wicks act like a hanging column of water in contact with the soil and so they can measure both saturated and unsaturated flow. However, to increase sampling effectiveness, the hydraulic properties of the wick should be matched as closely as possible to the hydraulic properties of the soil. Knutson and Selker (1994) derived relationships that could be used to predict saturated conductivity of the wick based on easily measured properties and presented a method for selecting wicks based on a given soil type and expected water flux. Rimmer et al. (1995) proposed an optimum wick length of about 30-40 cm for a sandy soil and >100 cm for finer textured soils. Placement of the wicks is critical to their functional use. One end of the wick is typically frayed and pushed against the bottom of the soil, consequently soil excavation is also needed for this method.

In addition to their design, placement of the solution samplers with respect to the flow pathways should be considered. If the samplers are located randomly, soil heterogeneity will require that a large number of solution samplers be installed. For example, Biggar and Nielsen (1976) found that 100 samplers would need to be sampled to come within ± 50% of the true field-scale solute flux. However, Kung and Donohue (1991) suggested that a protocol for placing the solution samplers could dramatically reduce the number of solution samplers needed. They proposed that the samplers be placed along preferential flow pathways (funnel flow in this study). The difficulty arises in how to predict the sampler location *a priori*. In sandy soils, ground-penetrating radar can be utilized to place solution samplers by first locating soil lenses/horizons that could restrict water flow (Kung and Donohue, 1991; Kung and Lu, 1993). However, ground-penetrating radar is less effective on finer-textured soils and no method, to date, exists for locating *a priori* preferential flow pathways resulting from soil voids and unstable wetting fronts.

B. Control

Recent studies indicate that preferential chemical transport occurs under a wide range of field conditions. However, the scope and magnitude of the preferential transport processes may vary dramatically depending upon hydrogeological setting, agricultural management practices, pesticide chemistry, and climate interactions. Consequently, our goal must be to minimize the detrimental impact of preferential flow on chemical transport. Two alternatives for minimizing detrimental impacts of preferential water flow are: enhancing diffusion transport, and development of management practices that maximize microbial activity.

1. Enhancing Diffusion Mechanisms

The scope and magnitude of preferential pesticide transport is a function of a number of variables (Helling and Gish, 1991). However, all preferential flow processes are associated with convection, the bulk transport of a chemical with the moving soil solution. Consequently, if the chemical can be forced from a convection mode into a diffusive transport mode, preferential transport should be reduced.

One method for increasing the influence of herbicide diffusion is to use controlled-release herbicide formulations. Wienhold and Gish (1992) observed that starch-encapsulated herbicides are released by a diffusion process that was governed by the solubility of the pesticide, temperature, water availability, and microbial activity. Diffusion-controlled herbicide release would result in only a small fraction of the herbicide being available for convective transport during a rain event. In addition, after the rain event has ceased, herbicide rates-of-release would remain high, allowing the herbicide to move into the soil around the starch granule. As the soil dries out, the herbicide would be drawn into the smaller pores of the soil matrix where it is less susceptible to preferential transport mechanisms. Field observations over several years confirmed this hypothesis, showing that a diffusion-controlled release process dramatically decreased herbicide mobility relative to commercial formulations (Gish et al., 1994; Figure 3). Unfortunately, a paradox can exist between controlled rates-of-release (which minimizes preferential transport) and weed control. Schreiber et al. (1994) resolved the weed control problems by reducing the size of the starch granules. The smaller starch granules gave better application coverage and faster rates-of-release. Williams et al. (1995) extended the work with the smaller granules and demonstrated that although reducing granule size increased mobility relative to larger starch granules, the mobilities were still much less than with the commercial formulation.

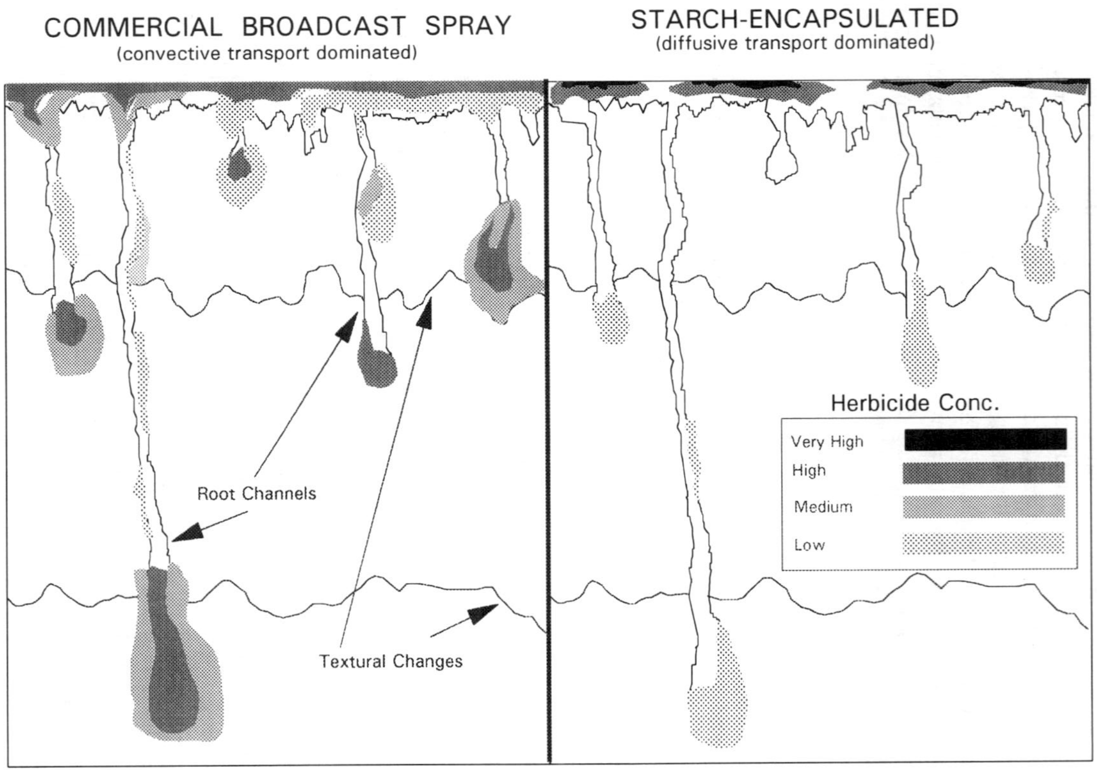

Figure 3. Convective versus diffusive atrazine transport. Convectively-dominated atrazine transport was surface broadcast-applied as a wettable powder,while diffusive-dominated transport was applied as a starch-encapsulated atrazine formulation.

Consequently, development of diffusion-limited herbicide formulations may be an effective means of reducing preferential pesticide transport while maintaining adequate weed control.

Rainfall amount and timing relative to application may also play a critical role in determining herbicide diffusion into the soil and subsequent transport. Although early rain events can result in rapid movement of pesticides (Isensee et al., 1990; Gish et al., 1991; Kladivko et al., 1991), very small rain events may actually be effective in moving solutes into the soil matrix where they will be less susceptible to preferential flow mechanisms. Shipitalo et al. (1990) observed that a 1-h 5-mm rain event preceding a large rain event reduced atrazine leachate concentrations by about 63% in 30-cm soil blocks.

2. Management Practices

Management systems modify the behavior of agricultural chemicals by altering the physical properties that affect surface and subsurface hydrology (Burwell et al., 1966; Allmaras et al., 1967; and Philips and Philips, 1984) and by influencing the amount and type of organic matter critical to herbicide degradation (Stevenson, 1972; Stearman et al., 1989). Consequently, a knowledge of how various management scenarios influence large-scale chemical behavior is critical to the development of best management practices.

Tillage reduces the amount of surface plant residue, weakens soil structure, and leads to the formation of soil crusts that may decrease water infiltration and enhance runoff (Rawls et al., 1980; Freebairn et al., 1989). Consequently, increased infiltration rates under no-tillage were thought to lead to enhanced pesticide leaching, relative to a tilled field. Over shallow perched water tables, no-tillage does appear to generate higher groundwater pesticide levels than tilled fields (Isensee et al., 1990; Gish et al., 1991; Richards and Steenhuis, 1988; Kladivko et al., 1991). However, few studies have been conducted on deep well-drained fields. Recently, Gish et al. (1995a) conducted a study on deep well drained fields in Maryland and observed that although atrazine initially moved deeper in the upper 0.6 m of soil under no-tillage, atrazine concentrations below the root zone (> 1.5 m) were always greater under tilled than under no-tillage. The attenuation of atrazine under no-tillage relative to a tilled field was attributed to the development of stable root channels that form in no-tillage fields. These stable root channels would have a microbial population and a carbon source, resulting from decomposing plant tissue, that would degrade pesticides. Consequently, pesticides that move preferentially through void root channels could be broken down by this microbial population that is preferentially located within the void root channels (Figure 4). In comparison, tilled fields are frequently cultivated. This soil disturbance prohibits the formation of stable void root channels through which water and solutes can move in the early spring. In support of this hypothesis, deethylatrazine (DEA), a biological metabolite of atrazine, collected over a four-year period was greater under no-tillage than with tillage (Figure 5). The averaged DEA concentrations

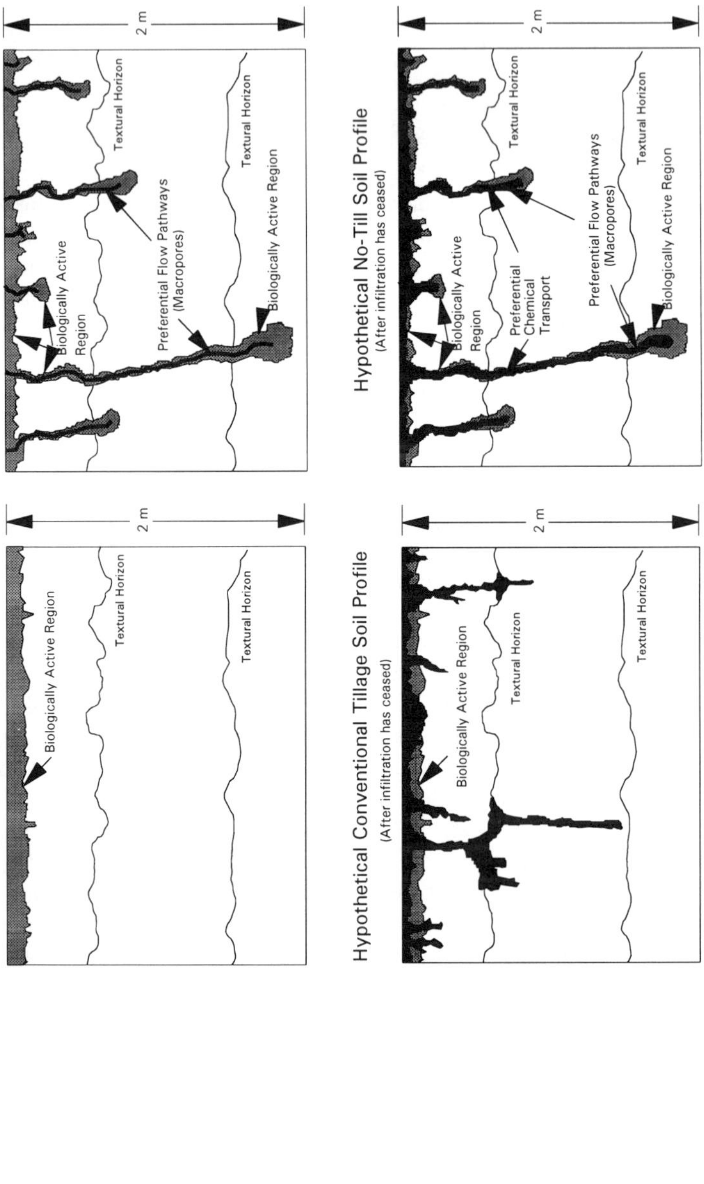

Figure 4. Hypothetical comparison of chemical transport before and after chemical infiltration on no-till and tilled fields.

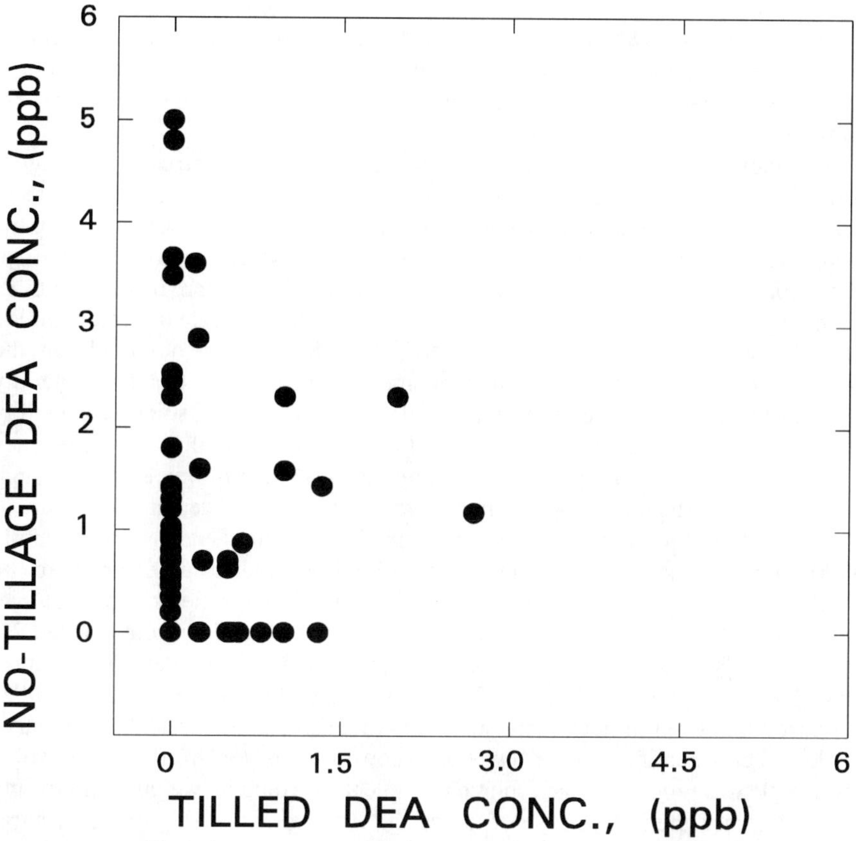

Figure 5. Deethylatrazine (DEA) concentrations below the root zone of no-till and tilled corn fields. Each dot represents field-averaged DEA concentrations for an individual sampling date. Solution samples collected from 1.5 and 1.8 m depths over a four-year period.

below the root zone over this period were 1.29 $\mu g\ L^{-1}$ under no-tillage (error of the mean,SE, 0.18), compared to 0.34 $\mu g\ L^{-1}$ (standard error of the mean 0.08) under tilled conditions. Consequently, the effectiveness of a management practice needs to be linked to the hydrogeological setting of the agricultural fields and there is a need to link preferential flow pathways with microbial activity so that the impact of microbial activity can be maximized for a given hydrogeological setting.

C. Heterogeneous Chemical Flux Index

A dimensionless index that compares the extent of preferential transport would be useful in characterizing not only the heterogeneity of a particular field site but also could be used in comparing chemical transit times among sites. Such an index would enhance our knowledge of soil fluid dynamics and aid regulatory agencies in identifying and ranking field sites that tend to be dominated by preferential flow mechanisms or are essentially homogeneous and are thereby dominated by matrix flow.

The movement of a chemically conservative tracer (e.g., bromide) through a large lysimeter containing relatively homogeneous soil is depicted in Figure 6. Simulated bromide breakthrough curves below the lysimeter schematic compare the transit time of bromide from the lysimeter drainage waters to the breakthrough curve for each suction lysimeter sampler. If the water fluxes are nearly identical, averaging the solution sampler concentrations alone may generate a single bromide breakthrough curve that is close to the "real" breakthrough curve observed from the bottom of the lysimeter. In the literature, chemical mobilities are generally evaluated by averaging the solution phase data. Unfortunately, soils are generally nonhomogeneous and simple averaging of the bromide breakthrough curves may not be representative of the field. For comparison, a set of bromide breakthrough curves from a lysimeter containing nonhomogeneous soil are depicted in Figure 7. Because of multiple water velocities through the nonhomogeneous soil, a simple averaging of the suction lysimeters data will not generate a curve similar to the "real" breakthrough curve measured at the bottom of the lysimeter. For example, Roth et al. (1991) observed that a fast moving, preferential flow front reached 2.2 m after only 31 mm of water had been applied while the slower velocities (matrix flow) had moved only 0.84 m after 853 mm of water had been applied. Ju (1993) proposed that even in heterogeneous soils, a realistic solute breakthrough curve could be generated if the water fluxes are known for the "localized" area in which the sample was removed. Each solution sample then would be weighted according to the flux measured at the time of sampling. Averaging the weighted concentrations would provide a more realistic breakthrough curve. Consequently, what is needed is a way to determine "localized" water fluxes. Once the "localized" water fluxes are calculated, the first moment or chemical center of mass can be calculated for each of the breakthrough curves by

$$M_{Center} = \frac{\int_0^\infty t\ C_t\ dt}{\int_0^\infty C_t}$$

where M_{Center} is the temporal center of mass for the breakthrough curve, and C_t is the herbicide concentration as a function of dimensionless time, t, (pore volumes). The difference between the flux-averaged center of mass and the concentration-averaged center of mass would give one index (ΔM) for characterizing the extent of homogeneity or heterogeneity for a given field site. Additionally, the peak height differential (ΔH) can be used to characterize preferential transport and could be utilized as a second index for characterizing the extent of preferential chemical

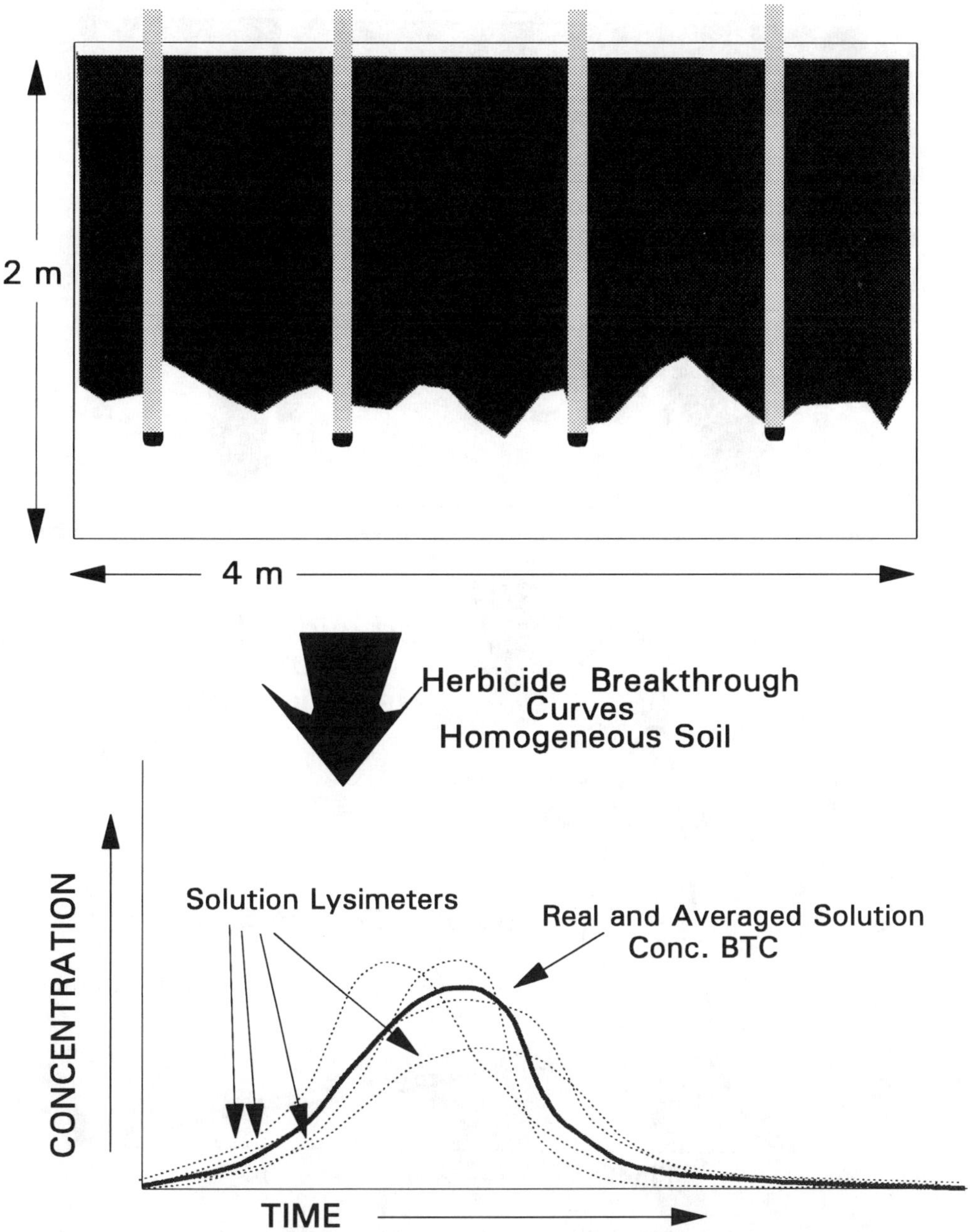

Figure 6. Schematic representation of chemical transport through a homogeneous soil profile with corresponding breakthrough curves. Dotted lines represent individual breakthrough curves (BTC) for each solution sampler, while the solid line represents a flux-averaged BTC.

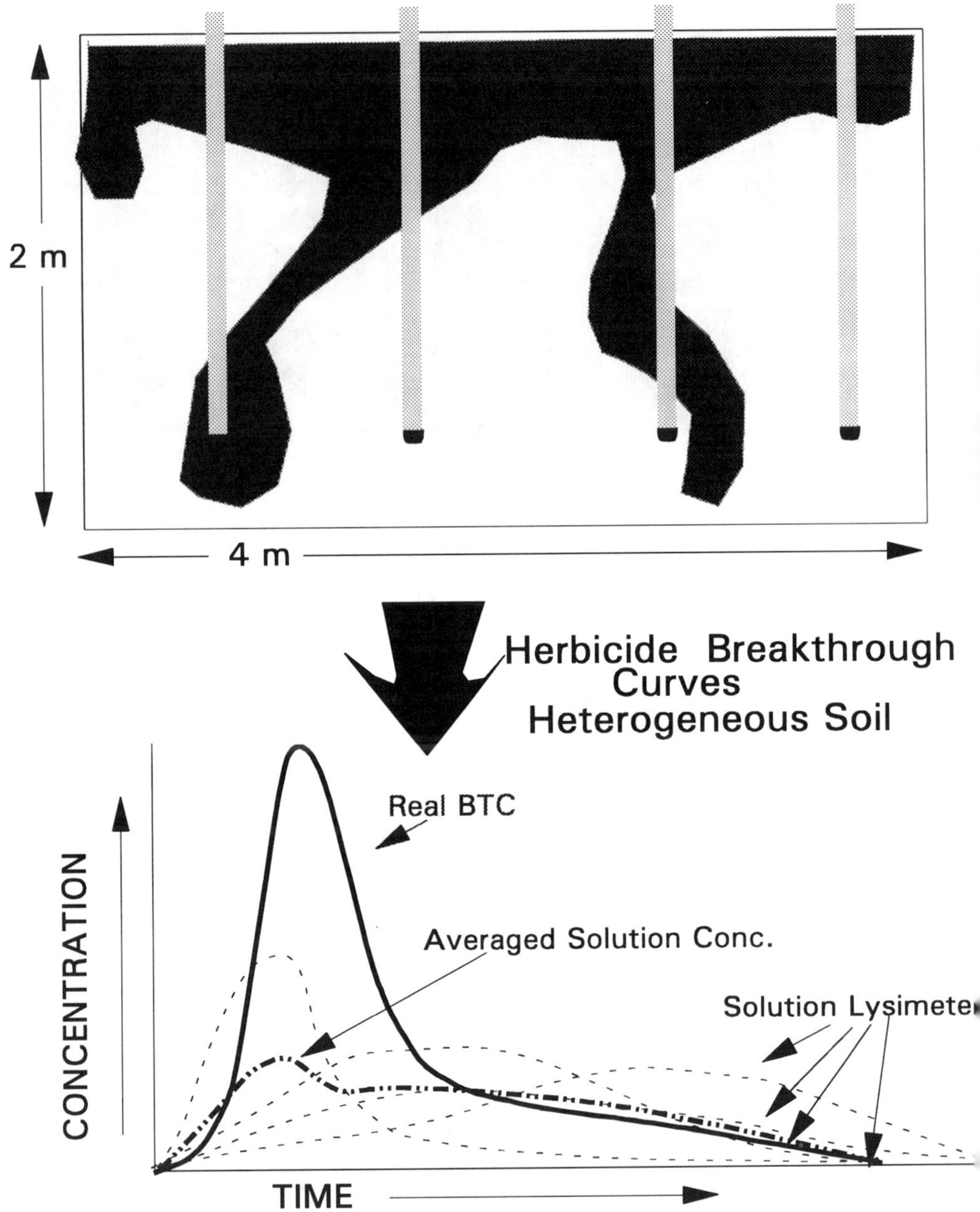

Figure 7. Schematic representation of chemical transport through a heterogeneous soil profile with corresponding breakthrough curves. Dotted lines represent individual breakthrough curves (BTC) for each solution sampler, while the solid line represents a flux-averaged BTC.

transport. Ju (1993) indicated that the difference between flux-averaged and the concentration-averaged peak heights can be as much as 400%. In a homogeneous soil, the flux-averaged and concentration-averaged breakthrough curves would be identical, resulting in ΔM and ΔH indexes that approach zero. As preferential flow becomes more dominant, the ΔM and ΔH indexes would increase (Figure 8). Subsequently, the magnitude of these indexes could be used to determine the extent of preferential transport *in situ*. Laboratory and large field lysimeters are presently being used to evaluate how the magnitude of these indexes correspond to the extent of preferential transport so that water quality sites can be characterized relative to their susceptibility to preferential chemical transport.

The new federal guidelines for small-scale groundwater monitoring studies require the application of a chemically conservative tracer (EPA, 1995). Consequently, the magnitude of ΔM and ΔH for these conservative tracers could be used to determine the relative heterogeneity of a field site, allowing transport data between sites to be grouped as a function of these indexes. Furthermore, ΔM and ΔH could be used to normalize transport data so that estimates of environmental impact could be determined.

III. Pesticide Volatilization

With regard to environmental loss pathways associated with pesticide use, most of the research has been focused on reducing runoff in surface water and leaching to ground water. Much less attention has been directed toward gaining a knowledge of pesticide volatilization and subsequent redeposition. Runoff and leaching losses of herbicides can be significant; however, frequently these loss pathways represent <5% of the applied herbicide while volatilization losses can often exceed 10%. Glotfelty et al. (1984) measured volatilization losses of 50% in <3 days, for heptachlor, chlordane, lindane, and trifluralin while Taylor (1978) reported volatilization losses of 90% for dieldrin, trifluralin, and heptachlor in 3 days. Wienhold and Gish (1994) measured alachlor losses of 14% of that applied to a tilled field under continuous corn production.

In some environments, volatilization and subsequent redeposition from the atmosphere is the major source of contamination (Eisenreich et al., 1981). Recently, Nations and Hallberg (1992) reported that several pesticides were found in rainwater; however, the volatilization processes and mode of transport were not well understood. Long-range atmospheric transport of pesticides to ecologically sensitive areas has also been made evident (Bidleman et al., 1989; McConnell et al., 1993). Consequently, methods for measuring and predicting, and eventually reducing pesticide volatilization and subsequent redeposition are critical research endeavors.

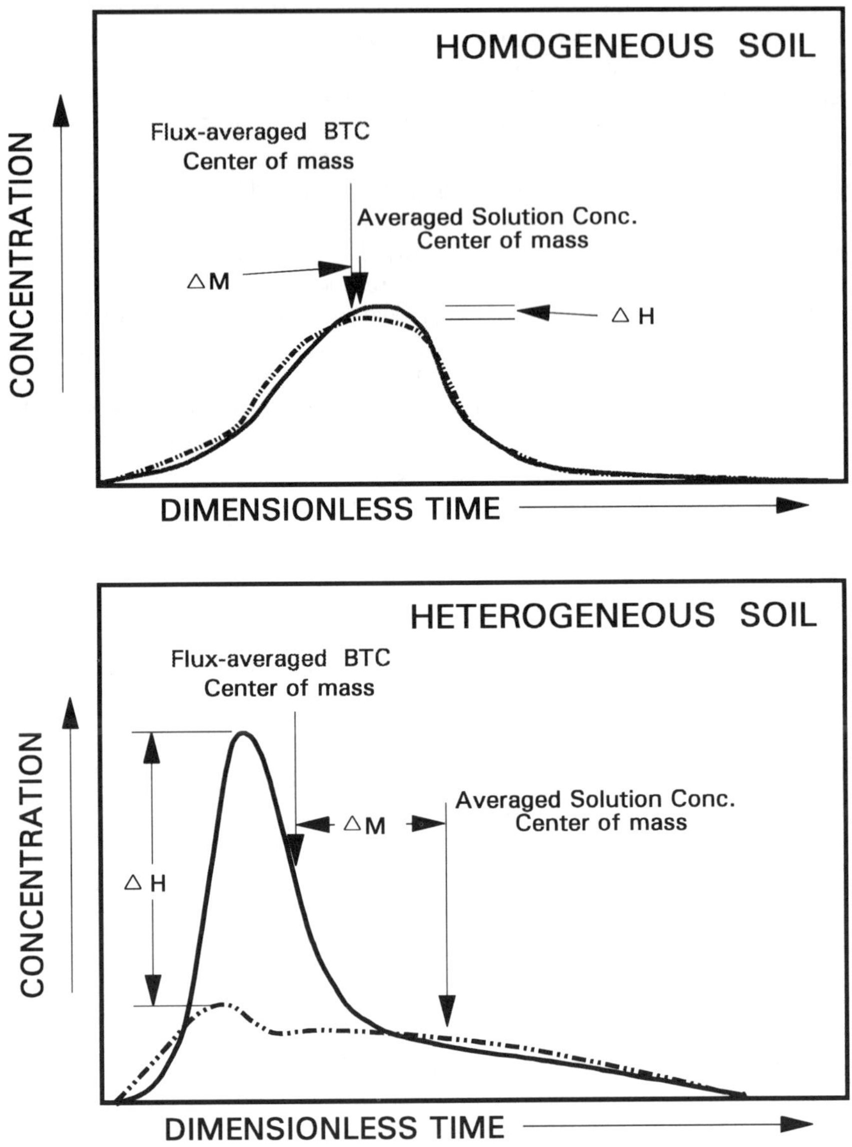

Figure 8. Comparison of flux- and concentration-averaged chemical breakthrough curves for a homogeneous and heterogeneous soil. Heterogeneous flux indices are first moment differential (△M) and peak height differential (△H) between flux- and concentration-averaged chemical breakthrough curves.

A. Measurement and Evaluation

Pesticides enter the atmosphere through volatilization from soil and plant residues, from drift during application, and from the physical transport of minute dust particles on which pesticides are attached. Consequently, pesticide concentrations associated with both air and soil particles are frequently measured (Junge, 1977; Duinker and Bouchertall, 1989; Ligocki and Pankow, 1989; Foreman and Bidleman, 1990). However, in evaluating pesticide redeposition, the predominant phase in which pesticides reside appears to be the vapor phase (Williams, 1986; Glotfelty et al., 1990). As a consequence, the primary processes governing pesticide volatilization can be grouped into two categories: 1) those that influence pesticide evaporation into the atmosphere immediately above soil and plant surfaces and, 2) those that influence pesticide mixing in the overlying atmosphere.

The saturated vapor pressure and solubility of the pesticide are the two most critical pesticide properties influencing pesticide volatility immediately above soil and its dispersal through the atmosphere. For dilute solutions, equilibrium between the pesticide vapor (a function of saturated vapor pressure) and the solution phase (a function of solubility) appears to follow Henry's Law (Moore, 1972). As a consequence, the Henry's Law constant, K_H, is often used as an index to characterize potential pesticide volatility. In an excellent review, Taylor (1995) lists the vapor pressure and Henry's Law constants for a number of pesticides. Nye (1979) and Hance (1980) propose that pesticides with $K_H > 10^{-4}$ diffuse primarily through the vapor phase and can therefore be classified as volatile compounds, while those pesticides where $K_H < 10^{-4}$ will diffuse primarily through the liquid phase and will therefore be much less volatile.

Along with the pesticide chemistry, soil moisture content and temperature are two variables that can have a dramatic impact on pesticide volatilization into the atmosphere immediately above soil and plant residues. Pesticide volatilization is greater from wet than dry soil, in part due to increased adsorption in dry soils (Spencer et al., 1969; Spencer and Cliath, 1974) and because less of the pesticide is in solution when the soil is dry. Under moist soil conditions, increases in soil temperature will also result in increased volatilization losses (Wienhold et al., 1993). However, as soils become increasingly dry, the temperature effects eventually become negligible. Parochetti and Warren (1966) observed that pesticide volatilization from wet soils increased with temperature, but temperature differential dampened out to near zero as the soil dried. The relationship between soil moisture content and herbicide volatilization is apparently reversible. If a dry soil is re-wetted, or exposed to 100% relative humidity, the pesticide volatilization rates will again increase dramatically (Spencer and Cliath, 1973; Prueger et al., 1993). Soil organic carbon can also influence pesticide volatilization by increasing soil adsorption (Chiou and Shoup, 1985). A doubling of the organic matter content essentially reduces herbicide volatilization by half (Spencer and Cliath, 1974).

Pesticide mixing into the atmosphere overlying agricultural lands is a function of several microclimatic variables. In the field, surface soil moisture will influence the relative humidity immediately above soil and plant residues. Thus, the relative humidity profile above the soil surface is important in evaluating or measuring pesticide volatilization (Spencer and Cliath, 1973, 1974; Harper et al., 1976; Glotfelty et al., 1984). Increasing wind speed and net solar radiation also increase pesticide volatilization (Burkhard and Guth, 1981; Glotfelty et al., 1984; Basile et al., 1986). In addition, the interaction of humidity, wind speed, and solar radiation is critical to evaluating pesticide volatilization rates, and as such requires frequent monitoring. Accurate measurements of herbicide volatilization and mixing into the atmosphere require a complex, expensive, labor and data intensive system where a complete energy balance can be constructed both day and night for several weeks. Since the cost associated with this type of study is prohibitive, alternative approaches for measuring pesticide vapor losses have been developed.

At present, there are three major approaches for determining pesticide volatilization into the atmosphere from a field: the energy balance, aerodynamic, and theoretical profile methods. In general, all three methods measure or calculate a pesticide vapor profile and subject that profile to microclimatic variables. The energy balance method measures pesticide vapor concentrations at various heights above the soil and links this pesticide vapor profile to measured temperature, solar radiation fluxes, and water vapor and wind speed profiles (Parmele et al., 1972; Majewski et al., 1990). Although the most theoretically accurate method, this approach is extremely labor intensive since all pesticide samples and microclimatic variables must be measured simultaneously on a short time scale, typically every 2 hours. In addition, a clear fetch of 100 times the height of the sampling tower is necessary in the upwind direction. Accordingly, the smallest treated area over which an acceptable pesticide concentration profile can be determined is approximately 1 ha (Glotfelty and Schomburg, 1989). The aerodynamic method for estimating pesticide vapor fluxes is probably the most common approach since data requirements and instrumentation needs are reduced relative to the energy flux method. The aerodynamic method utilizes the pesticide vapor and wind speed profiles above the field to calculate pesticide volatilization. Unfortunately, this method is likely to give unacceptable results if the field is subjected to strong surface heating or radiative cooling (Montieth, 1973). Additionally, the aerodynamic method is still subjected to the same field size limitations as the energy flux method. The third alternative, the theoretical profile method, requires wind speed and pesticide vapor sampling at one height above the soil and employs a two-dimensional theoretical trajectory model to predict vapor flux (Wilson et al., 1981). As a consequence, sample collection and analysis is reduced relative to the other two methods. Additionally, this latter approach is especially useful when field size is between 0.2 to 1 ha. Unfortunately, all vapor flux methods have considerable uncertainty associated with them. How are scale-dependent, spatially and temporally varying energy fluxes to be combined? How much uncertainty is associated with the calculated pesticide vapor flux? In a comparison of several vapor flux approaches, Majewski et al. (1990) concluded that there is considerable uncertainty with each method and that a better method for measuring pesticide vapor fluxes is needed.

Volatilization chambers are another alternative for estimating pesticide vapor losses. Much of the early work with pesticide volatilization (Spencer and Cliath, 1970, 1972, 1973; Nash et al., 1977) was performed with columns/chambers where air was passed over the soil surface and subsequently collected and analyzed. The chambers are useful in that they measure pesticide vapor losses from a finite soil area that is subjected to a given soil moisture content and wind speed. Since chambers monitor the pesticide vapor density close to the soil surface, and not necessarily the pesticide flux, they can only be used to obtain relative estimates of the pesticide vapor loss. However, this approach is inexpensive, does not require extensive sampling or micrometerological instrumentation, and is easy to use. Accordingly, Wienhold and Gish (1994) developed a mobile low-air flow volatilization chamber that is periodically moved to minimize localized temperature, soil moisture, and pesticide vapor density differences. The low air-flow minimizes negative pressure within the chamber that can affect volatilization rates. Additionally, pesticide vapor samples are collected as a function of days instead of minutes, reducing sample collection and analysis. This method is especially useful for relative comparisons of pesticide volatilization losses. Because chambers are insensitive to wind speed profile outside of the chamber, this method will under-predict watershed-scale volatilization losses on windy days and may over-predict on calm days.

In summary, although chambers and aerodynamic methods are easy to use, the energy balance technique is best equipped for generating the databases that will link pesticide volatilization to pesticide chemistry and climatic variables. However, due to the complexity and labor requirements for 24-hour monitoring, the energy-balance method is rarely utilized. As a result, no index has been developed that can accurately predict pesticide volatilization over a range of climatic conditions.

B. Control

Pesticide volatilization can be affected by various management practices such as the depth of pesticide incorporation, surface litter management, and pesticide formulation used. Glotfelty et al. (1984) observed that 75% of the applied heptachlor had volatilized from a moist soil surface whereas incorporating the same compound to a depth of 7.5-cm reduced volatilization losses to only 7% (Taylor et al., 1976). Because pesticide volatilization is reduced with soil incorporation, it has been proposed that increased usage of conservation tillage practices may lead to widespread losses of pesticide through volatilization. Whang et al. (1993) measured volatilization losses of three pesticides from side-by-side conventionally tilled and no-tillage fields. The higher volatilization losses observed under no-tillage could be a result of lower affinities of pesticide for vegetation than soil (Willis et al., 1983) or increased surface area and greater aerodynamic roughness, a consequence of plant residue and turbulent air flow. On the other hand, changes in climatic conditions dramatically alter pesticide volatilization and tillage interactions. Wienhold and Gish (1994) observed that small rain events that occurred 2 to 3 days after application could reduce pesticide volatilization losses from no-tillage relative to tilled fields. Atrazine and alachlor volatilization losses resulting from the early rain events were about 50% less with no-tillage than with tillage (Figure 9). Reduction in herbicide losses with no-tillage was attributed to pesticide washoff (Martin et al., 1978), placing the pesticide below the residue where soil temperatures and air exchange were reduced relative to the bare soil. However, this stagnant air-boundary resulting from the residue-precipitation interaction should also be a function of the Henry's Law constant, K_H, for each herbicide being applied (Gish et al., 1995b). In support of these findings, Spencer et al. (1988) found that volatilization of organics compounds with low K_H values are controlled by the air-boundary layer above the soil surface, while those with a higher K_H were less influenced by stagnant air. Consequently, interactions of pesticide application, plant residue, and climate need to be studied so that management practices can be developed that reduce pesticide volatilization for various climatic zones.

Increased awareness associated with detrimental impacts of pesticides has encouraged the development of new formulations that might reduce volatilization losses and thereby increase pesticide efficacy (Coffman and Gentner, 1980; and Coffman et al., 1984). Microencapsulation of chlorpropham in nylon capsules decreased volatilization losses five-fold relative to the emulsifiable concentrate (Turner et al., 1978). Schreiber et al. (1987) suggested that starch-encapsulation of herbicides should reduce volatilization losses by controlling the rate at which the chemical is released into the environment. Recently, Wienhold et al. (1993) showed that on a continuously moist soil, starch-encapsulation decreased atrazine volatilization, but increased or had no effect on alachlor. This study was extended to field conditions where starch-encapsulation reduced atrazine and alachlor volatilization on no-tillage soil (Figure 9; Wienhold and Gish, 1994). Consequently, controlled release formulations may be a viable alternative for reducing herbicide volatilization in the field, but their weed-control effectiveness also needs to be coupled with simultaneous measurements of runoff and leaching.

C. Prediction

With increased awareness associated with pesticide volatilization, deterministic pesticide volatilization models are becoming increasingly important in managing pesticide use and for estimating environmental impacts. To be effective for a wide range of soil and climatic conditions, these models must be able to account for pesticide evaporation from plant and soil material, determine the impact of microclimatic factors on pesticide volatilization rates (Basile et al., 1986), and predict the pesticide vapor concentrations as a function of height and distance downwind from an agricultural field.

To model pesticide vapor losses from soil and plant material, many soil-water-herbicide interactions must be characterized. Accordingly, Jury et al. (1983, 1984) developed a model for evaluating pesticide vapor losses from soil by combining

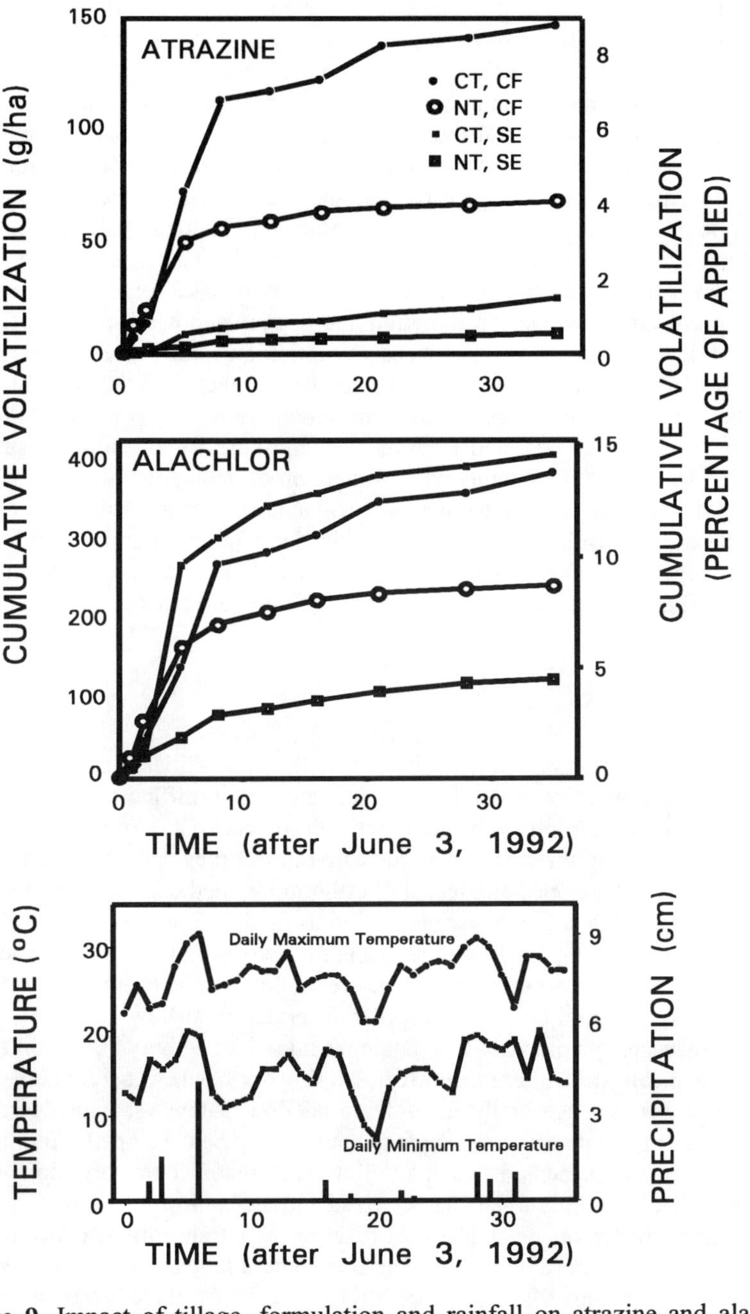

Figure 9. Impact of tillage, formulation and rainfall on atrazine and alachlor volatilization.

flow and continuity equations for pesticides in the sorbed, liquid, and gaseous phases in which molecular diffusion, convective transport, and degradation were accounted for. This Behavior Assessment Model (BAM) characterized many soil variables, but is a one-dimensional, isothermal, steady state model. Streile (1984) proposed a similar model but included temperature dependence. In an attempt to estimate off-site pesticide transport, Yates (1993) combined BAM (Jury et al., 1983) with wind speed profiles and the theoretical profile method (Wilson et al., 1981). In general, all these models can be used to categorize pesticide volatility as a function of a given set of environmental parameters. Although these models are useful in a screening mode, they are not useful for predicting pesticide volatilization losses that are cyclic due to periodic wetting and drying cycles, nor do they account for temporal variations in solar radiation interception and wind speed profiles.

In the development of a comprehensive pesticide volatilization model, several pesticide, soil, agricultural management, and climatic factors and their interactions must be accounted for. These factors and processes are: i) pesticide solubility, saturated vapor pressures, and Henry's Law constant; ii) soil bulk density, pH, pesticide soil adsorption isotherms as a function of depth, degradation rates as a function of water content and temperature, soil temperature, and soil water content; iii) mode and application rates of the pesticide, depth of incorporation, and surface litter management; iv) soil heat flux, sensible heat flux, latent heat flux, and wind speed profiles; and v) atmospheric pesticide degradation rates.

IV. Conclusions

Preferential flow and subsequent transport is the most critical pesticide leaching process. Although preferential pesticide transport is difficult to capture with solution samplers, it is impossible to detect with soil core data alone. Consequently, methods for determining "actual" field-scale transit times with solution lysimeter data need to be developed and tested. Furthermore, methods for determining the best location of the solution samplers need to be developed, especially on fine-textured soils. Two approaches for reducing the detrimental impact of preferential water flow and transport on a field-scale are: i) forcing the chemical into the soil matrix where it would be less susceptible to preferential flow processes, and ii) utilizing management practices that maximize microbial activity in the soil regions where chemicals are being transported. More work needs to be done on linking pesticide transport, especially in preferential flow pathways, with degradation processes so that more accurate simulation models can be developed. Additionally, an approach was presented that may allow preferential transport and field-site chemical flux variability to be characterized and indexed.

Climatic impact on pesticide volatilization and transport also needs to be investigated more thoroughly since there is some ambiguity associated with the impact of conservation tillage practices on field-scale herbicide behavior. Several approaches were discussed that would allow at least a first approximation of the pesticide vapor flux to be measured, regardless of field size. Furthermore, a need exists to more fully understand the impact of microclimatic variables on pesticide volatilization and redeposition so that it can be modeled and accurately predicted. Although an accurate volatilization index has not yet been developed, such an index must be linked to climatic variables. As a consequence, more emphasis should be given to measuring pesticide volatilization as a function of net energy balance.

References

Allmaras, R.R., R.E. Burwell, and R.R. Holt. 1967. Plow-layer and surface roughness from tillage as influenced by initial porosity and soil moisture of tillage time. *Soil Sci. Soc. Am. Proc.* 31:550-556.

Baker, R.S. and D. Hillel. 1990. Laboratory tests of a theory of fingering during infiltration into layered soils. *Soil Sci. Soc. Am. J.* 54:20-30.

Barbee, G.C. and K.W. Brown. 1986. Comparison between suction and free-drainage soil suction samplers. *Soil Sci.* 141:149-154.

Basile, M., N. Senesi, and F. Lamberti. 1986. A study of some factors affecting volatilization losses of 1,3-dichloropropene (1,3-D) from soil. *Agric. Ecosyst. Environ.* 17:269-279.

Bidleman, T.F., G.W. Patton, M.D. Walla, B.T. Hargrave, W.P. Vass, P. Ericson, B. Fowler, V. Scott, and D.J. Gregor. 1989. Toxaphene and other organochlorines in Arctic Ocean fauna: Evidence for atmospheric delivery. *Arctic* 42:307-313.

Biggar, J.W. and D.R. Nielsen. 1976. Spatial variability of the leaching characteristics of a field soil. *Water Resour. Res.* 12:78-84.

Bouma, J. 1981. Soil morphology and preferential flow along macropores. *Agric. Water Mgmt.* 3:235-250.

Brown, K.W., J.C. Thomas, and M.W. Holder. 1986. Development of a capillary wick unsaturated zone sampler. Coop. Agreement CR812316-01. U.S. Environ. Protection Agency Monit. Sys. Lab., Las Vegas, NV.

Burkhard, N. and J.A. Guth. 1981. Rate of volatilization of pesticides from soil surfaces; comparison of calculated results with those determined in a laboratory model system. *Pestic. Sci.* 12:37-44.

Burwell, R.E., R.R. Allamaras, and L.L. Sloneker. 1966. Structural alterations of soil surfaces by tillage and rainfall. *J. Soil Water Conserv.* 21:61-63.

Chiou, C.T. and T.D. Shoup. 1985. Soil sorption of organic vapors and effects of humidity on sorptive mechanism and capacity. *Environ. Sci. Technol.* 19:1196-1200.

Coffman, C.B. and W.A. Gentner. 1980. Persistence of several controlled release formulations of trifluralin in greenhouse and field. *Weed Sci.* 28:21-23.

Coffman, C.B., W.A. Gentner, and B.S. Shasha. 1984. Herbicide activity of controlled-release formulations of trifluralin. *Indian J. Agric. Sci.* 54:117-122.

Cushman, J.H. 1987. Development of stochastic partial differential equations for subsurface hydrology. *Stoch. Hydrol.* 1:241-262.

Dagan, G. and E. Bresler. 1979. Solute dispersion in unsaturated soil at field scale: I. Theory. *Soil Sci. Soc. Am. J.* 43:461-467.

Duinker, J.C. and F. Bouchertall. 1989. On the distribution of atmospheric PCB congeners between vapor phase, aerosols, and rain. *Environ. Sci. Technol.* 23:57.

Edwards, W.M., M.J. Shipitalo, W.A. Dick, and L.B. Owens. 1992. Rainfall intensity affects transport of water and chemicals through macropores in no-till soil. *Soil Sci. Soc. Am. J.* 56:52-58.

Eisenreich, S.J., B.B. Looney, and J.D. Thorton. 1981. Airborne organic contaminants in the Great Lakes ecosystem. *Environ. Sci. Technol.* 15:30-38.

Environmental Protection Agency. 1992. Another look: National survey of pesticides in drinking water wells. Phase II report. EPA 579/09-020. U.S. Environ. Protection Agency, Washington, D.C.

Environmental Protection Agency. 1995. Guidance for small-scale ground-water monitoring studies. Workshop draft, Environmental Fate and Effects Division, Office of Pesticide Programs, Mar. 16-17, Washington, D.C.

Foreman, W.T. and T.F. Bidleman. 1990. Semivolatile organic compounds in the ambient air of Denver, Colorado. *Atmos. Environ.* 24:2405-2416.

Freebairn, D.M., S.C. Gupta, C.A. Onstad, and W.J. Rawls. 1989. Antecedent rainfall and tillage effects on infiltration. *Soil Sci. Soc. Am. J.* 53:1183-1189.

Germann, P.F. and K. Beven. 1981. Water flow in soil macropores. I: An experimental approach. *J. Soil Sci.* 18:363-368.

Gish, T.J., A.R. Isensee, R.G. Nash, and C.S. Helling. 1991. Impact of pesticides on shallow groundwater quality. *Trans. Am. Soc. Agric. Eng.* 34:1745-1753.

Gish, T.J. and W.A. Jury. 1983. Effect of plant roots and root channels on solute transport. *Trans. Am. Soc. Agric. Eng.* 26:440-444, 451.

Gish, T.J., A. Sadeghi, and B.J. Wienhold. 1995b. Volatilization of atrazine and alachlor as influenced by surface litter. *Chemosphere* 31:2971-2982.

Gish, T.J., A. Shirmohammadi, R. Vyravipillai, and B.J. Wienhold. 1995a. Herbicide leaching under tilled and no-tillage fields. *Soil Sci. Soc. Am. J.* 59:895-901.

Gish, T.J., A. Shirmohammadi, and B.J. Wienhold. 1994. Field-scale mobility and persistence of commercial and starch-encapsulated atrazine and alachlor. *J. Environ. Qual.* 23:355-359.

Glass, R.J., J.-Y. Parlange, and T.S. Steenhuis. 1989. Wetting front instability: I. Theoretical discussion and dimensional analysis. *Water Resour. Res.* 25:1187-1194.

Glotfelty, D.E. and C.J. Schomburg. 1989. Volatilization of pesticides from soil. p. 181-207. In: B.L. Sawhney and K. Brown (eds.), *Reaction and Movement of Organic Chemicals in Soil.* Soil. Sci. Soc. Am. Spec. Publ. 22, Am. Soc. Agron. and Soil Sci. Soc. Am., Madison, WI.

Glotfelty, D.E., A.W. Taylor, B.C. Turner, and W.H. Zoller. 1984. Volatilization of surface-applied pesticides from fallow soil. *J. Agric. Food Chem.* 32:638-643.

Glotfelty, D.E., G.H. Williams, H.P. Freeman, and M.M. Leech. 1990. Regional atmospheric transport and deposition of pesticides in Maryland. p. 199-221. In: D.A. Kurtz (ed.), *Long Range Transport of Pesticides*. Lewis Publ. Inc., Chelsea, MI.

Hance, R.J. 1980. Transport in the vapor phase. p. 59-81. In: *Interactions of Herbicides with Soil*. Academic Press, London.

Harper, L.A., A.W. White, Jr., R.R. Bruce, A.W. Thomas, and R.A Leonard. 1976. Soil and microclimate effects on trifluralin volatilization. *J. Environ. Qual.* 5:236-242.

Helling, C.S. and T.J. Gish. 1991. Physical and chemical processes affecting preferential flow. p. 77-86. In: T.J. Gish and A. Shirmohammadi (eds.), *Preferential Flow*, Proc. Natl. Symp., December 16-17, 1991, Chicago, IL. Am. Soc. Agric. Eng., St. Joseph, MI.

Holder, M., K.W. Brown, J.C. Thomas, D. Zabick, and H.E. Murray. 1991. Capillary-wick unsaturated zone soil pore water samplers. *Soil Sci. Soc. Am. J.* 55:1195-1202.

Isensee, A.R., R.G. Nash, and C.S. Helling. 1990. Effect of conventional vs. no-tillage on pesticide leaching to groundwater. *J. Environ. Qual.* 19:434-440.

Ju, S-H. 1993. Properties of funnel-type preferential flow and its inference on contaminant transport, sampling, and modeling. 1993. Ph.D. thesis, Agric. Eng. Dept., Univ. Wisc., Madison, WI.

Ju, S-H. and K.-J.S. Kung. 1993. Simulating funnel-type preferential flow and overall flow property induced by multiple soil layers. *J. Environ. Qual.* 22:432-442.

Junge, C.E. 1977. Basic considerations about trace constituents in the atmosphere as related to the fate of global pollutants. In: I.H. Suffett (ed.), *Adv. Environ. Sci. Technol.*, vol. 8.

Jury, W.A., W.F. Spencer, and W.J. Farmer. 1983. Behavior assessment model for trace organics in soil: I. Model description. *J. Environ. Qual.* 12:558-564.

Jury, W.A., W.F. Spencer, and W.J. Farmer. 1984. Behavior assessment model for trace organics in soil: II. Chemical classification and parameter sensitivity. *J. Environ. Qual.* 13:567-572.

Kladivko, E.J., G.E. Van Scoyoc, E.J. Monke, K.M. Oates, and W. Pask. 1991. Pesticide and nutrient movement into subsurface tile drains on a silt loam soil in Indiana. *J. Environ. Qual.* 20:264-270.

Knutson, J.H. and J.S. Selker. 1994. Unsaturated hydraulic conductivities of fiberglass wicks in designing capillary porewater samplers. *Soil Sci. Soc. Am. J.* 58:721-729.

Kung, K.-J.S. 1990. Preferential flow in a sandy vadose zone: I: Field Observation. *Geoderma* 46:51-58.

Kung, K.-J.S. and S.V. Donohue. 1991. Improved solute-sampling protocol in a sandy vadose zone using ground-penetrating radar. *Soil Sci. Sci. Am. J.* 55:1453-1455.

Kung, K.-J.S. and Z-B. Lu. 1993. Using ground-penetrating radar to detect layers with abrupt discontinuity in dielectric constant from their surroundings. *Soil Sci. Soc. Am. J.* 57:335-340.

Libardi, P.L., K. Reichardt, D.R. Nelson, and J.W. Biggar. 1980. Simple field methods for estimating soil hydraulic conductivity. *Soil Sci. Soc. Am. J.* 44:3-6.

Ligocki, M.P. and J.F. Pankow. 1989. Measurements of gas-particle distributions of atmospheric organic compounds. *Environ. Sci. Technol.* 23:75-83.

Litaor, M.I. 1988. Review of soil solution samplers. *Water Resour. Res.* 24:727-733.

Luxmoore, R.J. 1991. On preferential flow and its measurement. In: T.J. Gish and A. Shirmohammadi (eds.), Preferential Flow. Proc. Natl. Symp., Chicago, IL, December 16-17, 1991. *Am. Soc. Agric. Eng.*, p. 113-121.

Majewski, M.S., D.E. Glotfelty, U. Kyaw Tha Paw, and J.N. Seiber. 1990. A field comparison of several methods for measuring pesticide evaporation rates from soil. *Environ Sci. Technol.* 24:1490-1496.

Martin, C.D., J.L. Baker, D.C. Erbach, and H.P. Johnson. 1978. Washoff of herbicides applied to corn residue. *Trans. Am. Soc. Agric. Eng.* 21:1164-1168.

McConnell, L.L., J.R. Kucklick, T.F. Bidleman, M.D. Walla. 1993. Long-range atmospheric transport of toxaphene to Lake Baikal. *Chemosphere* 27:2027-2036.

Montieth J.L. 1973. *Principles of Environmental Physics*. American Elsevier, New York.

Moore, W.J. 1972. *Physical Chemistry, 4th Ed.* Prentice Hall, Inc., Englewood Cliffs, New Jersey.

Nash, R.G., M.L. Beall, Jr., and W.G. Harris. 1977. Toxaphene and 1,1,1-trichloro-2,2-bis-(*p*-chlorophenyl)ethane (DDT) losses from cotton in an agroecosystem chamber. *J. Agric. Food Chem.* 25:336-341.

Nations, B.K. and G.R. Hallberg. 1992. Pesticides in Iowa precipitation. *J. Environ. Qual.* 21:486-492.

Nielsen, D.R., J.W. Biggar, and K.T. Erh. 1973. Spatial variability of field measured soil water properties. *Hilgardia* 2:215-259.

Nye, P.H. 1979. Diffusion of ions and uncharged solutes in soils and soil clays. *Adv. Agron.* 31:225-272.

Parmele, H., E.R. Lemon, and A.W. Taylor. 1972. Micrometerological measurements of pesticide vapor flux from bare soil and corn under field conditions. *Water Air Soil Pollu.* 1:433-451.

Parochetti, J.V. and G.F. Warren. 1966. Vapor losses of IPC and CIPC. *Weeds* 14:281-285.

Philips, R.E. and S.H. Philips (eds.). 1984. *No-Tillage Agriculture*. Van Nostrand Reinhold, New York. 301 pp.

Prueger, J.H., J.L. Hatfield, and R.L. Pfeiffer. 1993. Field measurements of metolachlor volatilization and surface energy components from a corn field using Bowen ratio and aerodynamic methods. *Agricultural Research to Protect Water Quality,* Conf. Proc., p. 149-151.

Quisenberry, V.L., R.E. Phillips, P.A. Nelson, and M.W. Hafield. 1991. Preferential flow in structured and non-structured soils. In: *Characterization of Transport Phenomena in the Vadose Zone*, Apr. 2-5, 1991, Tucson, AZ. p. 50-51.

Rawls, W.J., C.A. Onstad, and H.H. Richardson. 1980. Residue and tillage effects on SCS runoff curve numbers. *Trans. Am. Soc. Ag. Eng.* 23:357-361.

Richards, T.L. and T.S. Steenhuis. 1988. Tile drain sampling of preferential flow on a field scale. In: P.F. Germann (ed.), Rapid and far reaching hydrological processes in the vadose zone. *J. Contam. Hydrology* 3:307-325.

Rimmer, A., T.S. Steenhuis, and J.S. Selker. 1995. One-dimensional model to evaluate the performance of wick samplers in soils. *Soil Sci. Soc. Am. J.* 59:88-92.

Roth, K., W.A. Jury, H. Fluhler, and W. Attinger. 1991. Transport of chloride through an unsaturated field soil. *Water Resour. Res.* 27:2533-2541.

Schreiber, M.M., M.V. Hickman, and G.D. Vail. 1994. Efficacy of starch-encapsulated formulations of atrazine containing two or three herbicides in same granule. *Weed Technol.* 8:105-113.

Schreiber, M.M., B.S. Shasha, D. Trimnell, and M.D. White. 1987. Controlled release herbicides. p. 177-191. In: C.G. McWhorter and M. Gebhardt (eds.), *Methods of Applying Herbicides.* Weed Sci. Soc. Am. Monogr. Ser. WSSA, Champaign, IL.

Shipitalo, M.J., W.M. Edwards, W.A. Dick, and L.B. Owens. 1990. Initial storm effects on macropore transport surface-applied chemicals in no-till soil. *Soil Sci. Soc. Am. J.* 54:1530-1536.

Spencer, W.F. and M.M. Cliath. 1970. Desorption of lindane from soil as related to vapor density. *Soil Sci. Soc. Am. Proc.* 34:574-578.

Spencer, W.F. and M.M. Cliath. 1972. Volatility of DDT and related compounds. *J. Agric. Food Chem.* 20:645-649.

Spencer, W.F. and M.M. Cliath. 1973. Pesticide volatilization as related to water loss from soil. *J. Environ. Qual.* 2:284-289.

Spencer, W.F. and M.M. Cliath. 1974. Factors affecting vapor loss of trifluralin from soil. *J. Agric. Food Chem.* 22:987-991.

Spencer, W.F., M.M. Cliath, and W.J. Farmer. 1969. Vapor density of soil-applied dieldrin as related to soil-water content, temperature, and dieldrin concentration. *Soil Sci. Soc. Am. Proc.* 33:509-511.

Spencer, W.F., M.M. Claith, W.A. Jury, and L.-Z. Zhang. 1988. Volatilization of organic chemicals from soil as related to their Henry's law constant. *J. Environ. Qual.* 17:504-509.

Stearman, G.K., R.J. Lewis, L.J. Tortorelli, and D.D. Taylor. 1989. Herbicide reactivity of soil organic matter in no-tilled and tilled cotton. *Soil Sci. Soc. Am. J.* 53:1690-1694.

Steenhuis, T.S., J. Boll, E. Jolles, and J. Selker. 1996. Field evaluation of wicks and gravity pan samplers. In: L. Everett (ed.), *Vadose Zone Characterization and Monitoring: Principles, Methods, and Case Studies*. Lewis Publ., Chelsea, MI.

Stevenson, F.J. 1972. Organic matter reactions involving herbicides in soil. *J. Environ. Qual.* 1:333-343.

Streile, G.P. 1984. The Effect of Temperature on Pesticide Phase Partitioning, Transport, and Volatilization from Soil. Ph.D. dissertation. Soil and Environ. Sci. Dept., Univ. Calif., Riverside, CA.

Taylor, A.W. 1978. Post-application volatilization of pesticides under field conditions. *J. Air Pollu. Control Assoc.* 28:922-927.

Taylor, A.W. 1995. The volatility of pesticide residues. p. 258-306. In: T.R. Roberts and P.C. Kearney (eds.), *Environmental Behavior of Agrochemicals*.

Taylor, A.W., D.E. Glotfelty, B.L. Glass, H.P. Freeman, and W.M. Edwards. 1976. Volatilization of dieldrin and heptachlor from a maize field. *J. Agric. Food Chem*. 24:625-634.

Turner, B.C., D.E. Glotfelty, A.W. Taylor, and D.R. Watson. 1978. Volatilization of microencapsulated and conventionally applied chlorpropham in the field. *Agron. J.* 70:933-937.

van Genuchten, M.Th. and R.J. Wagenet. 1989. Two-site/two-region models for pesticide transport and degradation: Theoretical developments and analytical solutions. *Soil Sci. Soc. Am. J.* 53:1303-1310.

van Schilfgaarde, J. 1991. Water resource issues for the 21st century: An Agricultural Research Service perspective. p. 20-26. In: W.H. Blackburn and J.G. King (eds.), *Water Resource Challenges and Opportunities for the 21st Century, Proc. First USDA Water Resource Research and Technology Workshop*, Aug. 26-30, 1991, Denver, CO.

Whang, J.M., C.J. Schomburg, D.E. Glotfelty, and A.W. Taylor. 1993. Volatilization of fonofos, chlorpyrifos and atrazine from conventional and no-till surface soils in the field. *J. Environ. Qual.* 22:173-180.

Wienhold, B.J. and T.J. Gish. 1992. Effect of water potential, temperature, and soil microbial activity on release of starch-encapsulated atrazine and alachlor. *J. Environ. Qual.* 21:382-386.

Wienhold, B.J. and T.J. Gish. 1994. Effect of formulation and tillage practice on volatilization of atrazine and alachlor. *J. Environ. Qual.* 23:292-298.

Wienhold, B.J., A.M. Sadeghi, and T.J. Gish. 1993. Effect of starch encapsulation and temperature on volatilization of atrazine and alachlor. *J. Environ. Qual.* 22:162-166.

Williams, G.H. 1986. Field Measurement of Pesticide Washout in Rain Near Beltsville, Maryland. MS Thesis, Univ. Maryland, College Park, MD.

Williams, C.F., S.D. Nelson, and T.J. Gish. 1995. Release and mobility of starch-encapsulated atrazine in calcareous soils. *Clean Water-Clean Environment-21st. Century, Conf. Proc.* Am Soc. Agric. Eng. 1:173-174. March 5-8, 1995, Kansas City, MO.

Willis, G.H., L.L. McDowell, L.A. Harper, L.M. Southwick, and S. Smith. 1983. Seasonal disappearance and volatilization of toxaphene and DDT from a cotton field. *J. Environ. Qual.* 12:80-85.

Wilson, J.D., G.W. Thurtell, and G.E. Kidd. 1981. *Boundary Layer Meteorol.* 21:295-313.

Yates, S.R. 1993. Determining off-site concentrations of volatile pesticides using the trajectory-simulation model. *J. Environ. Qual.* 22:481-486.

Yeh, J.T.-C., L.W. Gelhar, and A.L. Gutjahr. 1985. Stochastic analysis of unsaturated flow in heterogeneous soils. I. Statistical isotropic media. *Water Resour. Res.* 21:447-456.

Integrated Crop Management and Off-Site Movement of Nutrients and Pesticides

J.L. Baker, S.K. Mickelson, and W.G. Crumpton

I. Introduction

A. Water Quality Concerns

Losses of agricultural chemicals, including herbicides used for weed control, with surface runoff water, eroded soil, and leaching water have the potential to pollute surface and groundwater resources. With respect to nutrients, concern usually is focused on the inorganic ions ammonium-nitrogen (NH_4-N), nitrate-nitrogen (NO_3-N), and phosphate-phosphorus (PO_4-P). For NH_4-N, the concern relates mainly to the toxicity of ammonia (NH_3) to fishes. A concentration of 0.02 mg L^{-1} NH_3

represents a toxic level to several species, and at normal temperatures and near neutral pH, this concentration would result from a concentration of about 2 mg L^{-1} NH_4-N. Hence the 2 mg L^{-1} NH_4-N concentration often is quoted as a level of concern (USEPA, 1976). For NO_3-N, the concern relates to drinking water and the conversion of nitrate to nitrite in the body, and the reaction of nitrite with hemoglobin in the blood to reduce its oxygen transport ability. This is particularly a concern in infants under three months of age. The enforceable drinking water standard, or maximum contaminant level (MCL, the maximum permissible level of a contaminant in water that is delivered to any user of a public water system), for NO_3-N is 10 mg L^{-1} (USEPA, 1976). For PO_4-P, the concern relates to the surface water ecosystem and the development of biological nuisances and accelerated eutrophication. Although a standard as such does not exist, it is believed that PO_4-P levels should not exceed 0.050 mg L^{-1} in any stream where it enters a lake, or 0.025 mg L^{-1} within the lake (USEPA, 1976).

With respect to pesticides used in row-crop production in the central U.S., concern usually is focused first on herbicides and then on insecticides. Human health concerns are expressed by drinking water standards (MCL) or health advisory (HA) levels as discussed by Benson (1989). These are based on the toxicity of the pesticides assessed by epidemiological and animal tests. Because of a large degree of uncertainty, a safety factor of 10 to 1000 often is used. For the 20 herbicides and six insecticides listed as significantly used on corn and soybeans in Tables 1 and 2, respectively, many MCL values are yet to be determined. Of the pesticides listed, MCL values currently exist only for atrazine (3 µg L^{-1}), alachlor (2 µg L^{-1}), 2,4-D (70 µg L^{-1}) and glyphosate (700 µg L^{-1}). Lifetime HA levels exist for 50% of the pesticides listed. Pesticides also are classified by the USEPA as to whether they are a human carcinogen, probable human carcinogen, possible human carcinogen, or noncarcinogen for humans.

Concerns also exist for the impact of pesticides on the aquatic ecosystem. Current aquatic testing requirements for pesticide registration consist of acute studies of pesticide concentrations in water on three organisms (two fishes and one daphid). Additional testing, as discussed in the Society of Environmental Toxicology and Chemistry (SETAC, 1994) report on aquatic risk assessment and mitigation, would include more species as well as chronic tests of pesticide exposure in water. As an example of data cited in this report, chlorpyrifos with lethal concentration to 50% of the test species (LC50) of 203 and 1.3 mg L^{-1} for fathead minnow and bluegill, respectively, caused 0.8% mortality for fathead minnows in a field study at a concentration of 32 mg L^{-1}, and at 6.3 mg L^{-1} caused 38% mortality for bluegill. There is a lot of uncertainty in this area as well, and, as stated by Minshall (1995), our "nation sometimes spends large sums of money removing contaminants that are not affecting aquatic organisms. However, the reliance on chemical criteria or laboratory-derived toxicological information, out of context of the entire environmental milieu to which organisms are exposed in nature, also has resulted in permitted levels of toxicants that exceed an aquatic population's (or community's) ability to survive."

Table 1. Agricultural chemicals used on corn in 1994 by major producing statea[a]

Chemical	Area Treated[b] %	Rate kg ha^{-1}	Total Applied metric tons
Fertilizer			
nitrogen	97	145	3,563,500
phosphate	83	65	1,357,200
potash	72	90	1,634,300
Herbicides[c]			
2,4-D	14	0.48	1,647
acetochlor	7	2.01	3,378
alachlor	17	2.22	9,673
atrazine	68	1.20	20,599
bromoxynil	10	0.27	656
cyanazine	21	2.35	12,560
dicamba	29	0.39	2,868
metolachlor	32	2.17	17,787
nicosulfuron	13	0.03	113
TOTALS	98	--	77,212
Insecticides[d]			
chlorpyrifos	8	1.21	2,361
fonofos	2	1.19	703
methyl parathion	2	0.59	335
permethrin	3	0.12	101
tefluthrin	4	0.12	110
terbufos	6	1.20	1,937
TOTALS	27	--	6,212

[a]Source: USDA-NASS, 1995. Agricultural Chemical Usage 1994 - Field Crops Summary. Economic Research Service, Washington, D.C.

[b]In 1994, 25.3 million ha planted to corn in ten major producing states.

[c]Only herbicides used on at least 7% of area are listed; most fields receive treatment with more than one herbicide.

[d]Only insecticides used on at least 2% of area are listed.

Table 2. Agricultural chemicals used on soybeans in 1994 by major producing states[a]

Chemical	Area Treated[b] %	Rate kg ha-1	Total Applied metric tons
Fertilizer			
nitrogen	13	24	65,500
phosphate	20	53	189,900
potash	25	92	414,000
Herbicides[c]			
2,4-D	10	0.45	811
acifluorfen	10	0.25	413
alachlor	7	2.25	2,732
bentazon	14	0.77	1,915
chlorimuron-ethyl	15	0.02	60
fluazifop-P-butyl	8	0.07	102
glyphosate	15	0.58	1,577
imazaquin	18	0.11	346
imazethapyr	42	0.07	474
metolachlor	8	2.23	3,185
metribuzin	10	0.34	620
pendimethalin	25	1.08	4,697
thifensulfuron	14	0.003	9
trifluralin	24	0.92	3,859
TOTALS	98	--	22,401

[a]Source: USDA-NASS, 1995. Agricultural Chemical Usage 1994 - Field Crops Summary. Economic Research Service, Washington, D.C.

[b]In 1994, 17.8 million ha planted to soybeans in nine major producing states.

[c]Only herbicides used on at least 7% of area are listed; most fields receive treatment with more than one herbicide.

B. Agricultural Chemical Use and Benefits

Intensive use is made of fertilizer and pesticides in corn and soybean row-crop production as shown in Tables 1 and 2, respectively. Of the corn grown in major producing states, 97% received nitrogen (N) fertilizer in 1994. At an average rate of 145 kg ha^{-1}, the amount of N applied to corn totalled over 3.5 million metric tons and at \$0.33 kg^{-1} (\$0.15 lb^{-1}) cost about \$1.2 billion. Phosphate was applied to at least 50% and potash to at least 90% of the corn and soybeans grown. Herbicides (often more than one) were used on 98% of the corn and soybeans grown, and 27% of the corn grown received an insecticide treatment. A total of over 105,000 metric tons of pesticides were used on corn and soybeans, and, at an average of \$20 kg^{-1} (\$9 lb^{-1}), cost about \$2.1 billion.

The economic benefits derived from agricultural chemical use generally are much greater than their costs, which explains their intensive use. In a five-year study of continuous corn production in Iowa (IDALS, 1994), the optimum rate of N was about 200 kg ha^{-1}, based on N at \$0.33 kg^{-1} and corn at \$98 metric ton^{-1} (\$2.50 bu^{-1}). At that rate, the increase in corn production over no N applied was about 3.6 metric tons ha^{-1}. Allowing \$7 ha^{-1} for application costs, the rate of return was \$353 divided by \$73, or 484% (in less than four months). In this same study, the best herbicide combination was atrazine (1.68 kg ha^{-1}) and metolachlor (2.80 kg ha^{-1}) applied preemergent. Corn yields for this treatment, with mechanical cultivation, averaged about 3.6 metric tons ha^{-1} greater than the no-herbicide treatment with mechanical cultivation. Again at \$98 metric ton^{-1} of corn and herbicide costs of \$77 ha (\$7 ha^{-1} application costs and \$70 ha^{-1} for product), the rate of return was \$353 divided by \$77, or 458%. In addition, in the fifth year of the study the estimated weed count on the atrazine/metolachlor plots was 300 ha^{-1}, while it was 3,000,000 ha^{-1} on the no-herbicide/mechanical-cultivation plots. Similar economic and weed count data were obtained for soybeans in rotation with corn with and without herbicide use.

C. Ground and Surface Water Surveys

In the recent past, several monitoring surveys have been performed to assess the contamination of groundwater by agricultural chemicals at national, regional, and state levels. At the national level, the USEPA (1990) monitored 564 community and 783 rural domestic wells across the U.S. for NO_3-N and 126 pesticides and pesticide metabolites. They found that about 7% of the well samples contained detectable pesticide(s), but less than 0.8% exceeded health concerns for any pesticide. About 2% of the wells exceeded the 10 mg L^{-1} MCL for NO_3-N. At the regional level, Monsanto Agricultural Company (1990) surveyed 1430 wells in a 26-state area of extensive herbicide and fertilizer use for alachlor as well as atrazine, simazine, metolachlor, and NO_3-N. They found that about 13% of the well samples contained detectable pesticide(s), but less than 0.1% exceeded MCL or HA levels for any pesticide. About 5% of the wells exceeded the 10 mg L^{-1} MCL for NO_3-N.

At the state level, surveys have been performed in several states, including Iowa (Kross et al., 1990) and Ohio (Baker et al., 1994). In Iowa, 686 rural domestic wells across the state were sampled for 16 pesticides and pesticide metabolites and NO_3-N in 1988-89. They found that about 14% of the well samples contained detectable pesticides, with about 1% exceeding MCL or HA levels for any pesticide. About 18% of the wells exceeded the 10 mg L^{-1} MCL for NO_3-N. It also was determined that shallow wells (<15 m) were more susceptible to contamination. In Ohio, over 16,000 wells were tested from 1987 to 1993. Of those tested for alachlor, atrazine, and cyanazine (610 wells), about 1% exceeded a MCL or HA level. For all wells, about 3% exceeded the 10 mg L^{-1} MCL for NO_3-N, with the likelihood of contamination decreasing with well depth below 15 m.

For surface water resources, the USEPA through Section 305 (b) of the Clean Water Act requires states to provide a biennial assessment of the quality of their rivers, lakes, and coastal waters. Part of this assessment must include "a description of the nature and extent of nonpoint sources of pollutants." In the most recent report (USEPA, 1994), data from 1992 were summarized and showed that agriculture was the leading source of impairment for rivers and lakes. A comparison with the 1988 data (USEPA, 1990) indicated that the overall quality of rivers and lakes may be declining with the impact of agriculture increasing while the impact from municipal/industrial and other sources was decreasing. In 1992, siltation, nutrients, and pesticides were involved in the impairment for 45, 37, and 26%, respectively, of the river miles assessed (this adds to greater than 100% because of multiple impairments); corresponding values for lake areas were 22, 40, and 9%. Overall, 40% of U.S. rivers, lakes, and estuaries were still not suitable for fishing and swimming.

As an example of more specific agricultural nonpoint source pollution, data are available for a five-year study in east-central Iowa of the quality of water in surface runoff, subsurface drainage, and stream flow for a 50-km^2 agricultural watershed (Johnson and Baker, 1982, 1984). Concentrations and losses for the herbicides, alachlor, atrazine, cyanazine, metribuzin, and propachlor, were found to be the greatest right after spring application. In-stream concentrations as high as 100 µg L^{-1} were measured for individual herbicides at times of peak surface runoff. In-stream herbicide concentrations in the periods between surface runoff events and over the winter were generally below detection (0.1 µg L^{-1}), with the exception of atrazine for which concentrations often ranged between 0.1 and 2.0 µg L^{-1}. Annual herbicide losses with stream flow were $\leq$ 2% of that applied in the watershed. Losses from treated areas at the field boundary showed much higher concentrations (with no dilution water with runoff from untreated areas as exists in the stream) and higher percent annual losses, up to 7.2% in a very wet year (with no field-to-stream attenuation as exists in a mixed-cover watershed).

In-stream NO_3-N concentrations for the same study area averaged about 9 mg L^{-1} and generally exceeded the 10 mg L^{-1} MCL when stream flow was dominated by the combination of base flow and subsurface drainage water flow through tile lines (tile drainage water samples averaged 12 mg L^{-1}). Annual NO_3-N losses from the whole watershed over the five-year study averaged 17 kg ha^{-1}. In-stream NH_4-N concentrations averaged about 0.5 mg L^{-1}, with the highest concentrations during snowmelt.

Concentrations in field surface runoff and tile drainage water generally averaged about 0.5 and 0.1 mg L^{-1}, respectively. One exception was field surface runoff with over 2 mg L^{-1} for an event 24 h after N fertilizer application. Annual NH_4-N losses from both fields and the whole watershed were usually $\leq$ 1 kg ha^{-1}. In-stream PO_4-P concentrations averaged about 0.1 mg L^{-1}. Concentrations in field surface runoff ranged from about 0.1 to over 1.0 mg L^{-1}, with the higher concentrations associated with surface P fertilization of row-cropped fields and with the cattle-grazed pasture site. Concentrations in the tile drainage water were generally $\leq$ 0.1 mg L^{-1}. Annual PO_4-P losses from both fields and the whole watershed were almost always $\leq$ 1 kg ha^{-1}.

On a larger scale (the Mississippi River watershed at three points above Baton Rouge, LA) and more recently (April 1991 to September 1992), the U.S. Geological Survey (Coupe et al., 1995) measured herbicides and NO_3-N in river water. Atrazine was detected in almost all samples with highest concentrations at all locations in May, June, and July (4.4 µg L^{-1} was the maximum for all samples) and an overall average concentration of less than 1 µg L^{-1}. For the next three most commonly detected herbicides, alachlor, metolachlor, and cyanazine, concentrations were also highest in May, June, and July. However, their maximum concentrations were somewhat lower being 0.9, 2.3, and 3.2 µg L^{-1}, respectively. In addition, there were extended periods when alachlor, metolachlor, and/or cyanazine concentrations were below the limit of detection. Concentrations of NO_3-N in river water ranged from about 1 to 5 mg L^{-1}, being highest during some high flow periods when evapotranspiration was the lowest (i.e., times other than summer). Leaching of NO_3-N through the surface soil profile with subsequent lateral movement with tile drainage water and base flow to streams feeding the Mississippi River could explain this trend.

D. Approach to the Problem

As noted by Puckett (1995) more than $540 billion has been spent since the 1972 Clean Water Act was passed, and yet in 1992 more than 40% of U.S. surface water resources did not fully support the uses of these waters as designated by the states (USEPA, 1994). However, because of the Act's point-source focus, nearly 90% of dollars spent on water pollution prevention have been spent on point-source controls. Less has been spent on nonpoint-source controls for a number of reasons, one of the major ones being it is not always evident how important nonpoint sources are in causing pollution problems compared to easily identifiable point-sources, such as sewage treatment plants and factory outlets. Generally, it is believed that prevention of nonpoint source pollution rather than cleanup, particularly for groundwater resources, is preferable.

A recent Office of Technology Assessment report (1995) addressed program strategies for reducing environmental impacts from agriculture, including soil and water quality problems. That report emphasized the need to target priority areas and apply low-cost solutions. Use of education and technical assistance, and minimal one-time subsidies, were cited as the methods of choice to promote adoption of technologies that simultaneously "enhance environmental quality and maintain acceptable profits." More costly management technologies could be encouraged through economic incentives or disincentives. Caution was expressed for use of costly programs used to conserve land through long-term retirement. In attempting to identify water quality priorities, the importance of the watershed scale and managing whole agroecosystems emerges, and the use and siting of both in-field (e.g., tillage, cropping, and chemical management) and off-site (e.g., vegetated buffer strips and wetlands) practices for reducing or preventing water quality problems becomes critical.

II. Factors Affecting Pollutant Transport

A. Transport Mechanisms

Transport of agricultural chemicals to water resources can take place adsorbed to sediment and dissolved in water in surface runoff and/or dissolved in leaching water from the bottom of the root zone. Surface runoff and leaching water can transport chemicals to surface water resources, the latter through base flow or artificial subsurface drainage (Figure 1). Groundwater resources are mostly impacted through leaching water. In some instances, such as alluvial groundwater systems, there can be groundwater recharge directly from surface waters. In addition, special situations involving losing streams, sink holes, and agricultural drainage wells, further illustrate why surface and groundwater must be considered together or as parts of a watershed system when trying to protect water quality.

B. Chemical Factors

Four chemical properties largely determine the fate and possible off-site transport of agricultural chemicals applied to fields: persistence (resistance to transformation or degradation), vapor pressure, solubility, and soil adsorption. Solubility and adsorption are related, with adsorption generally increasing with decreasing solubility, although the degree of soil adsorption is more important in determining concentrations in water. This is particularly true for exceptions such as the cationic herbicide paraquat which, while very soluble, is so strongly adsorbed that concentrations in water are usually below the limit of detection.

For both nutrients and pesticides, the longer they exist in the field, the greater the chance for off-site transport. For nutrients, the rates of transformations, including those involved with plant uptake, are important. For example, the rate of nitrification of NH_4-N, either that naturally present in the soil or added as manure or fertilizer, affects the potential for NH_4 loss with water; the longer it is present or the greater mass that is present, the greater the potential loss. For pesticides, the greater the resistance to decomposition through chemical, photochemical (sunlight), or microbiological actions, the greater the persistence and the potential to be lost with water. The banning of chlorinated hydrocarbon insecticides, e.g., DDT and dieldrin, was due mainly to their long persistence and resulting existence throughout the biosphere. Similarly, one

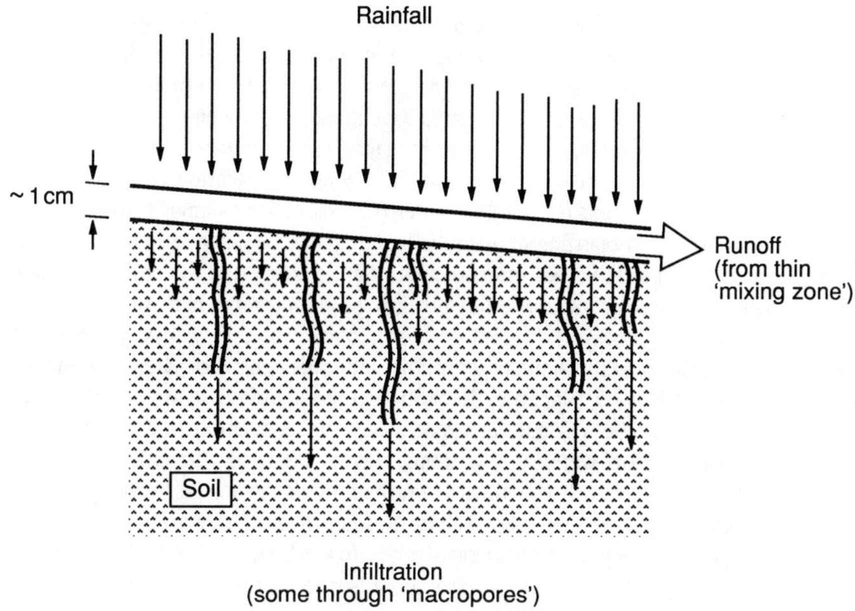

Figure 1. Mechanisms of chemical transport to water resources.

of the main reasons atrazine detection in water resources is widespread, besides its extensive use, is its relatively long persistence.

Vapor pressure, as a measure of the tendency of a chemical to evaporate, is important in determining how much NH_3 or pesticide might be lost to the atmosphere. Losses can take place both during and after application, with generally the higher vapor pressure, the higher the loss. However, other factors are also important, e.g., temperature and drop-size for application losses of sprayed pesticides. For post-application losses, the interaction with or adsorption to soil is particularly important in reducing volatilization. For ammoniacal fertilizers the degree of soil interaction is influenced by pH with greater adsorption of NH_3 (or NH_4^+) occurring at lower pH. In one study (Carter and Allison, 1961), loss of NH_4^+ applied to a sandy loam soil averaged about 5% at a pH of 5.9, but averaged about 22% when the pH was increased to 6.6 with calcium carbonate. Furthermore, at both pH levels, the losses were increased at higher NH_4^+ concentrations (e.g., up to 7 meq/100 g), when competition for adsorption sites could be occurring.

For pesticides, in one study (Tremwell, 1985) of volatilization in a covered system, about 2% of the herbicide propachlor applied to soil escaped into the air in the first 5 days after application. In the same time interval, less than 0.1% of the applied cyanazine evaporated, largely because its vapor pressure is 140,000 times less than propachlor. Furthermore, when the propachlor was applied to corn residue covered soil, volatilization loss was 12%. In a study of washoff of herbicides applied directly to corn residue open to the atmosphere (Baker and Shiers, 1989), a mass balance indicated that more than 60% of the applied propachlor was lost, presumably by volatilization, in the time (1 day) between application and rainfall simulation.

Solubility and soil adsorption determine the division between being dissolved in soil water or being adsorbed to soil particles for chemicals. The empirical Freundlich isotherm equation often is used to describe how this division changes with concentration at constant temperature. In the special case where the relationship between the concentrations in sediment (C_s) and in water (C_w) is, or is assumed to be, linear, a partition or distribution coefficient, usually designated K, is defined as C_s/C_w. Thus, chemicals with higher K values interact with or are adsorbed to the soil to a greater degree, with the degree of interaction determining the major mechanism of loss. For chemicals with K values 50 to 100 or more, placed in the "strongly-adsorbed" category, the major loss mechanism would be with sediment. For chemicals with K values from about 1 to 50, placed in the "moderately-adsorbed" category, the major loss mechanism would be with surface runoff water (dependent to some degree on the relative masses of runoff water and sediment leaving the field; this ratio is usually > 100:1). For chemicals with K values from 0 to 1, placed in the "weakly to non-adsorbed" category, the major loss mechanism likely would be with leaching water. As examples, in the five-year study in east-central Iowa discussed earlier (Johnson and Baker, 1982, 1984), field losses of the strongly adsorbed cationic herbicide, paraquat (K>1000), were totally with sediment, averaging about 1.5% of that applied; field losses of the moderately adsorbed herbicides, alachlor, propachlor, metribuzin, and cyanazine (Ks about 5 to 15), were dominated by that in surface runoff water (about 4/5 of the total losses were with water), averaging about 1.3% of that applied; and watershed losses of the nonadsorbed NO_3-N (K=0) were dominated by that in base flow and/or subsurface drainage (about 7/8 of the total), equivalent to about 20% of the fertilizer N applied in the watershed.

C. Hydrologic Factors

The most important hydrologic factor affecting the transport of chemicals with surface runoff is the soil's infiltration rate relative to rainfall intensity. The relative infiltration and rainfall rates determine the timing and volume of runoff, which in turn affects not only the masses of runoff water and sediment, but also potentially affects chemical concentrations in water and sediments. For transfer of chemicals from soils to overland flow, it is believed that there is a thin "mixing zone" of soil at or near the surface (Figure 2) from which chemicals are released (Heathman et al., 1986). The amount of chemical within this mixing zone, and thus concentrations in surface runoff, generally decrease with time as infiltrating water leaches chemical to a deeper depth. The rate of leaching and thus concentration decrease is dependent both on the volume of infiltrating water and the degree of chemical soil adsorption. Several studies (e.g., Barisas et al., 1978; Baker and Laflen, 1982; Baker et al., 1978; Baker and Laflen, 1979; Baker et al., 1982) have shown that nutrient and pesticide concentrations in runoff decrease with time during a rainfall-runoff event, presumably a result of decreasing amounts remaining in the mixing zone. Based on herbicide concentration decreases of 75% or greater during storms when runoff losses were less than 5 to 10%,

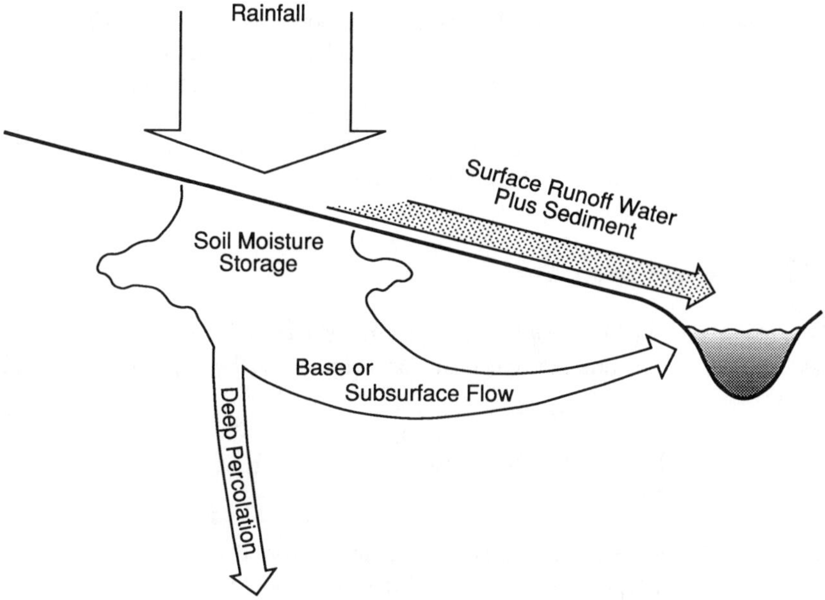

Figure 2. Schematic illustrating the thin mixing zone and macropores.

the dominant process for decreasing the amount in the mixing zone must be localized leaching and not losses with surface runoff.

Further evidence of the importance of infiltration rates on both volumes of carriers and concentrations are found in rainfall simulation studies involving compaction (Baker and Laflen, 1979) and surface crop residue (Baker et al., 1982). Soil compaction from tractor traffic on disked plots caused runoff to begin sooner and be greater, by 8%, than for similar disked plots without tractor traffic. Furthermore, average concentrations of the three surface-applied herbicides studied were more than tripled in runoff from the compacted plots resulting in losses about four times those from the uncompacted plots. Crop residue on the surface of disked plots protected the soil against surface sealing, and delayed and reduced surface runoff compared to disked plots without crop residue. The residue reduced runoff by 72%, and, because concentrations also were reduced (by about a factor of two), measured herbicide losses were reduced by 86%.

The transport of chemicals with leaching water also is affected by the soil's infiltration rate. In a study of NO_3-N losses with subsurface drainage from continuous corn plots (Kanwar et al., 1995), increased infiltration and drainage volume for no-till relative to the moldboard plow tillage system resulted in larger losses, even though NO_3-N concentrations were lower for no-till. In addition to the effect on leaching losses of the rate or volume of leaching water, the route of water moving through the soil profile can affect chemical concentrations and therefore losses. Figure 2 illustrates the possible presence of preferential flow paths or macropores. In some studies of NO_3-N leaching (Kanwar et al., 1985; Juo and Lal, 1979), it was believed that the movement of water quickly through macropores allowed water to bypass NO_3-N that was within soil aggregates and reduced leaching losses. In other NO_3-N (Tyler and Thomas, 1977) and herbicide (Kanwar et al., 1993) leaching studies, the increased existence of macropores all the way to the soil surface with conservation tillage was believed responsible for higher chemical concentrations in water percolating from the root zone compared to the moldboard plow system.

D. Management Factors

Total chemical loss from cropland is equal to the summation of the products of the masses of the carriers times the chemical concentrations in the respective carriers, i.e.:

$$\text{Total loss} = \sum_{n=1}^{3} \text{mass}_n \text{ x concentration}_n$$

where n=1 through 3 represents sediment, surface runoff water, and leaching water. Thus, a management practice that reduces carrier mass and/or concentration reduces loss. When a single management practice is not sufficient to provide the desired level of control, a system of practices is needed. To devise a single or a system of management practices that is efficient in reducing chemical transport to water resources, knowledge of the major mechanism of transport is needed. This requires information on the chemical's properties, the source(s) of the chemical, and the soil and climatic conditions that exist. A system may include a combination of in-field and off-site practices.

III. In-Field Management Practices

Although it may be possible to devise several different management practices to use in designing a system to reduce nonpoint source pollution, practical and economic realities considerably limit the number of practices currently feasible. In this section, emphasis will be given to the major management factors of cropping, tillage, and the rate, method, and timing of chemical application.

A. Cropping Practices

Crops planted on agricultural lands affect the surface and subsurface hydrology, in addition to chemical inputs, and thus affect the volumes of carriers and chemical concentrations in the carriers from those lands. Although economics plays a large role in crop selection (often dictating use of row-crops), close-grown or solid-seeded crops such as perennial grasses and legumes generally result in both lower runoff volumes and sediment concentrations as well as lower chemical concentrations in water and sediment, compared to row-crops. NO_3-N concentrations in subsurface drainage and groundwater under crops such as oats and perennial grass are generally much less than

10 mg L^{-1} (Baker, 1980). Strip intercropping, where narrow crop strips are grown contiguously in the same field, has the potential to maintain or even increase crop yields while providing surface and/or subsurface water quality benefits (Cruse et al., 1995). In a three-crop system of corn, soybeans, and oats/clover, where the corn was planted into the previous oats/clover strips, the NO_3-N concentrations for these systems were below 10 mg L^{-1} and less than those for a corn or soybeans monocrop system (Kanwar et al., 1995).

Overall, cropping practices that reduce soil losses and drainage volumes, and require lower chemical inputs such as herbicides, should reduce nonpoint source pollution. Good examples would be alfalfa, oats, and permanent bromegrass.

B. Tillage Systems

Tillage systems, as they affect hydrology, erosion, and chemical application, can affect chemical losses. Several reviews of these effects have been published (Logan et al., 1987; D'Itri, 1985; Fawcett et al., 1994; Baker and Laflen, 1983), where comparisons are made between the various forms of conservation tillage and the moldboard plow system. To be classified as conservation tillage by the Natural Resources Conservation Service (NRCS), at least 30% of the soil surface after planting must be covered with residue from the previous crop; no-till is the extreme form of conservation tillage with no soil or residue disturbance between harvest and planting. In general, soil losses decrease exponentially with residue cover (e.g., for a given soil, soil loss may decrease by half for each additional increment, such as 15%, in residue cover) such that for no-till, where residue after a corn crop usually ranges from 50 to 90%, soil losses are normally only 10% of those for moldboard plow. As reviewed by Baker (1987), data for annual surface runoff volumes show that conservation tillage generally decreases surface runoff in the range of 0 to 25% relative to the moldboard plow, although the reduction is much greater in certain situations (Edwards et al., 1989). However, due to increased roughness and porosity caused by tillage, generally runoff is less for the tilled area versus the untilled area for the first rain or two after any tillage (Johnson and Baker, 1982, 1984). This does have water quality consequences since chemical runoff losses generally are the greatest for the first runoff event after application, and then decrease with time during the growing season as the chemical dissipates and moves downward out of the thin surface mixing zone.

Another complicating factor with conservation tillage is the conflict between the water quality benefits of soil incorporation of chemicals and the desire to leave crop residue on the soil surface to protect against wind and water erosion. Surface runoff losses of surface-applied nutrients (Timmons et al., 1973) and herbicides (Baker and Laflen, 1979) are much higher compared to those soil-incorporated. In addition, with herbicides, there is concern for the fate of those broadcast sprayed that are intercepted by surface crop residue and prevented from reaching the soil surface. Volatilization or evaporation from the crop residue is one concern that has been noted previously (Baker and Shiers, 1989), with an increased amount volatilized from residue relative to soil (Tremwell, 1985). Herbicides also have been observed to wash off crop residue fairly quickly with rain (Martin et al., 1978; Baker and Shiers, 1989), and if that wash-off water becomes part of runoff, herbicide concentrations may be higher for higher residue coverage with conservation tillage. This has been observed in some studies (e.g., Baker et al., 1978).

Overall, conservation tillage, by leaving at least 30% of the soil surface covered with residue, significantly reduces erosion and usually reduces annual runoff volumes. This generally results in a reduction in surface runoff losses of herbicides and other agricultural chemicals. Exceptions to this usually are connected with unusual runoff events caused by severe storms timed right after chemical application. The increased infiltration with conservation tillage may increase herbicide and NO_3-N leaching, but this increased leaching, and the exceptions noted above, are seldom severe enough to outweigh the total environmental benefits of conservation tillage. The other criticism of conservation tillage increasing the reliance on herbicides must be weighed against the fact that almost all producers, whether using conservation tillage or not, are using everything in their means, including herbicides, to maintain good weed control.

C. Rate of Application

The rate of chemical application has a major effect on the concentrations and losses of chemicals in both surface runoff and subsurface drainage. For example, in a natural rainfall study of atrazine application rates ranging from 0.6 to 9.0 kg ha^{-1}, Hall et al. (1972) found that concentrations and losses in surface runoff water and sediment were roughly proportional to the amount applied, with losses averaging about 3% for all rates. Barnett et al. (1967) found that doubling the application rate of 2,4-D about doubled runoff losses. In studies of herbicide banding, where the areas treated and field rates were reduced by a factor of two or three, leaching losses of atrazine (Baker et al., 1995) and cyanazine (Baker and Melvin, 1994) with subsurface drainage were proportionately reduced.

Surface runoff and subsurface drainage losses of N and P also generally increase with the rate applied, although not always in direct proportion. This is due at least in part to: 1) the "buffering capacity" of the soil, and 2) the natural occurrence of N and P. Both solution N and P runoff losses from corn were increased about 30% when the N rate was increased from 168 to 448 kg ha^{-1} and the P rate was increased from 39 to 97 kg ha^{-1} (Schuman et al., 1973a,b). Romkens and Nelson (1974) showed that P losses in solution and with sediment were increased 85 and 62%, respectively, when the P rate was doubled from 56 to 112 kg ha^{-1}.

Concentrations and losses of NO_3-N with subsurface drainage as a function of N applied were found in one study (Gast et al., 1978) to double, going from 112 to 224 kg $ha^{-1,}$ and then doubled again going from 224 to 448 kg ha^{-1}. Baker and Johnson (1981) also found that concentrations and losses in tile drainage water doubled when the N rate was doubled in corn rotated with oats or soybeans. In a more recent five-year study of both continuous corn and a corn-soybean rotation (Baker and Melvin, 1994), average NO_3-N concentrations in subsurface drainage increased nearly linearly with N application rate beginning at 112 kg/ha^{-1} for continuous corn and 56 kg ha^{-1} for the corn-soybean rotation.

Overall, any practice that can reduce the rates of herbicide or other agricultural chemical application, reduces the source of nonpoint pollution. Source reduction, whether by herbicide banding or other means, should be one of the first things considered when a water quality problem exists.

D. Method of Application

The method of application in terms of formulation, additives, or placement can affect the "availability" of the applied chemical to move or be lost with surface runoff or subsurface drainage. In a review of pesticide loss, Wauchope (1978) reported that losses of pesticides formulated as wettable powders were generally greater than for other formulations. Use of microencapsulation, such as for alachlor, or starch to produce a "slow-release" herbicide, has been promoted because of the potential reduction in the availability of a pesticide at any one point in time to be lost with surface runoff or subsurface drainage. With ammonium-based N fertilizers, the use of additives in the form of nitrification inhibitors is expected to reduce N leaching losses by maintaining the applied N in the soil-adsorbed NH_4^+ cation form as opposed to the nonadsorbed and very mobile NO_3^- anion form.

The placement of agricultural chemicals relative to the thin mixing zone illustrated in Figure 2 is important in determining surface runoff losses. In a study of herbicide incorporation, Baker and Laflen (1979) found that atrazine, alachlor, and propachlor losses with runoff water and sediment were reduced by a factor of three for incorporation with a disk versus a surface broadcast application without incorporation. Furthermore, losses were twelve times lower when the herbicides were incorporated compared to surface application to plots with compaction from tractor traffic. Others (e.g., Hall et al., 1983) also have found that soil incorporation reduces pesticide runoff losses. One problem that does exist with incorporation is that incorporation with tillage also can destroy soil-protecting surface crop residue. Some new tillage tools do have the potential for herbicide incorporation without significant crop residue incorporation (Mickelson et al., 1995).

There also have been several studies showing reduction in surface runoff losses of N and P if N and P fertilizers are incorporated. For example, Timmons et al. (1973) showed in a rainfall simulation study that losses of N and P in solution were reduced 49 and 77%, respectively, when N and P fertilizers were disk incorporated relative to a surface application to fallow soil. Corresponding reductions were 80 and 87% when the fertilizers were incorporated with the moldboard plow. In addition, Baker and Laflen (1982), in a rainfall simulation study, showed that N and P losses in solution were reduced 60 and 86%, respectively, when liquid N and P fertilizers were point-injected versus a surface broadcast application. The point-injector fertilizer applicator was developed primarily to incorporate liquid fertilizers without incorporating crop residue with conservation tillage systems (Baker et al., 1989).

With respect to NO_3-N leaching and N fertilizer placement, work is ongoing to determine the feasibility of manipulating the soil during N application to force the major portion of infiltrating water to move in zones remote from the zone of N fertilization. In large column studies (Kiuchi et al., 1994; Baker et al., 1996), the leaching of NO_3-N and other anion tracers has been shown to be reduced by compacting the soil or placing barriers above the zone of chemical application. In another study (Ressler et al., 1995), a field-scale machine has been built and tested that places liquid N fertilizer in a line, cutting and smearing any nearby macropores, compacting the soil above the line, and finally doming a small amount of soil at the surface over the line. Results from preliminary measurements show reduced leaching with this type of microscale soil management.

Overall, when surface runoff losses of herbicides or other agricultural chemicals are a problem, the possible use of soil incorporation to reduce the contact of the chemical with rainfall and overland flow should be considered. This is one of the more efficient best management practices to reduce nonpoint source pollution and should receive more attention.

E. Timing of Application

Timing of chemical application relative to storm events is important because increased time intervals between chemical applications and runoff or subsurface drainage events can decrease chemical losses significantly if rapid dissipation (mainly for pesticides) or crop uptake (mainly for nutrients) occurs. Because of the limits of current weather prediction and practical considerations of the time available for and need to apply chemicals, timing options may be restricted somewhat. However, timing relative to expected storm events and weather trends should be considered. In a rainfall simulation study, White et al. (1967) found that atrazine runoff losses were reduced by over one-half for a rainfall 96 h after application versus 1 h. Wauchope (1978) has summarized several other studies showing a similar effect of timing of application on pesticide runoff losses. One of the potential advantages proposed for fall application of metolachlor in the northern U.S. is the likelihood of lower surface runoff losses compared to spring application when intense rain storms are more likely (Tierney et al., 1996).

In addition to the timing of a chemical application relative to runoff or subsurface drainage events, timing of nutrient applications with respect to crop needs also is important. Timmons and Dylla (1981) measured about a 20% reduction of NO_3-N concentrations in subsurface drainage when 240 N ha^{-1} was applied to corn (with irrigation water) four times during the growing season compared to all at once in the spring. In field (Kanwar et al., 1988) and lysimeter (Baker and Timmons, 1994) studies, improved N management systems of lower total rate/split applications (versus a higher rate/single application) reduced NO_3-N concentrations in subsurface drainage by roughly one-third.

Overall, when feasible, the timing of herbicide and other chemical applications should be managed to reduce the time available to be lost off-site. The rate of degradation, or crop uptake for nutrients, relative to expected excess precipitation patterns needs to be considered in the decision process.

IV. Off-Site Landscape Modification Practices

Nonpoint source agricultural pollution is a watershed-scale problem whose solution likely will require the reconfiguration of agricultural landscapes through a combination of in-field and off-site approaches. In-field approaches, as just discussed, include reduced amounts and improved methods and timing of nutrient and pesticide applications; management practices to reduce soil erosion and chemical transport; and improved cropping systems. However, ongoing research continues to show that the reduction of some nonpoint source contaminants to acceptable levels in agricultural landscapes likely will not be accomplished solely by the adoption of better in-field farming practices and improved cropping systems. Off-site approaches also likely will be needed, such as vegetated buffer strips and restored or constructed wetlands.

A. Vegetated Buffer Strips

Vegetated buffer strips can be situated as riparian zones, on field borders, or within fields on the contour or as grassed waterways. They have the potential of influencing the volume and quality of surface runoff and shallow subsurface drainage from the field-buffer strip system. The vegetation and rooting system slows overland flow, and the generally better soil structure and lower moisture content than for the field draining to the buffer strip increases infiltration and removal of chemicals dissolved in that flow. Surface roughness and reduced flow velocity within the buffer strip reduce the carrying capacity for sediment, and sediment and chemicals adsorbed to it are deposited. In addition, depending on their chemical and physical properties, some chemicals are removed from overland flow through adsorption to in-place soil and/or living and dead vegetation. The differences in reduction of chemical transport between riparian zones, contour buffer strips, and grassed waterways will be manifested in the effects of differences in relative areas of drainage to the vegetated buffer strip, the length of travel through the vegetated buffer strip, and the degree of concentration (or depth) of flow. Therefore, the topography and relative geometry of the source area and the vegetated area will be important.

In a study of the effects of vegetated buffer strips on runoff, sediment transport, and herbicide removal, Wauchope et al. (1990) reported that one-third less rain was needed for bare plots to produce the same runoff as for grassed plots in a rainfall simulation study. Asmussen et al. (1977) conducted a study on the reduction of 2,4-D load in surface runoff flowing through a grassed waterway. By using simulated rainfall on 24.4 m long waterways, incoming suspended sediment was reduced 98 and 94% for dry and wet antecedent conditions, respectively. The total losses of 2,4-D from the plots were 2.5 and 10.3% for the dry and wet plots, respectively. Only about 30% of the 2,4-D that entered the top of the waterway reached the bottom. In a similar study, Rhode et al. (1980) determined percent surface runoff losses of trifluralin broadcast applied to fields and how vegetated buffer strips receiving that runoff reduced transport. Annual losses as percentages of that applied were low, 0.17 and 0.03%, for the two years of measurement under natural rainfall. When runoff caused by rainfall simulation was directed onto vegetated buffer strips, trifluralin losses were reduced 96% if the buffer strip was initially dry and 86% if it was pre-wetted. Over one-half of this reduction was attributed to adsorption on vegetation, organic matter, and soil. Hall et al. (1983) determined atrazine surface-runoff losses from treated corn fields with and without an oat strip at the slope base, and also as affected by preplant herbicide incorporation. Atrazine loss without the oat strip, or incorporation, was 3.5% of that applied. With the oat strip, loss was reduced to 0.33%. Preplant incorporation reduced losses even further over a preemergence application.

In a rainfall simulation study on the effect of vegetative buffer strip length on atrazine transport, reductions in herbicide transport of 31.7 and 55.4% were observed for 4.6 and 9.1 m lengths, respectively, with no significant difference in reduction whether the runoff contained sediment or not (Mickelson and Baker, 1993). Arora et al. (1993) reported that for the first field-runoff event after herbicide application, an average of 12.5% atrazine, 27.3% metolachlor, and 21.1% cyanazine was retained for a 15:1 area ratio (drainage area to vegetated buffer strip area); corresponding values for a 30:1 area ratio were 9.3, 15.3, and 7.2%. Dillaha et al. (1989) conducted a study to investigate the performance of vegetative buffer strips as an agricultural nonpoint source pollution-control measure. They concluded that vegetative buffer strips are effective on steeper hill slopes; that they become ineffective under concentrated flow conditions with time; and that when sediment was deposited higher than the adjacent field, flow parallel to the strips took place that reduced their effectiveness.

The National Council of the Paper Industry for Air and Stream Improvement (NCASL, 1992) published a technical bulletin on the effectiveness of buffer strips in reducing sediment, nutrient, and pesticide transport and losses occurring from silvicultural operations. Velocity of water flow, size distribution of incoming sediment, slope and slope length before reaching the buffer strips, slope and slope length of buffer strips, vegetation characteristics, water depth and vegetation height, pesticide properties, and water quality were stated as critical factors responsible for determining the effectiveness of buffer strips. It was concluded that due to the variety of ongoing chemical, physical, and biological processes, buffer strips have high potential to reduce the transport of nutrients, pesticides, and sediment with surface runoff. They pointed out that specific site factors (e.g., soil type, topography, vegetation in buffer areas, and the nature of the surface water being protected) should be considered when designing buffer strips.

Overall, buffer strips can reduce overland transport of herbicides and other agricultural chemicals. However, the degree of reduction is highly variable and buffer strips should be viewed as a supplementary practice to in-field best management practices and not as an "answer-all" solution to water quality problems.

B. Restored/Constructed Wetlands

One of the more promising strategies for reducing agricultural chemical contamination of surface and groundwater is the construction or restoration of wetlands in agricultural watersheds specifically as sinks for agricultural chemical contaminants (van der Valk and Jolly, 1992). Wetlands are areas of intense biological activity, and there is considerable opportunity for chemical transformation and loss as pollutants dissolved in water move through these systems (Crumpton et al., 1993). Studies suggest that wetlands may act as sinks for a variety of compounds (e.g., Howard-Williams, 1985), and wetlands may be especially effective as sinks for NO_3-N loads from row-cropped fields. In particular, constructed wetlands that are integrated into new or existing drainage systems may have considerable potential to remove NO_3-N from shallow subsurface drainage enroute from the field to a surface water resource. Publications dealing with wetlands and water quality note the probable importance of denitrification, with resulting gaseous loss of nitrous oxide (N_2O) and N_2. In fact, with rare exception, denitrification is cited as the primary reason wetlands may serve as N sinks (Crumpton et al., 1993). However, N transformations in wetlands involve complex spatial and temporal patterns, and wetlands can differ greatly in their efficiency as NO_3-N sinks, even when depth, area, and residence time are taken into account. In particular, seasonal and diurnal changes in factors such as temperature, NO_3-N concentration, and oxygen concentration dramatically affect N transformation in sediments (Anderson et al., 1984).

There is considerable interest in the potential for using constructed wetlands for agricultural wastewater treatment. The USDA Natural Resources Conservation Service has technical requirements for reconstructed wetlands, but they consciously have not written a "Conservation Practice Standard" because "knowledge of the technology, performance, and utility of constructed wetlands for agricultural wastewater treatment is still very limited" (Krider and Boyd, 1992). The wastewaters usually being considered, such as animal or processing wastes, have high levels of NH_4-N, organic matter, solids, and oxygen demand, which present special problems such as oxidation of NH_4-N and organic matter (Payne et al., 1992).

A major research initiative at Iowa State University has specifically addressed the water quality functions of watersheds (Crumpton et al., 1993, 1995; Crumpton and van der Valk, 1994). The studies have confirmed that wetlands have considerable capacity to remove NO_3-N (Crumpton et al., 1993), and that denitrification is the dominant loss process for externally loaded NO_3-N in the study systems (Isenhart, 1992). Even under highly aerobic conditions, NO_3-N concentrations declined rapidly in water overlying wetland sediments in all of the experiments. When mesocosms with residence times of approximately one week were loaded with 3 to 15 mg L^{-1} of NO_3-N, percent NO_3-N removal exceeded 80%. Estimates of loss rates range from 0.5 to 5 g m^{-2} per day of NO_3-N in the overlying water. These studies suggest that a mature, 1-ha wetland could remove significant amounts of the NO_3-N lost in water draining approximately 100 ha of corn at moderately high N application rates (Crumpton et al., 1993; Isenhart, 1992). These were chosen as starting conditions for model simulations where it was predicted that the wetland would remove approximately 60% of the annual NO_3-N load (Crumpton and Baker, 1993). At least as important as percent NO_3-N removal, of course, is the concentration of NO_3-N in water exiting the wetland. Although inlet concentrations averaged 18 mg L^{-1}, predicted outlet concentrations fell below the drinking water standard of 10 mg L^{-1}. These results are particularly encouraging given the very conservative nature of the hydrologic model used, which underestimates residence times in the wetland. The model also was used to predict the effect on NO_3-N concentrations of mature wetlands receiving subsurface drainage water from watersheds with areas of 33 and 300 times the area of the wetland. For these simulations, as could be expected, concentrations were lower for wetlands draining less acreage and thus receiving lower flows. Predicted NO_3-N removal for a 1-ha "mature" wetland ranged from 35% for a 300-ha watershed to 82% for a 33-ha watershed, whereas NO_3-N removal for a "new" wetlands ranged from 20% for a 300-ha watershed to 76% for a 33-ha watershed.

Overall, the potential for wetlands to reduce the transport of herbicides to surface water resources is not well documented; whereas, it is well established that wetlands can reduce NO_3-N in drainage waters by denitrification. The potential for the biomass in wetlands to adsorb herbicides or enhance their degradation still needs to be determined.

V. Summary

How to integrate off-farm environmental quality considerations into on-farm decision making is one of the major problems facing agriculture and society (Committee on Long-Range Soil and Water Conservation, 1993). In terms of scale, management decisions can be made on field-, farm-, and watershed-scales. At the field-scale, only in-field management practices are considered. Currently there is increased interest in management planning at the farm-scale, which includes off-site practices. However, to effectively place or locate off-site practices for water quality enhancement, the watershed-scale should be used. Watershed-scale planning would allow for targeting to obtain the greatest improvement for the effort. Watershed-scale planning is also more complicated than field- or farm-scale planning, and requires cooperation of all the producers/land owners in the watershed, and may impact some more than others.

In addition to the organizational problems of watershed-scale planning, the technical problems of obtaining/managing watershed data and its use in decision making still need work. Agricultural landscapes generally are mosaics of cropland, pastures, grassed waterways/buffer strips, woodlands, streams, and wetlands. Each of these landscape elements will function differently with respect to water quality depending on the specific characteristics of that element. For example, pesticide loss from a field will depend on various physical attributes of the field, the cropping system, and management practices. The functioning of some elements, particularly fields, can be altered significantly by altering in-field practices. The functioning of other elements, e.g., perennial pasture, is more constant seasonally and between years. A GIS-based system model would have the capability to link output from water quality models for different elements (Tim and Jolly, 1994). Development of such a model would require refinement of existing water quality models for each element, integration of these models with GIS databases to form a comprehensive watershed model, and calibration of the integrated model using observed data from watersheds. However, use of such a model would allow a prediction of how the interactions of various combinations of in-field and off-site practices would impact water quality for policy/implementation purposes.

With herbicide contamination of water resources, particularly surface water resources, being a known water quality problem, any improvements in efficiency of weed control and/or herbicide use should provide water quality benefits. Practices that replace or reduce the rate of herbicide use and still maintain weed control need to be developed and implemented. Two practices with potential for reducing rate involve "precision or site-specific" technologies and weed identification/vision systems, both of which need further research and development. Other research needs that could result in elimination of herbicide use on all or part of a field involve alternative weed control methods (e.g., flaming) and identification and treatment of spatially variable weed populations with respect to known economic thresholds. There is also a need for new chemical products and formulations that have properties that maintain weed control but reduce the potential for off-site movement. More information on potential benefits of formulations such as starch- or micro-encapsulation is still needed. The equipment industry also has research opportunities with respect to improved methods of application and/or soil incorporation. Use of point-injection technology, with the potential of reducing spray-drift losses while allowing herbicide incorporation without the incorporation of surface crop residue, needs to be refined. With respect to herbicide transport from the field to a surface water resource, one void in our understanding is the mechanism and magnitude of removal that could generally be expected upon passage of agricultural drainage through a wetland, given all the wetland variables, such as hydraulic retention time, that would affect removal.

References

Anderson, T.K., M.H. Jensen, and J. Sorensen. 1984. Diurnal variation of nitrogen cycling in coastal marine sediments. 1. Denitrification. *Marine Biology* 83:171-176.

Arora, K., J.L. Baker, S.K. Mickelson, and D.P. Tierney. 1993. Evaluating herbicide removal by buffer strips under natural rainfall. ASAE Paper No. 93-2593, St. Joseph, MI, presented at ASAE International Winter Mtg., Chicago, IL.

Asmussen, L.E., A.W. White, E.W. Hauser, and J.M. Sheridan. 1977. Reduction of 2,4-D load in surface runoff down a grassed waterway. *J. Environ. Qual.* 6:159-162.

Baker, J.L. 1980. Agricultural areas as nonpoint sources of pollution. In: M.R. Overcash and J.M. Davidson (eds.), *Environmental Impact of Nonpoint Source Pollution,* Ann Arbor Science Publishers, Inc., Ann Arbor, MI, p. 275-310.

Baker, J.L. 1987. Hydrologic effects of conservation tillage and their importance relative to water quality. In: T.J. Logan, J.M. Davidson, J.L. Baker, and M.R. Overcash (eds.), *Effects of Conservation Tillage and Groundwater Quality*, Lewis Publishers, Inc., Chelsea, MI, Chapter 6.

Baker, J.L. and H.P. Johnson. 1981. Nitrate-nitrogen in tile drainage as affected by fertilization. *J. Environ. Qual.* 10:519-522.

Baker, J.L. and J.M. Laflen. 1979. Runoff losses of surface-applied herbicides as affected by wheel tracks and incorporation. *J. Environ. Qual.* 8:602-607.

Baker, J.L. and J.M. Laflen. 1982. Effects of corn residue and fertilizer management on soluble nutrient runoff losses. *Trans. ASAE* 25:344-348.

Baker, J.L. and J.M. Laflen. 1983. Water quality consequences of conservation tillage. *J. Soil and Water Conserv*. 38:186-193.

Baker, J.L. and S.W. Melvin. 1994. Chemical management - overview. p. 27-60. In: 1994 Annual Report on Agricultural Drainage Well Research and Demonstration Project, Iowa Dept. Ag. and Land Stewardship, Des Moines, IA.

Baker, J.L. and L.E. Shiers. 1989. Effects of herbicide formulation and application method on washoff from corn residue. *Trans. ASAE* 32:830-833.

Baker, J.L. and D.R. Timmons. 1994. Effect of fertilizer management on leaching of labeled nitrogen for no-till corn in field lysimeters. *J. Environ. Qual.* 23:305-310.

Baker, J.L., J.M. Laflen, and R.D. Hartwig. 1982. Effects of corn residue and herbicide placement on herbicide runoff losses. *Trans. ASAE* 25:340-343.

Baker, J.L., J.M. Laflen, and H.P. Johnson. 1978. Effect of tillage systems on runoff losses of pesticides, a rainfall simulation study. *Trans. ASAE* 21:886-892.

Baker, J.L., J.M. Laflen, and M.M. Schreber. 1996. Potential use of localized compaction to reduce anion leaching. *J. Environ. Qual.* 25.

Baker, D.B., L.K. Wallrabenstein, and R.P. Richards. 1994. "Well Vulnerability and Agricultural Contamination: Assessments From a Voluntary Well Testing Program." New Directions in Pesticide Research, Development, Management, and Policy: Proceedings of the 4th National Conference on Pesticides. Blacksburg, VA: Virginia Polytechnic Institute. p. 470-494.

Baker, J.L., T.S. Colvin, D.C. Erbach, and R.S. Kanwar. 1995. Potential water quality and production efficiency benefits from reduced herbicide inputs through bonding. p. 5.6-5.9. In: Comprehensive Report, Integrated Farm Management Demonstration Program, rep. no. IFM 16, Iowa State Ext., Ames, IA.

Baker, J.L., T.S. Colvin, S.J. Marley, and M. Dawelbeit. 1989. A point-injector applicator to improve fertilizer management. *Applied Eng. in Agric.* 5:334-338.

Barisas, S.G., J.L. Baker, H.P. Johnson, and J.M. Laflen. 1978. Effect of tillage systems on runoff losses of nutrients, a rainfall simulation study. *Trans. ASAE* 21:893-897.

Barnett, A.P., E.W. Hauser, A.W. White, and J.H. Holladay. 1967. Loss of 2,4-D in washoff from cultivated fallow land. *Weeds* 15:133-137.

Benson, R.W. 1989. EPA drinking water standards. In: Proceedings of the 5th National Domestic Water Quality Symposium, Dec. 11-12, 1989, New Orleans, LA, p. 67-76.

Carter, J.N. and F.E. Allison. 1961. The effects of rates of application of ammonium sulfate on gaseous losses of nitrogen from soils. *Soil Sci. Soc. Am. Proc.* 25:404-408.

Committee on Long-Range Soil and Water Conservation. 1993. Soil and Water Quality. National Academy Press, Washington, D.C. 516 p.

Coupe, R.H., D.A. Goolsby, J.L. Iverson, D.J. Markovchick, and S.D. Zaugg. 1995. Pesticide, nutrient, water-discharge and physical-property data for the Mississippi River and some of its tributaries, April 1991-September 1992. U.S. Geol. Surv. open file rept. 93-657, Denver, CO.

Crumpton, W.G. and J.L. Baker. 1993. Integrating wetlands into agricultural drainage systems: Predictions of nitrate loading and loss in wetlands receiving agricultural subsurface drainage. In: Proceedings, International Symposium on Integrated Resource Management & Landscape Modification for Environmental Protection, American Society of Agricultural Engineers, Dec. 13-14, Chicago, IL, p. 118-126.

Crumpton, W.G. and A.G. van der Valk. 1994. Transformation and fate of nitrate and atrazine in freshwater wetlands. Leopold Ctr. Prog. Rept. 3:29-31. Leopold Center for Sustainable Agric., Ames, IA.

Crumpton, W.G., T.M. Isenhart, and S. Fisher. 1993. Transformation and fate of nitrate in wetlands receiving nonpoint source agricultural inputs. In: G.A. Moshiri (ed.), *Constructed Wetlands for Water Quality Improvement*, Lewis Publishers, Chelsea, MI.

Crumpton, W.G., J.L. Baker, J. Owens, C. Rose, and J. Stenback. 1995. Wetlands and streams as off-site sinks for agricultural chemicals. In: *Proceedings, Clean Water-Clean Environment-21st Century*, Vol. I, p. 49-52.

Cruse, R.M., M. Ghaffarzadeh, and J.B. Swan. 1995. Strip intercropping. p. 73-76. In: Proc. Clean Water-Clean Environment-21st Century, Vol. III.

Dillaha, T.A., J.H. Renueau, S. Mostaghimi, and D. Lee. 1989. Vegetative filter strips for agricultural nonpoint source pollution control. *Trans. ASAE* 32(2):513-519.

D'Itri, F.M. (ed.). 1985. *A Systems Approach to Conservation Tillage*, Lewis Publishers, Inc., Chelsea, MI.

Edwards, W.M., M.J. Shipitals, L.B. Owens, and L.D. Norton. 1989. Water and nitrate movement in earthworm burrows within long-term no-till cornfields. *J. Soil and Water Conserv.* 44:240-243.

Fawcett, R.S., B. Christensen, D.P. Tierney. 1994. The impact of conservation tillage on pesticide runoff into surface water. *J. Soil and Water Conserv.* 49:126-135.

Gast, R.G., W.W. Nelson, and G.W. Randall. 1978. Nitrate accumulation in soils and loss in tile drainage following nitrogen applications to continuous corn. *J. Environ. Qual.* 7:258-261.

Hall, J.K., N.L. Hartwig, and L.D. Hoffman. 1983. Application mode and alternate cropping effects on atrazine losses from a hillside. *J. Environ. Qual.* 12:336-340.

Hall, J.K., M. Pawlus, and E.R. Higgins. 1972. Losses of atrazine in runoff water and soil sediment. *J. Environ. Qual.* 1:172-176.

Heathman, G.C., L.R. Ahuja, and J.L. Baker. 1986. Test of a non-uniform mixing model for transfer of herbicides to surface runoff. *Trans. ASAE* 29:450-455, 461.

Howard-Williams, C. 1985. Cycling and retention of nitrogen and phosphorus in wetlands: a theoretical and applied perspective. *Freshwater Biology* 15:391-431.

IDALS. 1994. Agricultural drainage well research and demonstration project, annual rept., Iowa Dept. of Agric. and Land Stewardship and Iowa State Univ., 63 p.

Isenhart, T.M. 1992. Transformation and fate of nitrate in northern prairie wetlands. Ph.D. Dissertation. Iowa State University, Ames, IA. 93 p.

Johnson, H.P. and J.L. Baker. 1982. Field-to-stream transport of agricultural chemicals and sediment in an Iowa watershed: Part I. Data base for model testing (1976-1978). U.S. Dept. Agr., Washington, D.C.

Johnson, H.P. and J.L. Baker. 1984. Field-to-stream transport of agricultural chemicals and sediment in an Iowa watershed: Part II. Database for model testing (1979-1980). Report No. EPA-600/S3-84-055, Environmental Research Laboratory, Athens, GA.

Juo, A.S.R. and R. Lal. 1979. Nutrient profile in a tropical Alfisol under conventional and no-till systems. *Soil Sci.* 127:168-173.

Kanwar, R.S., J.L. Baker, and J.M. Laflen. 1985. Nitrate movement through the soil profile in relation to tillage system and fertilizer application method. *Trans. ASAE* 28:1802-1807.

Kanwar, R.S., J.L. Baker, and D.G. Baker. 1988. Tillage and split N-fertilization effects on subsurface drainage water quality and corn yield. *Trans. ASAE* 31:453-461.

Kanwar, R.S., T.S. Colvin, and D.L. Karlen. 1995. Tillage and crop rotation effects on drainage water quality. p. 163-166. In: *Proceedings, Clean Water-Clean-Environment-21st Century*, Vol. III.

Kanwar, R.S., D.E. Stoltenberg, R. Pfeiffer, D.L. Karlen, T.S. Colvin, and W.W. Simpkins. 1993. Transport of nitrate and pesticides to shallow groundwater systems affected by tillage and crop rotation practices. In: *Proceedings of National Conference on Agricultural Research to Protect Water Quality*, p. 270-273.

Kiuchi, M., R. Horton, and T.C. Kaspar. 1994. Leaching characteristics of repacked soil columns as influenced by subsurface flow barriers. *Soil Sci. Soc. Am. J.* 50:1212-1218.

Krider, J.N. and W.A. Boyd. 1992. SCS technical requirements for constructed wetlands for agricultural wastewater treatment. Paper 92-4523 presented at ASAE Winter Mtg., Nashville, TN, Dec. 15-18, 1992.

Kross, B.C., G.R. Hallberg, R.D. Libra, L.F. Burmeister, and K.L. Cherryholmes. 1990. The Iowa State-Wide Rural Well-Water Survey: Water Quality Data: Initial Analysis. Technical Information Series 19. Des Moines, IA: Iowa Department of Natural Resources.

Logan, T.J., J.M. Davidson, J.L. Baker, and M.R. Overcash (eds.). 1987. *Conservation Tillage and Groundwater Quality*, Lewis Publishers, Inc., Chelsea, MI.

Martin, C.D., J.L. Baker, D.C. Erbach, and H.P. Johnson. 1978. Washoff of herbicides applied to corn residue. *Trans. ASAE* 21:1164-1168.

Mickelson, S.K. and J.L. Baker. 1993. Buffer strips for controlling herbicide runoff losses. ASAE Paper No. 93-2084, St. Joseph, MI, presented at ASAE International Summer Mtg., Spokane, WA.

Mickelson, S.K., J.L. Baker, J.A. Baldauf, and P.M. Boyd. 1995. Tillage and herbicide incorporation effects on runoff, erosion and herbicide loss. ASAE Paper No. 95-2695, St. Joseph, MI, presented at ASAE International Annual Mtg., Chicago, IL.

Minshall, G.W. 1995. Reemergence of the importance of biology in water quality assessments. *Water Resources Update*, No. 98, p. 21-216.

Monsanto Agricultural Company. 1990. The National Alachlor Well Water Survey (NAWWS): Data Summary. Monsanto Technical Bulletin. St. Louis, MO: Monsanto Agricultural Company.

NCASL, Technical Bulletin #631. 1992. The effectiveness of buffer strips for ameliorating offsite transport of sediments, nutrients, and pesticides from silvicultural operations.

Office of Technology Assessment. 1995. Targeting environmental priorities in agriculture: Reforming program strategies, OTA-ENV-640, U.S. Govt. Printing Office, Washington, D.C.

Payne, V.W.E., T.A. McCaskey, and J.T. Eason. 1992. Constructed wetland for treating swine lagoon effluent. Paper No. 92-4526, presented at ASAE Winter Mtg., Nashville, TN, Dec. 15-18, 1992.

Puckett, J.J. 1995. Identifying the major sources of nutrient water pollution. *Environ. Sci. Tech.* 29(9):408A-414A.

Ressler, D.E., J.L. Baker, T.C. Kaspar, R. Horton, and J. Green. 1995. Localized compaction and doming to increase N-use efficiency and reduce leaching. p. 215-218. In: *Proceedings, Clean Water-Clean Environment-21st Century*, Vol. III.

Rhode, W.A., L.E. Asmussen, E.W. Hauser, R.D. Wauchope, and H.D. Allison. 1980. Trifluralin movement in runoff from a small agricultural watershed. *J. Environ. Qual.* 9:37-42.

Romkens, M.J.M. and D.W. Nelson. 1974. Phosphorus relationships in runoff from fertilized plots. *J. Environ. Qual.* 3:10-13.

Schuman, G.E., R.E. Burwell, R.F. Piest, and R.G. Spomer. 1973a. Nitrogen losses in surface runoff from agricultural watersheds on Missouri Valley losses. *J. Environ. Qual.* 2:299-302.

Schuman, G.E., R.G. Spomer, and R.F. Piest. 1973b. Phosphorus losses from four agricultural watersheds on Missouri Valley loess. *Soil Sci. Soc. Am. Proc.* 37:424-427.

SETAC. 1994. Aquatic Risk Assessment and Mitigation, Final report (dialogue group), Pensacola, FL, 220 p.

Tierney, D., B.P. Chistensen, K. Golden, M. Colletine, M. Blanchard, R. Fawcett, and P.A. Banks. 1996. Metolachlor spring and fall application: Comparative surface runoff and groundwater infiltration potential. WSSA Abstracts, p. 61.

Tim, U.S. and R.W. Jolly. 1994. Evaluating agricultural nonpoint source pollution using integrated GIS and hydrological water quality models. *J. Environ. Qual.* 23:25-35.

Timmons, D.R. and A.S. Dylla. 1981. Nitrogen leaching as influenced by nitrogen management and supplemental irrigation level. *J. Environ. Qual.* 10:421-426.

Timmons, D.R., R.E. Burwell, and R.F. Holt. 1973. Nitrogen and phosphorus losses in surface runoff from agricultural land as influenced by placement of broadcast fertilizer. *Water Resources Res.* 9:658-667.

Tremwell, T.K. 1985. Fate of broadcast herbicides used with consumption tillage systems. M.S. Thesis. Iowa State University, Ames, IA, 94 p.

Tyler, D.D. and G.W. Thomas. 1977. Lysimeter measurements of nitrate and chloride losses from soil under conventional and no-tillage corn. *J. Environ. Qual.* 6:63-66.

USDA-NASS. 1995. Agricultural Chemical Usage 1994 - Field Crops Summary. Economic Research Service, Washington, D.C.

USEPA. 1976. *Quality Criteria for Water.* Washington, D.C., 256 p.

USEPA. 1990. National Pesticide Survey Phase 1 Report. EPA 570/9-90-003, Sept., Washington, D.C.

USEPA. 1994. National Water Quality Inventory: 1992 Report to Congress. EPA 841-R94-001, Mar., Washington, D.C.

van der Valk, A.G. and R.W. Jolly. 1992. Recommendations for research to develop guidelines for the use of wetlands to control rural NPS pollution. *Ecological Engineering* 1:115-134.

Wauchope, R.D. 1978. The pesticide content of surface water draining from agricultural fields - a review. *J. Environ. Qual.* 7:459-472.

Wauchope, R.D., G.W. Randall, and L.R. Marti. 1990. Runoff of sulfometuron-methyl and cyanazine from small plots: effects of formulation and grass cover. *J. Environ. Qual.* 19:119-125.

White, A.W., A.P. Barnett, B.G. Wright, and J.H. Holladay. 1967. Atrazine losses from fallow land caused by runoff and erosion. *Environ. Sci. Tech.* 1:740-744.

Microbial Interactions with Weed Seeds and Seedlings and Its Potential for Weed Management

R.J. Kremer

I. Introduction

Soil microorganisms are responsible for many key processes including nutrient cycling, plant nutrition, xenobiotic transformations, and interacting (either beneficially or antagonistically) with other organisms inhabiting the soil. Recent concerns about possible detrimental effects on the environment by agrichemicals have generated interest in developing alternative pest management strategies including biological control using microorganisms occurring in nature. Some success has been achieved using insects and fungi for controlling specific weeds in certain agroecosystems (Harley and Forno, 1992). However, development of microbial agents for widespread use in managing economically important weeds in cropping and pasture/rangeland systems has not advanced as rapidly as expected. Causes for delays in establishing successful strategies for biological control of these weeds include variable (inconsistent) levels of efficacy, host specificity constraints, and lack of persistence in the field of the selected microorganisms (Kremer and Kennedy, 1996).

At least two important limitations impact the development of environmentally safe, alternative weed management strategies. First, such strategies require detailed quantitative understanding of the ecological and biological factors affecting the dynamics of weed growth and populations (Lindquist et al., 1995). Secondly, knowledge of the ecology and biology of microorganisms that interact with weeds at various stages of a weed's life cycle is needed to properly fit biological approaches in a weed management system. Little research has been conducted to quantify the relationships between soil and rhizosphere microorganisms and weed seeds and seedlings. These relationships are complex and likely are unique to each weed family, if not species, and must be described so that fundamental hypotheses can be made and applied in developing practical weed management systems integrated with biological control strategies. The purpose of this paper is to assess the relationships between soil and rhizosphere microorganisms and weed seeds and seedlings and describe potential weed management strategies that can be developed based on an understanding of these relationships.

II. Weed Seeds and Seedlings and Microorganisms

A. Soil Microbial Diversity

The biological component of soil has largely been ignored as an important aspect in weed seed and seedling growth dynamics and as a resource for potential nonchemical control systems. Soil microorganisms have been implicated in weed seed depletion or seedling growth inhibition, but these effects have not been extensively quantified (Kremer, 1993). When the biological diversity or the variety of microbial species occurring in an ecosystem is considered, the tremendous potential for discovery of organisms with significant impact for managing weeds or other agricultural pests is readily apparent. Kennedy and Smith (1995) estimate that one gram of soil contains one million to ten billion bacteria, which may comprise up to 10,000 different species. Of the approximately one million microbial species on earth, only about 110,000 are known and most of these have yet to be fully described. It is concluded that only a minuscule portion of the total soil microbial potential is currently known, and a wealth of bioactivity and genetic information is waiting to be discovered (Kennedy and Smith, 1995).

Despite the presence of vast numbers of diverse microorganisms, seeds of most weed species persist viably in soil for many years with only a small portion succumbing to microbial attack. Seeds retrieved from soil often possess intimate microbial associations. This is illustrated in Figure 1, which depicts the extent of colonization of giant foxtail (*Setaria faberii* Herrm.) and redroot pigweed (*Amaranthus retroflexus* L.) seeds from the upper 10 cm of a silt loam in central Missouri. High proportions of all seed viability categories [nondormant (imbibed or germinating), viable (dormant, nongerminating), and dead] revealed fungal associations when surface-sterilized seeds were cultured using standard microbiological procedures (Kremer et al., 1984). Bacterial associations were also detected to a lesser extent in all seed categories for both weed species. Microbial associa-

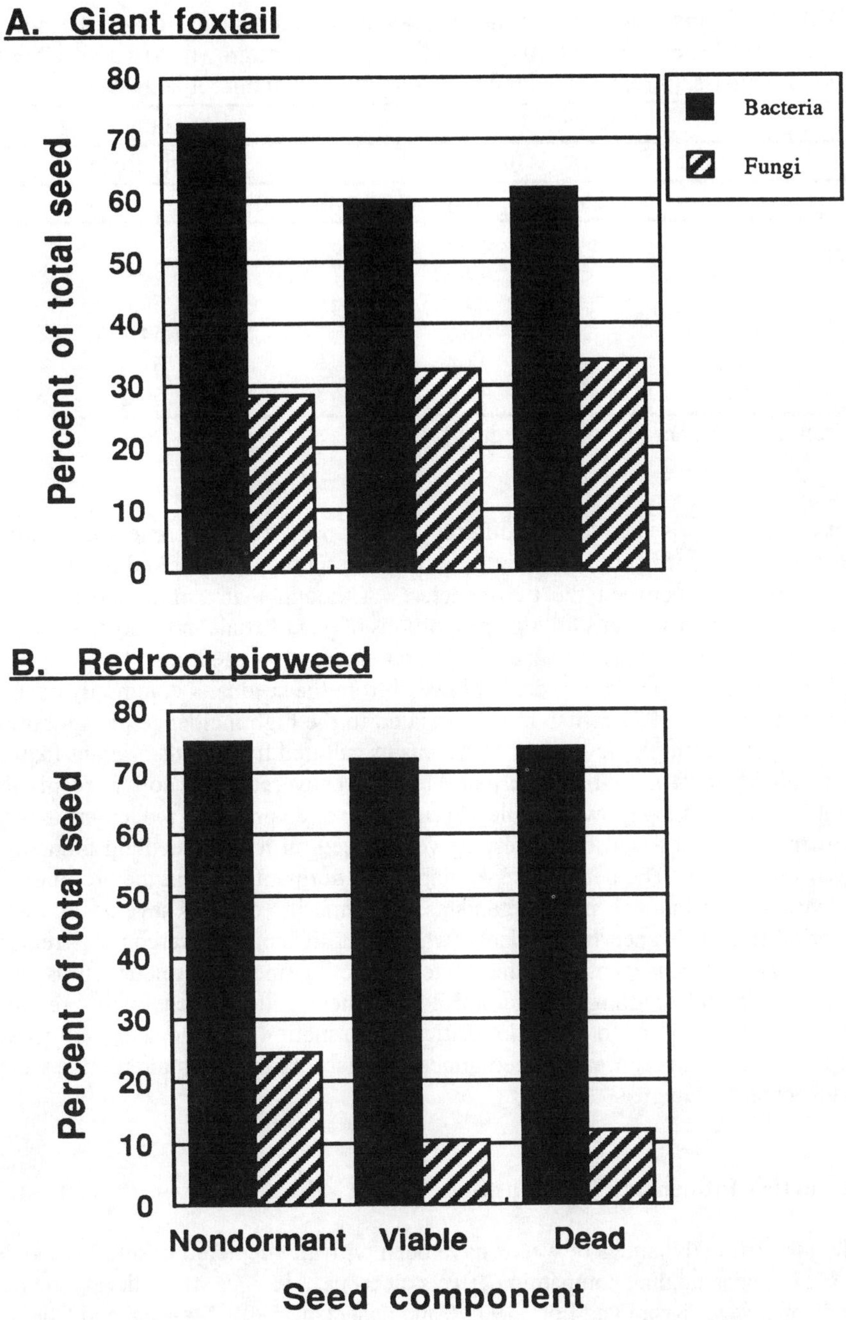

Figure 1. Microorganisms associated with different components of weed seeds retrieved from the surface 10 cm of a Mexico silt loam in central Missouri in 1993. (From R.J. Kremer, unpublished data.)

Table 1. Phytoxicity of microbial isolates from weed seeds retrieved from a Mexico silt loam in central Missouri; reduction in root growth based on lettuce seedling bioassays (each value represents an individual microbial isolate)

Bacteria		Fungi
Velvetleaf	Giant foxtail	Giant foxtail
% reduction in root growth		
23.0	15.4	7.7
30.0	40.0	36.4
39.5	53.4	54.0
67.4	67.0	54.5
78.8	73.4	61.5
80.0	84.6	69.2

(From R.J. Kremer, unpublished data.)

tions often comprised up to three different species per seed. Differences in relative proportions of seeds colonized by fungi or bacteria may be due to differences in seed structures between the two species. Although many of the seeds were colonized by microorganisms, high proportions of giant foxtail and redroot pigweed seeds in the seedbank were also viable and emerged as seedlings the following spring. Poor emergence of redroot pigweed from the seedbank commonly occurs (Forcella et al., 1992), which may be related to the high incidence of seedborne bacteria. Interestingly, several microorganisms cultured from seeds of giant foxtail and velvetleaf (*Abutilon theophrasti* Medic.) recovered from soil were highly inhibitory to seedling growth (Table 1) causing up to 85 and 69% reductions in root growth for bacteria and fungi, respectively, based on lettuce seedling bioassays (Alstrom, 1987). The persistence of viable and dormant seeds in the presence of potentially detrimental microorganisms is primarily due to several inherent deterioration resistance mechanisms, which are difficult to overcome (Kremer, 1993). These mechanisms and other factors affecting microbial relationships with weed seeds and seedlings in the soil environment and agroecosystems require further investigation to provide critical information for devising effective approaches for incorporating weed-attacking microorganisms in alternative weed management strategies.

B. Factors Influencing Relationships

The population dynamics of weeds have been typically depicted in simple models to aid in understanding competition effects on crops at various stages during the life cycle of a weed (Kropf and Lotz, 1992; Lindquist et al., 1995; Maxwell and Ghersa, 1992). Occurrence of microbial associations at stages of the weed's life cycle and factors influencing these associations can be superimposed on a generic model of an annual weed to indicate the complex microorganism-weed plant relationships

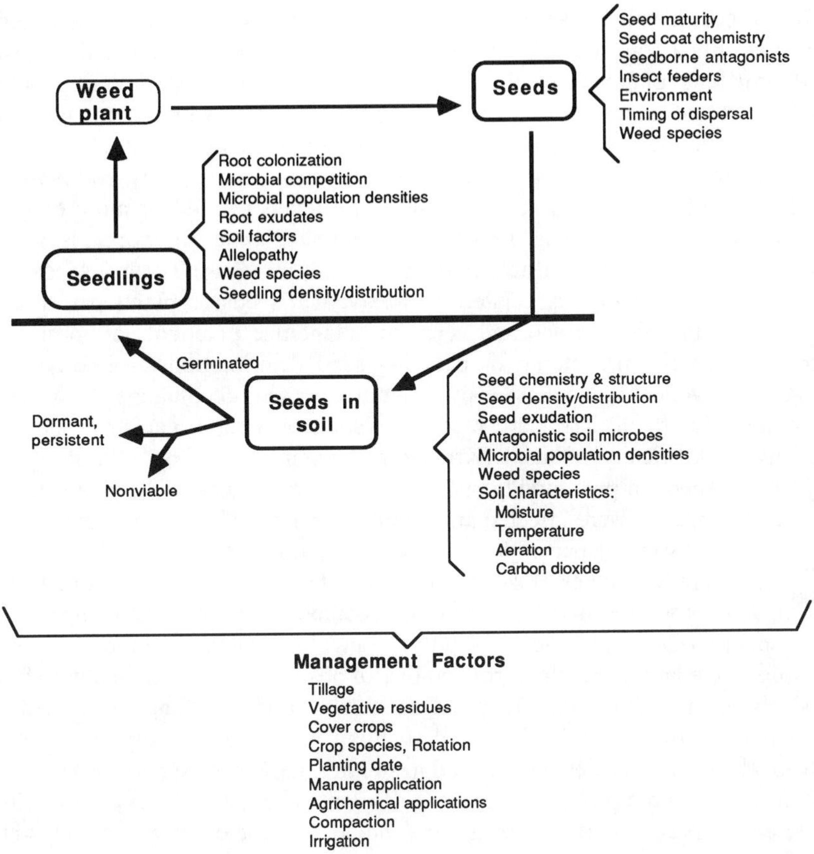

Figure 2. Factors influencing interactions of microorganisms with weed seeds and seedlings.

(Figure 2). Interactions of pathogens with actively growing weeds have been described previously and serve as bases for development of classical and mycoherbicide biocontrol strategies. The focus of the present paper is on interactions of weed seeds and seedlings with soil and rhizosphere microorganisms. Overviews on the development and deployment of pathogens for biological control of weeds are available in previous publications (Charudattan, 1991; TeBeest et al., 1992; Watson, 1991).

Several factors affect interactions of soil and rhizosphere microorganisms associated with seeds and seedlings at the various demographic stages of an annual weed (Figure 2). For example, the extent and composition of bacteria and fungi inhabiting the seed coat surface and interior seed structures while seeds develop on the plant are affected by weed species, seed maturity, and several seed coat characteristics (Kirkpatrick and Bazzaz, 1979; Kremer, 1987; Kremer et al., 1984).

Weed seeds in soil, the seedbank, are exposed to multitudes of soil microorganisms, yet only a few studies are available on seed-microbial relationships in soil. Seedborne fungi of green foxtail [*S. viridis* (L.) Beauv.] and giant foxtail persisted as colonists of seeds along with two soilborne fungi with phytopathogenic properties after the seeds were incorporated into soil in the field (Pitty et al., 1987). Seed germination of wild oat (*Avena fatua* L.) was depressed by both soil and seedborne microorganisms in soils at moisture levels of 50% or more of water holding capacity (Kiewnick, 1964). Soil and seedborne microorganisms also affected germination of bull thistle [*Cirsium vulgare* (Savi) Tenore] seeds (van Leeuwan, 1981). Fungi associated with velvetleaf seeds on plants persist after dispersal to the soil surface and seem to antagonize potential soilborne seed pathogens thereby preventing or delaying seed deterioration (Kremer, 1986). However, velvetleaf seeds previously penetrated by seed-feeding insects while on the plant prior to dispersal are readily invaded by microorganisms, and seed viability is drastically reduced (Kremer and Spencer, 1989b). Similarly, the majority of seeds of puncturevine (*Tribulus terrestris* L.) attacked by a selective seed-feeding weevil were infected and viability destroyed by soil microorganisms when the seeds were dispersed to soil (Goeden and Ricker, 1973).

Based on limited studies, rhizospheres of weed seedlings support high numbers of diverse microorganisms similar to those documented for crop and horticultural plant species (Curl and Truelove, 1986). Fungal associations reported on weed seedlings have largely resulted from empirical observations of detrimental effects of soilborne fungi on specific weed species. Soilborne fungi examined for biological control potential inhibited seedling emergence or suppressed plant growth when the fungi were established in soil at relatively high densities (approx. 100 million or more propagules per m^2) (Boyette et al., 1984; Grey et al., 1995; Jones et al., 1988). Different soils were included as factors in one study, which revealed that the fungal-weed seedling association was not affected by soil texture (Boyette et al., 1984). Similarly, actinomycetes able to produce compounds phytotoxic to barnyardgrass (*Echinochloa crusgalli* Beauv.) seedling root growth were isolated from a diverse collection of soils with no clear indication of particular edaphic factors favoring development of inhibitory isolates (Heisey et al., 1985). The abundance and composition of bacteria colonizing the rhizospheres of weed seedlings vary among weed species (Kremer et al., 1990). Rhizobacteria are those components of rhizosphere bacteria able to colonize root surfaces, some of which detrimentally affect plant growth and vigor and are often host-specific. It was suggested that the distinctive rhizobacteria-weed species associations were influenced by exudation of specific root compounds or controlled by certain genes in the plant. A subsequent study demonstrated that specific bacteria were attracted toward the seeds or seedlings in response to exudates from both imbibed seeds and seedlings of velvetleaf (chemotaxis) in soil (Begonia and Kremer, 1994). Currently, several research projects are investigating deleterious rhizobacteria (DRB) originating from weed seeds or seedlings for potential biological control of at least 18 weed species (Kremer and Kennedy, 1996). Few studies, however, have investigated factors that influence efficacy of DRB in soils and rhizospheres. Among these is a report indicating that cool, moist soils are most conducive to colonization of downy brome (*Bromus tectorum* L.) seedlings by the DRB *Pseudomonas fluorescens* D7 (Johnson et al., 1993).

Microbial relationships between weed seeds and seedlings are further impacted by several management factors imposed in agroecosystems (Figure 2). Nearly all research on the influence of management on soil microorganisms has dealt with those associated with crop plants. Soil continuously cropped to potato (*Solanum tuberosum* L.) caused a shift in the rhizosphere microbial equilibrium toward increasing activities of the soilborne pathogens *Verticillium dahliae*, *Rhizoctonia solani* and *Streptomyces* spp. detrimental to root development (Schippers et al., 1986). More phytotoxic rhizosphere bacteria were isolated from corn (*Zea mays* L.) grown continuously compared to corn in a corn-soybean [*Glycine max* L. (Merr.)] rotation (Turco et al., 1990). A similar situation may develop in soils under cultivation with continuous or annual velvetleaf infestations in which populations of wilt pathogens (*Verticillium* spp.) increase to cause significant reductions in velvetleaf seedling growth, competition, and seed production (Lindquist et al., 1995). Agrichemicals including herbicides and insecticides often increase rhizosphere microbial populations primarily due to stimulation of root exudates by affected plants (Curl and Truelove, 1986). Crop residues remaining on the soil surface in minimum-tillage systems can serve as substrates for specific bacteria able to produce metabolites inhibitory to seedlings (Stroo et al., 1988). Soil surface layers in long-term no-till fields contain accumulated organic and inorganic substances that may provide optimum environments for intense biological activity ideal for proliferation of weed seed predators and pathogens causing shifts in weed compositions (Cardina et al., 1991).

Development of approaches to exploit the interactions for practical application in weed management is only now being pursued by a few research projects, most of which are in the exploratory or preliminary testing stages (Kremer and Kennedy, 1996). The work summarized above illustrating microorganism-weed seed/seedling interactions and related effects of various environmental and management factors involved a limited number of weed species or was based on results from studies with crop species. It is precarious to suggest that similar relationships would occur with other weed species under the same set of factors. A more complete understanding of the ecology and biology of microorganisms that interact with specific weeds is needed not only to define their usefulness as potential biological control agents, but also to develop practices to manipulate the environment to favor either the development of naturally-occurring weed-suppressive agents or those agents introduced into the agroecosystem as part of alternative weed management approaches.

III. Potential of Soil and Rhizosphere Microorganisms in Weed Management

A recent review of research on weed seedbanks (Kremer, 1993) noted that weed infestations continue to occur despite the use of advanced weed management technologies, and that complete control of weed growth practiced over several years would not eliminate weeds in the field. Subsequent weed infestations are due to a small but highly persistent proportion of the seedbank that is not affected by conventional practices nor associated microorganisms (Figure 1). Some progress has been made in promoting the decline in weed seed numbers in soil through various cultural practices that enhance weed seed-decaying microorganisms. Weed management involving biological control may ultimately rely on several organisms each of which possesses unique mechanisms of action for controlling weeds at one or more demographic stage of development.

Although our understanding of the factors influencing the relationships between microorganisms and weed seeds and seedlings (Figure 2) is limited, enough information is available to devise strategies for exploiting these relationships for consideration in alternative weed management. Some strategies have previously been proposed (Kremer, 1993; Kremer and Kennedy, 1996) and will be summarized and amended with recent developments in this area.

A. Application of Selected Microorganisms

In studies conducted to date, selected microorganisms have typically been applied directly to soil or vegetative residues to attack germinating seeds and emerging seedlings and for eventual suppression of weed growth. This strategy seeks to regulate development of specific weeds before or coincident with emergence of crop plants. Thus, the problem weed is not eradicated, but early growth is significantly suppressed to allow the developing crop plants to effectively compete for growth requirements with the weakened weed seedlings. This strategy is most effective when weed growth coincides with environmental factors conducive to microbial growth and plant-suppressive activity, as illustrated by *P. fluorescens* D7 selected for biocontrol of downy brome in winter wheat (Kennedy et al., 1991). Selection of seed and seedling microorganisms possessing certain key properties greatly improves chances for effectiveness in soil. An ideal combination includes selective attraction of microorganisms (chemotaxis) to weed seeds and seedlings by exudates diffused from seeds released during germination or from seedling roots, rapid seed and seedling root colonizing ability, and toxin production. Chemotaxis has been demonstrated for bacterial isolates toward velvetleaf seeds and seedlings, which was related to effective seedling growth suppression (Begonia and Kremer, 1994). Aggressive root colonization was a major factor in establishing and sustaining growth suppression of downy brome by *P. fluorescens* D7 through the winter wheat growing season (Kennedy et al., 1991). Little is known about the colonizing ability of other microbial weed biocontrol agents, thus it is suggested that this important trait be assessed in screening programs searching for effective agents (Kremer and Kennedy, 1996).

B. Integration of Multiple Agents for Weed Management

Several recent studies indicate that combinations of biocontrol microorganisms are more effective in controlling target pests in soils, spermospheres, and rhizospheres than if inocula are comprised of individual microorganisms (Fukui et al., 1994; Pierson and Weller, 1994). Increased biocontrol of soilborne diseases by combinations of biocontrol bacteria is likely due to greater diversity of introduced phenotypes able to more thoroughly colonize roots and survive the biological, chemical, and physical changes that occur in soils and rhizospheres. Thus, multiple strains of biocontrol microorganisms result in a greater variety of traits for pest suppression, which can be expressed over a wide range of environmental conditions and a broad range of microhabitats. Likewise, potential problems of inconsistent control of weeds from site to site might be overcome by devising different strain combinations for different sites to account for differences in soil properties, weed biotypes or management system. This approach has been proposed for improving the performance of bacterial control agents for take-all disease in wheat (*Triticum aestivum* L.) (Cook, 1993). The multiple strain or organism approach will aid in managing a greater diversity of weed species and biotypes, involve a wider array of biocontrol mechanisms, and be effective under a broader range of environments. These approaches may be adaptable for systems designed for site-specific management of spatially distributed aggregates or clumps of weeds, typical of many fields in row-crop production (Mortensen et al., 1993). As suggested for biocontrol of take-all disease in winter wheat (Pierson and Weller, 1994), biocontrol of weeds with DRB may ultimately require the use of many (≥10) "core strains" in a mixture based on weed species composition, soil type, crop cultivars, and tillage system. Preliminary research has indicated that multiple DRB strains increased suppression of downy brome in winter wheat compared with single strains (A.C. Kennedy, personal communication).

Enhancement of detrimental activity of seedborne fungi by the selective seed-feeding insect *Niesthrea louisianica* on velvetleaf seed viability has been described (Kremer and Spencer, 1989a,b). This combined biocontrol agent strategy for preventing or reducing seed production by weeds escaping early-season control could be integrated as part of a total weed management program for optimizing use of both chemical and biological approaches (Kropf and Lotz, 1992). The most practical application of DRB and insect combinations would be in situations where the insect agent feeds on roots or crowns of target weeds. Indeed, it has been suggested that leafy spurge (*Euphorbia esula* L.) control resulting from feeding by root-boring larvae of flea beetles (*Aphthona* spp.) may be enhanced due to secondary invasion by plant pathogens naturally present in soils (Rees and Spencer, 1991). Exploitation of flea beetle larvae as vectors of DRB selective for suppression of leafy spurge could contribute an additional strategy for control of this noxious range weed and serve as a model for integration of root insect-DRB combinations on other weeds.

C. Phytotoxins

Deleterious activity toward weed seed viability and seedling growth by most microorganisms under study for biological control is due to the production of phytotoxins. The fungus *Gliocladium virens* applied to soil on a rice (*Oryza sativa* L.) grain substrate produced the herbicidal metabolite viridiol that prevented pigweed emergence without harming emerging cotton (*Gossypium hirsutum* L.) seedlings (Howell and Stipanovic, 1984). Rhizobacteria for biological control of weeds likely metabolize phytotoxins at root surfaces where they are readily absorbed by the plant. It is not known how widespread phytotoxin production is among weed biocontrol rhizobacteria, but evidence is accumulating showing that phytotoxins play a causal role in deleterious activity (Souissi and Kremer, 1994; Tranel et al., 1993). There is currently some question as to whether phytotoxins produced in culture and applied as a bioherbicide are as effective in controlling weeds compared to application of the intact organism. Durbin (1983) points out that some bacterial pathogens are unable to produce phytotoxins in culture but only produce them *in planta*, possibly due to the requirement of specific seed or seedling exudate as substrates for the inhibitory compounds. Therefore, before considerable research efforts are devoted to *in vitro* production of 'natural products' for weed control, a complete understanding of the conditions required for optimum and effective phytotoxin production is necessary. It is likely that successful establishment of rhizobacterial inocula that produce high levels of phytotoxin in the rhizosphere would be more economical than chemical synthesis of the compound (Arshad and Frankenberger, 1991).

Known metabolites produced in the rhizosphere of plants that can be phytotoxic at higher than physiologic concentrations include the auxins and hydrogen cyanide. Rhizosphere-inhabiting microorganisms likely synthesize and release auxins as secondary metabolites because of rich supplies of substrates exuded from plant roots. Some microorganisms produce auxins in the presence of a suitable precursor such as tryptophan (TRP) (Arshad and Frankenberger, 1991). The prospect of using rhizobacteria that produce excessive amounts of growth-regulating substances in the rhizosphere to suppress growth of weed seedlings has received little attention. Such rhizobacteria could selectively colonize weed seedling roots, localize excess auxin production, and minimize potential deleterious effects on crop growth. In a preliminary study, different plant seedlings reflected a wide range in response (measured as root length) to application of a selected DRB alone and in combination with TRP (Table 2). Root length inhibition ranged from 90.5% for field bindweed (*Convolvulus arvensis* L.) to 24.8% for soybean. Inoculation with the isolate alone resulted in strong inhibition of root growth in field bindweed, green foxtail, and morningglory (*Ipomoea* sp.) while lesser inhibition was observed for seedlings of wheat and soybean. The enterobacterial strain used as inoculum in this study is representative of others typically found in plant rhizospheres, which readily

Table 2. Effect of tryptophan (TRP; 10 μM) and isolate 3.8.12.7 applied alone or combined on seedling root growth of different plant species in agar bioassay

Treatment	Wheat	FBW	VL	RRP	GFT	MG	SB
Control	57.8d[a]	47.5d	59.7d	31.2c	30.9c	48.2b	25.8b
TRP alone	48.6c	38.8c	53.2c	31.5c	28.4c	49.6b	29.3b
Isolate 3.8.12.7	33.5b	10.0b	36.6b	14.6b	8.5b	13.9a	25.4b
TRP + isolate 3.8.12.7	19.1a	4.5a	17.1a	7.3a	3.8a	9.1a	19.4a

[a]Mean values within columns followed by the same letter are not significantly different at $p < 0.05$; plant species codes: FWB = field bindweed; VL = velvetleaf; RRP = redroot pigweed; GFT = green foxtail; MG = morningglory; SB = soybean. (Modified from Sarwar and Kremer, 1995.)

colonize plant roots and actively produce auxins (Sarwar and Kremer, 1995). This finding demonstrated that live inocula delivered to the rhizosphere of weeds produced effective concentrations of auxins causing phytotoxicity. Efficacy of the biocontrol organisms was further enhanced when TRP was provided as a precursor for producing phytotoxic concentrations *in situ*.

D. Formulations and Delivery of Microbial Agents

Formulation and delivery systems that promote survival and colonization of weed seeds and seedlings by DRB in the field are critical in attaining an acceptable level of efficacy. Formulations that enhanced bioactivity of biocontrol microbes against soilborne pathogens (Roberts and Fravel, 1993) and those developed for plant growth-promoting rhizobacteria (Caesar and Burr, 1991) can be adapted for use in developing prototype inoculants for weed biocontrol agents. Zorner et al. (1993) stressed the importance of utilizing formulations that alter the soil or rhizosphere environment to favor establishment and optimum activity of biological control agents. The use of exotic carbon sources can specifically enhance establishment and phytotoxic activity in the weed rhizosphere. Inoculum containing a unique carbon source that is not widespread in nature will enhance the establishment of selected agents in soil or rhizosphere and out-compete native bacteria, simultaneously producing phytotoxins to be absorbed by the weed seedling. Inocula can be prepared to contain slow-release formulations such as starch-encapsulated alginate granules (Daigle and Connick, 1990) so that the exotic carbon source will be released over time thereby extending the biological activity. Formulations can also be used to provide precursors of known phytotoxic compounds produced by the agent. For example, addition of L-TRP with the excess auxin-producing rhizobacterium *E. taylorae* 3.8.12.7 in a starch formulation applied to soil caused significant suppression of green foxtail root growth but had no effect on soybean growth (Table 3). Although preliminary, these results are encouraging and indicate the potential for manipulating weed rhizospheres by providing suitable precursors

Table 3. Effect of L-tryptophan (L-TRP) and selected rhizobacteria on root growth of soybean and green foxtail planted together in soil

		Root length (mm)	
Isolate	L-TRP (μM)	Soybean	Green foxtail
None	0	26.1	9.3
3.8.12.7	0	25.4	8.6
None	10	24.4	9.6
3.8.12.7	10	24.2	6.4[a]

[a]Significantly different ($p < 0.05$) from other treatment combinations within the column based on LSD.
(From R.J. Kremer and M. Sarwar, unpublished data.)

for production and delivery of phytotoxic compounds in soil and in the presence of crop roots.

A unique approach for delivery of microbial agents to soil infested with weed seeds is by either direct inoculation of crop seeds with the agents or by promoting colonization of crop roots by the agents in formulations applied at planting. Preliminary studies with a starch-based formulation (Connick et al., 1993) containing rhizobacteria for controlling giant foxtail showed high root colonization was achieved not only on giant foxtail but also on soybean (Figure 3). Giant foxtail growth was suppressed, but soybeans were not affected (R.J. Kremer, unpublished data). Based on this study, crop roots not only may deliver microbial agents to adjacent roots of weeds but also maintain or even enhance the agent's numbers for attack of seedlings emerging later in the season.

E. Integration with Cultural Practices

1. Agrichemicals

Integration of herbicides with biocontrol agents may be necessary to reduce adverse impacts of weeds on the crops and to establish the weed-attacking microorganisms for long-term control. Only limited examples of agrichemicals used in combination with seed and seedling microbial agents are available. Seedling emergence and seed viability of velvetleaf in soil were significantly reduced when the fungus *Fusarium oxysporum* was combined with several agrichemicals (Kremer and Schulte, 1989). Reduction in emergence of Texas gourd [*Cucurbita texana* (Scheele) Gray] seedlings by *Fusarium solani* f. sp. *cucurbitae* was enhanced when the fungus was combined with trifluralin (Weidemann and Templeton, 1988). Greaves and Sargent (1986) found that colonization of wheat roots by *Pseudomonas* spp. was greatly enhanced and resulted in extensive cellular and tissue damage when plants were treated with mecoprop. They suggested that exploitation of plant root-herbicide-microorganism interactions potentially could be an effective strategy

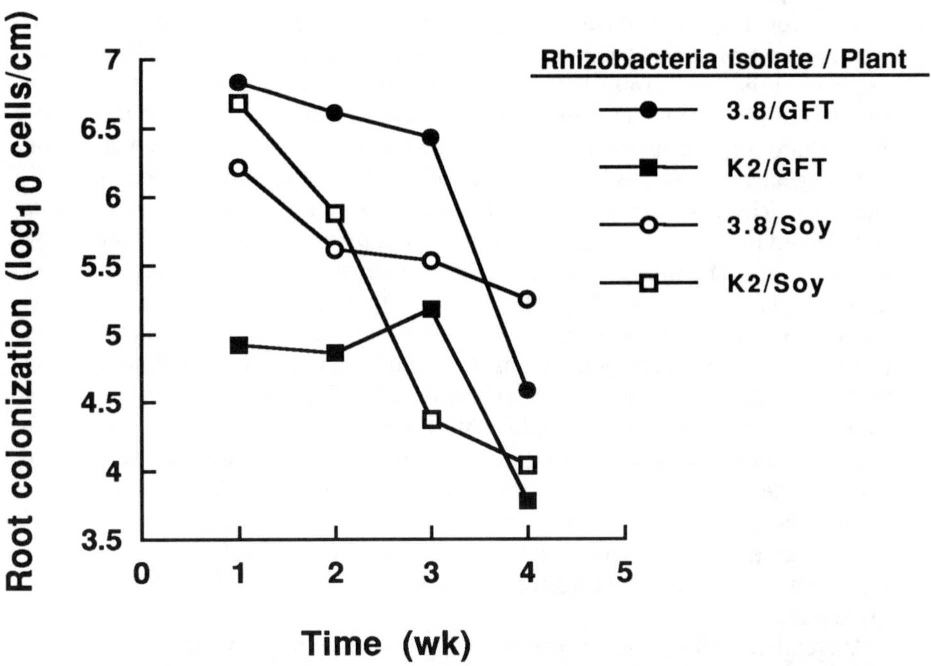

Figure 3. Root colonization of giant foxtail (GFT) and soybean (Soy) by rhizobacteria in starch-base inocula applied to a Mexico silt loam.
(From R.J. Kremer, unpublished data.)

for biological control of weeds. Rhizobacteria inhibitory to downy brome and jointed goatgrass (*Aegilops cylindrica* Host.) exhibited higher growth-suppressive activity in soil when combined with herbicides at reduced rates of application (Kremer and Kennedy, 1996). Growth suppression by some DRB combined with metribuzin and diclofop applied at sublethal rates was additive. Diclofop plus bacteria increased root growth suppression of downy brome by 0 to 12% over diclofop alone. Further research into the mechanisms of herbicide-rhizobacteria interactions is needed to develop strategies where DRB selected for activity toward a weed can be paired with a specific chemical that increases susceptibility of the weed to the DRB. Successful development of this integrated strategy will increase efficacy of DRB agents, reduce amounts of herbicides required for weed control, and decrease potential environmental contamination.

2. Cover Crops, Mulches, and Allelopathy

Cover crops and mulches as components of alternative management systems may be used for integrating biocontrol agents by delivering the agents on seeds and promoting their establishment in soils for attack of weed seeds and seedlings prior to planting the main crop. Previous research demonstrating that certain legume cover crops promoted populations of soilborne plant pathogens of cotton (Rothrock et al., 1995) suggests that cover crops may be useful in establishing weed-attacking microorganisms in soil well in advance of weed seedling emergence. A rhizobacterium deleterious toward giant foxtail was coated on oat (*Avena sativa* L.) seeds and planted in soil containing giant foxtail seeds prior to planting soybeans as the primary crop. The rhizobacterium readily colonized oat seedling roots as well as those of giant foxtail seedlings (Figure 4). The vegetative growth of oats was removed after eight weeks to accommodate soybean planting. The roots continued to support rhizobacterial populations that established in soil and colonized giant foxtail seedlings emerging with soybeans (R.J. Kremer, unpublished data). Combined effects of potential allelopathic activity by the oats plus the rhizobacterium may also impact weed growth suppression. An apparent interaction between allelopathic substances from kochia [*Kochia scoparia* (L.) Shrad.] and the fungus *Rhizopus* sp. on inhibition of sugarbeet (*Beta vulgaris* L.) germination and seedling growth (Wiley et al., 1985) suggests the potential for discovery of similar relationships for microbial agents and crop seeds that may be useful in weed management.

Vegetative residues at or near the soil surface could serve as substrates for production of weed-suppressive agents by DRB applied directly to the residues. As previously demonstrated, numbers of wheat-inhibitory DRB increased dramatically when applied on crop residues, which promoted production of toxins inhibitory to wheat (Stroo et al., 1988). An approach is envisioned in which DRB applied on surface residues for producing phytotoxins might suppress weed growth prior to planting the crop, similar to a preemergence herbicide tactic.

3. Cropping Practices

Cultural practices used for nonchemical weed control offer convenient application methods for integrating microorganisms for biological control of weeds in cropping systems. Tillage can influence the frequency of inhibitory bacteria occurring in soil and their growth-suppressive activity. Downy brome and jointed goatgrass were suppressed by rhizobacteria under either conventional or minimum tillage suggesting that application of selected DRB during tillage may be effective in integrated weed management (Kremer and Kennedy, 1996).

Crop rotation is a practice that may also be manipulated to encourage development of specific inhibitory bacteria on weed roots. Previous work reporting a rotation effect in corn was due partly to certain rhizobacteria specifically associated with corn roots illustrates the potential for using DRB to achieve suppression of weeds in crop rotation systems (Turco et al., 1990).

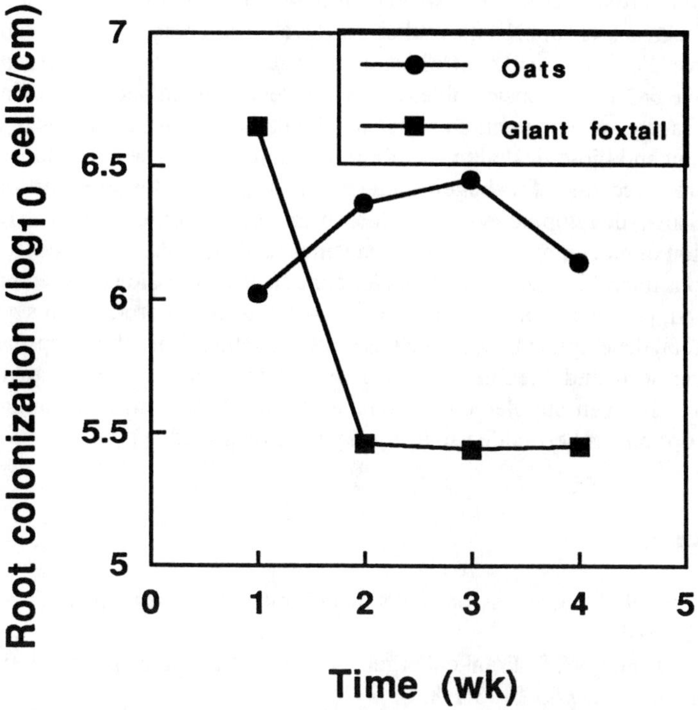

Figure 4. Root colonization of giant foxtail and oats by rhizobacterial isolate 3.8.12.7 inoculated on oat seeds and planted in a Mexico silt loam. (From R.J. Kremer, unpublished data.)

4. Competition

Efficacy of DRB in field studies has been attributed partly to reduced competitiveness of weeds due to growth suppression by DRB (Kennedy et al., 1991). Increasing crop interference in the field by manipulating row spacing, seeding rates, and other cultural practices to suppress early weed growth has been proposed as a viable component of integrated weed management (Jordan, 1993). Highly competitive varieties selected through crop breeding could be used in conjunction with weed-suppressive microorganisms and further enhance effectiveness and acceptance of integrated weed management systems emphasizing nonchemical control.

IV. Conclusions and Research Needs

An urgent need exists to develop alternative weed management systems that will reduce herbicide use, provide optimum crop yields, and protect the environment.

The need to provide efficacious weed control with low inputs of agrichemicals will determine research approaches taken to develop alternative weed management programs. For biological control components to be seriously considered in these alternative programs, considerable research in development and implementation of effective agents will be required. Intensive fundamental investigations in the areas of ecology and biological activity of microorganism-weed seed/seedling relationships are necessary for better understanding and exploitation of biocontrol mechanisms, development of suitable formulations for use in the field, and integration of biocontrol into weed management and agricultural practices. As more efficacious microbial-based strategies for consistent suppression of weed growth are developed, prospects for acceptance of these strategies in production systems and commercial development should increase considerably. Microbial approaches that target the seed and seedling stages of weeds are additional biological control strategies that can supplement current mycoherbicides and offer augmentative control options as herbicide use becomes more restricted.

References

Alstrom, S. 1987. Factors associated with detrimental effects on plant growth. *Plant Soil* 102:3-9.

Arshad, M. and W.T. Frankenberger, Jr. 1991. Microbial production of plant hormones. *Plant Soil* 133:1-8.

Begonia, M.F.T. and R.J. Kremer. 1994. Chemotaxis of deleterious rhizobacteria to velvetleaf (*Abutilon theophrasti*) seeds and seedlings. *FEMS Micobiol. Ecol.* 15:227-236.

Boyette, C.D., G.E. Templeton, and L.R. Oliver. 1984. Texas gourd (*Cucurbita texana*) control with *Fusarium solani* f.sp. *cucurbitae. Weed Sci.* 32:649-655.

Caesar, A.J. and T.J. Burr. 1991. Effect of conditioning, betaine, and sucrose on survival of rhizobacteria in powder formulations. *Appl. Environ. Microbiol.* 57:168-172.

Cardina, J., E. Regnier, and K. Harrison. 1991. Long-term tillage effects on seed banks in three Ohio soils. *Weed Sci.* 39:186-194.

Charudattan, R. 1991. The mycoherbicide approach with plant pathogens. p. 24-57. In: D.O. TeBeest (ed.), *Microbial Control of Weeds*. Chapman and Hall, New York.

Connick, W.J., W.R. Nickle, and B.T. Vinyard. 1993. "Pesta": New granular formulations for *Steinernema carpocapsae. J. Nematol.* 25:198-203.

Cook, R.J. 1993. Making greater use of introduced microorganisms for biological control of plant pathogens. *Ann. Rev. Phytopathol.* 31:53-80.

Curl, E.A. and B. Truelove. 1986. *The Rhizosphere*. Springer-Verlag, New York.

Daigle, D.J. and W.J. Connick. 1990. Formulation and application technology for microbial weed control. p. 288-304. In: R.E. Hoagland (ed.), *Microbes and Microbial Products as Herbicides*. Am. Chem. Soc., Washington, D.C.

Durbin, R. 1983. The biochemistry of fungal and bacterial toxins and their modes of action. p. 137-162. In: J. Callow (ed.), *Biochemical Plant Pathology*. John Wiley & Sons, New York.

Forcella, F., R.G. Wilson, K.A. Renner, J. Dekker, R.G. Harvey, D.A. Alm, D.D. Buhler, and J.A. Cardina. 1992. Weed seedbanks of the U.S. Cornbelt: magnitude, variation, emergence, and application. *Weed Sci.* 40:636-640.

Fukui, R., M.N. Schroth, M. Hendson, and J.G. Hancock. 1994. Interactions between strains of pseudomonads in sugarbeet spermospheres and their relationship to pericarp colonization by *Pythium ultimum* in soil. *Phytopathology* 84:1322-1330.

Goeden, R.D. and D.W. Ricker. 1973. A soil profile analysis for puncturevine fruit and seed. *Weed Sci.* 21:504-507.

Greaves, M.P. and J.A. Sargent. 1986. Herbicide-induced microbial invasion of plant roots. *Weed Sci.* 34 (Suppl. 1):50-53.

Grey, W.E., P.C. Quimby, Jr., D.E. Mathre, and J.E. Young. 1995. Potential for biological control of downy brome (*Bromus tectorum*) and medusahead (*Taeniatherum caput-medusae*) with crown and root rot fungi. *Weed Technol.* 9:362-365.

Harley, K.L.S. and I.W. Forno. 1992. *Biological Control of Weeds: A Handbook for Practitioners and Students*, Inkata Press, Melbourne.

Heisey, R.M., J. DeFrank, and A.R. Putnam. 1985. A survey of soil microorganisms for herbicidal activity. p. 337-349. In: A.C. Thompson (ed.), *the Chemistry of Allelopathy*. Am. Chem. Soc., Washington, D.C.

Howell, C.R. and R.D. Stipanovic. 1984. Phytotoxicity to crop plants and herbicidal effects on weeds of viridiol produced by *Gliocladium virens*. *Phytopathology* 74:1346-1349.

Johnson, B.N., A.C. Kennedy, and A.G. Ogg, Jr. 1993. Suppression of downy brome growth by a rhizobacterium in controlled environments. *Soil Sci. Soc. Am. J.* 57:73-77.

Jones, R.W., W.T. Lanini, and J.G. Hancock. 1988. Plant growth response to the phytotoxin viridiol produced by the fungus *Gliocladium virens*. *Weed Sci.* 36:683-687.

Jordan, N. 1993. Prospects for weed control through crop interference. *Ecol. Appl.* 3:84-91.

Kennedy, A.C. and K.L. Smith. 1995. Soil microbial diversity and the sustainability of agricultural soils. *Plant Soil* 170:75-86.

Kennedy, A.C., L.F. Elliott, F.L. Young, and C.L. Douglas. 1991. Rhizobacteria suppressive to the weed downy brome. *Soil Sci. Soc. Am. J.* 55:722-727.

Kiewnick, L. 1964. Experiments on the influence of seedborne and soilborne microflora on the viability of wild oat seeds (*Avena fatua* L.) II. Experiments on the influences of microflora on the viability of seeds in the soil. *Weed Res.* 4:31-43.

Kirkpatrick, B.L. and F.A. Bazzaz. 1979. Influence of certain fungi on seed germination and seedling survival of four colonizing annuals. *J. Appl. Ecol.* 16:515-527.

Kremer, R.J. 1986. Microorganisms associated with velvetleaf (*Abutilon theophrasti*) seeds on the soil surface. *Weed Sci.* 34:233-236.

Kremer, R.J. 1987. Identity and properties of bacteria inhabiting seeds of selected broadleaf weed species. *Microb. Ecol.* 14:29-37.

Kremer, R.J. 1993. Management of weed seed banks with microorganisms. *Ecol. Appl.* 3:42-52.

Kremer, R.J. and A.C. Kennedy. 1996. Rhizobacteria as biological control agents of weeds. *Weed Technol.* 10:(In press).

Kremer, R.J. and L.K. Schulte. 1989. Influence of chemical treatment and *Fusarium oxysporum* on velvetleaf (*Abutilon theophrasti*). *Weed Technol.* 3:369-374.

Kremer, R.J. and N.R. Spencer. 1989a. Impact of a seed-feeding insect and microorganisms on velvetleaf (*Abutilon theophrasti*) seed viability. *Weed Sci.* 37:211-216.

Kremer, R.J. and N.R. Spencer. 1989b. Interaction of insects, fungi, and burial on velvetleaf (*Abutilon theophrasti*) seed viability. *Weed Technol.* 3:322-328.

Kremer, R.J., L.B. Hughes, Jr., and R.J. Aldrich. 1984. Examination of microorganisms and deterioration resistance mechanisms associated with velvetleaf seed. *Agronomy J.* 76:745-749.

Kremer, R.J., M.F.T. Begonia, L. Stanley, and E.T. Lanham. 1990. Characterization of rhizobacteria associated with weed seedlings. *Appl. Environ. Microbiol.* 56:1649-1655.

Kropf, M.J. and L.A.P. Lotz. 1992. Optimization of weed management systems: the role of ecological models of interplant competition. *Weed Technol.* 6:462-470.

Lindquist, J.L., B.D. Maxwell, D.D. Buhler, and J.L. Gunsolus. 1995. Velvetleaf (*Abutilon theophrasti*) recruitment, survival, seed production, and interference in soybean (*Glycine max*). *Weed Sci.* 43:226-232.

Maxwell, B.D. and C. Ghersa. 1992. The influence of weed seed dispersion versus the effect of competition on crop yield. *Weed Technol.* 6:196-204.

Mortensen, D.A., G.A. Johnson, and L.J. Young. 1993. Weed distribution in agricultural fields. p. 113-124. In: P.C. Robert, R.H. Rust, and W.E. Larson (eds.), *Soil Specific Crop Management*. Am. Soc. Agron., Madison, WI.

Pierson, E.A. and D.M. Weller. 1994. Use of mixtures of fluorescent pseudomonads to suppress take-all and improve growth of wheat. *Phytopathology* 84:940-947.

Pitty, A., D.W. Staniforth, and L.H. Tiffany. 1987. Fungi associated with caryopses of *Setaria* species from field-harvested seeds and from soil under two tillage systems. *Weed Sci.* 35:319-323.

Rees, N.E. and N.R. Spencer. 1991. Biological control of leafy spurge. p. 182-192. In: L.F. James (ed.), *Noxious Range Weeds*. Westview Press, Boulder, CO.

Roberts, D.P. and D.R. Fravel. 1993. Strategies and techniques for improving biocontrol of soilborne plant pathogens. p. 323-331. In: S.O. Duke, J.J. Menn, and J.R. Plimmer (eds.), *Pest Control with Enhanced Environmental Safety*. Am. Chem. Soc., Washington, D.C.

Rothrock, C.S., T.L. Kirkpatrick, R.E. Frans, and H.D. Scott. 1995. The influence of winter legume cover crops on soilborne plant pathogens and cotton seedling diseases. *Plant Dis.* 79:167-171.

Sarwar, M. and R.J. Kremer. 1995. Enhanced suppression of plant growth through production of L-tryptophan-derived compounds by deleterious rhizobacteria. *Plant Soil* 172:261-269.

Schippers, B., A.W. Bakker, P.A.H.M. Bakker, P.J. Weisbeek, and B. Lugtenberg. 1986. Plant-growth-inhibiting and stimulating rhizosphere micro-organisms. p. 35-48. In: V. Jensen, A. Kjoller, and L.H. Sorensen (eds.), *Microbial Communities in Soil*. Elsevier, London.

Souissi, T. and R.J. Kremer. 1994. Leafy spurge (*Euphorbia esula*) cell cultures for screening deleterious rhizobacteria. *Weed Sci.* 42:310-315.

Stroo, H.F., L.F. Elliott, and R.I. Papendick. 1988. Growth, survival and toxin production of root-inhibitory pseudomonads on crop residues. *Soil Biol. Biochem*. 20:201-207.

TeBeest, D.O., X.B. Chang, and C.R. Cisar. 1992. The status of biological control of weeds with fungal pathogens. *Ann. Rev. Phytopathol.* 30:637-657.

Tranel, P.J., D.R. Gealy, and A.C. Kennedy. 1993. Inhibition of downy brome (*Bromus tectorum*) root growth by a phytotoxin from *Pseudomonas fluorescens* strain D7. *Weed Technol*. 7:134-139.

Turco, R.F., M. Bischoff, D.P. Breakwell, and D.R. Griffith. 1990. Contribution of soil-borne bacteria to the rotation effect in corn. *Plant Soil* 122:115-120.

van Leeuwan, B.H. 1981. Influence of microorganisms on the germination of the monocarpic *Cirsium vulgare* in relation to disturbance. *Oecologia (Berlin)* 48:112-115.

Watson, A.K. 1991. The classical approach with plant pathogens. p. 3-23. In: D.O. TeBeest (ed.), *Microbial Control of Weeds*. Chapman and Hall, New York.

Weidemann, G.J. and G.E. Templeton. 1988. Control of Texas gourd, *Cucurbita texana*, with *Fusarium solani* f. sp. *cucurbitae. Weed Technol*. 2:271-274.

Wiley, R.B., E.E. Schweizer, and E.G. Ruppel. 1985. Interaction of kochia (*Kochia scoparia*) and *Rhizopus* sp. on sugarbeet (*Beta vulgaris*) germination. *Weed Sci.* 33:275-279.

Zorner, P.S., S.L. Evans, and S.D. Savage. 1993. Perspectives on providing a realistic technical foundation for the commercialization of bioherbicides. p. 79-86. In: S.O. Duke, J.J. Menn, and J.R. Plimmer (eds.), *Pest Control with Enhanced Environmental Safety*. Am. Chem. Soc., Washington, D.C.

Crop Rotation and Legume Residue Effects on Weed Emergence and Growth: Applications for Weed Management

Matt Liebman and Tsutomu Ohno

I. Introduction

One of the dominant trends in American agriculture since the Second World War has been a reduction in crop diversity at both farm and regional levels. As crop production has become more capital-intensive, isolated from livestock production, and linked to federal financial policy, farms and regions have become increasingly specialized in the crops they produce (Power and Follett, 1987; NRC, 1989; Bullock, 1992). This trend is exemplified by changes in cropping patterns in Illinois between 1938 and 1970. Marked reductions during this period in oat (*Avena sativa* L.), wheat (*Triticum aestivum* L.), hay, and pasture acreage were accompanied by large increases in soybean (*Glycine max* [L.] Merr.) and corn (*Zea mays* L.) acreage (USDA, 1973). Corn and soybean occupied 45% of Illinois cropland in 1958 (USDA, 1973), but almost 85% of the state's cropland in 1982 (Power and Follett, 1987).

In many areas of the U.S., the spatial homogeneity of cropland is accompanied by a low level of crop diversity in the temporal dimension. A substantial portion of American field crop production occurs in monoculture, in which a particular crop is planted repeatedly on the same land (Power and Follett, 1987). A 1991 survey of U.S. field crops indicated that 61% of the land planted with cotton (*Gossypium hirsutum* L.), 29% of the land planted with wheat, 24% of the land planted with corn, and 10% of the land planted with soybean were occupied by those same crops in both 1990 and 1989 (ERS, 1992). Examined on a more local scale, the incidence of monoculture may increase dramatically. In 1991, for example, production of the same crop on the same field for at least three consecutive years occurred on 82% of the land used for cotton in Louisiana, 86% of the land used for wheat in Oklahoma, 55% of the land used for corn in Nebraska, and 57% of the land used for soybean in Mississippi (ERS, 1992).

Even when monoculture is not practiced, crop rotation sequences are generally short and diversity is limited. For example, about 50% of the U.S. corn crop and 60% of the U.S. soybean crop are grown in rotation sequences that include only those crops (e.g., corn-soybean, corn-corn-soybean; ERS, 1992). Longer rotations for corn that include hay, pasture, and green manure crops have generally been abandoned (Bullock, 1992). About 27% of the U.S. wheat crop is now grown in "rotation" sequences that include only wheat and extended fallow periods (ERS, 1992).

Concomitant with the loss of crop diversity in U.S. agriculture at farm and regional levels has been the increased use of synthetic fertilizers and pesticides, particularly nitrogen (N) and herbicides. Since 1960, applications of N fertilizer and herbicides to U.S. cropland have increased more than three-fold (NRC, 1989; Power, 1990). Nitrogen fertilizer use on U.S. cropland now exceeds 9×10^6 Mg year^{-1} (Power, 1990), and agricultural pesticide use now exceeds 350×10^6 kg of active ingredients annually, with herbicides comprising $>200 \times 10^6$ kg (>60%) of this total (Aspelin, 1994). Nitrogen fertilizers and herbicides have substantially increased yield per unit land area and per unit labor, but have been detected with increasing frequency as contaminants of ground and surface water (Hallberg, 1989; NRC, 1989; Goolsby et al., 1991; Thurman et al., 1991; USEPA, 1992; Goolsby et al., 1993; Nelson and Jones, 1994).

During the past decade, concerns over environmental contamination with fertilizers and pesticides, coupled with questions over the social, economic, and health-related impacts of conventional agricultural systems, have led to efforts to improve agricultural sustainability through research, education, and policy initiatives (NRC, 1991; Liebman, 1992). As a result, there is growing interest in the design and management of agroecosystems that are primarily reliant on manipulations of ecological interactions, rather than on applications of agrichemicals, to maintain soil productivity and manage pest populations (Stinner and House, 1989). Observations of economically viable, agronomically productive U.S. farms that have greatly reduced or eliminated use of synthetic fertilizers and pesticides suggest that their ability to produce acceptable yields and profits is linked to crop diversification, especially through rotation sequences that include forage legumes (Lockeretz et al., 1981; Reganold et al., 1990; Taylor et al., 1992).

This chapter focuses on the use of crop rotation as a means of improving weed management with reduced reliance on herbicides. Because we believe that best use of crop rotation for weed management is dependent on improved knowledge of mechanisms, we have sought to identify how crop rotation may affect biological, chemical, and physical factors regulating weed survival, germination, growth, reproduction, and competitive interactions. Because crop rotations often include legume species to alter soil N dynamics and help satisfy crop fertility requirements, and because both crop and weed species can be highly responsive to N conditions, we have emphasized potential interactions between weed management and use of legumes as a source of fertility. Finally, because crop residues, including those of legume species, may function as sources of allelopathic compounds, we have examined the challenges and potential benefits of managing the biochemistry of crop residues for weed suppression.

II. Potential Benefits of Crop Rotation Systems

Crop rotation involves growing different crops in recurring sequence on the same unit of land, in contrast to monoculture, in which a particular crop is grown in the same place repeatedly. Rotation may involve crop sequences extending over several years (e.g., a three-year sequence of oat-clover [*Trifolium* spp.]/grass hay-potato [*Solanum tuberosum* L.] [Hepler et al., 1984, 1985]), or sequences compressed within a single year (e.g., winter wheat-soybean [Lewis and Phillips, 1976]). Rotation may also involve the use of nonharvested crops grown for soil improvement (green manures), soil conservation (cover crops), and nutrient retention (catch crops) during seasons when cash or feed crops are not in the field. Crop sequences exemplifying these practices include corn-hairy vetch (*Vicia villosa* Roth), corn-winter rye (*Secale cereale* L.), and lettuce (*Lactuca sativa* L.)-phacelia (*Phacelia tanacetifolia* Benth.) (Smith et al., 1987; Shipley et al., 1992; Jackson et al., 1993). For the purposes of our discussion, we have restricted our attention to rotation systems in temperate regions, although we recognize rotation systems are also used in the tropics (see, e.g., Beets, 1982; Ashby and Pachico, 1987; Kass et al., 1993).

Depending on the specific crops used and their duration, potential benefits of rotation may include:

- improvements in soil physical characteristics, e.g., decreased bulk density, improved aggregate stability, improved water infiltration (Sumner, 1982; MacRae and Mehuys, 1985; Smith et al., 1987; Bullock, 1992; Karlen et al., 1994);

- reductions in soil erosion (Hepler et al., 1984, 1985; Gantzer et al., 1991);

● increases in N available to subsequent crops following N_2 fixation by legume species (Heichel and Barnes, 1984; Voss and Shrader, 1984; Fox and Pielielek, 1988; Frye et al. 1988; Hesterman, 1988);

● reductions in nitrate leaching (Shipley et al., 1992; Jackson et al., 1993);

● reductions in the incidence and severity of certain soil-borne crop diseases (Maloy and Burkholder, 1959; Emmond and Ledingham, 1972; Sturz and Bernier, 1987; Dabney et al., 1988; McEwen et al., 1989);

● reductions in densities of certain insect pests (Lashomb and Ng, 1984; Wright, 1984; Brust and King, 1994); and

● reductions in weed density, whether herbicides are used (Schreiber, 1992; Martin and Felton, 1993; Blackshaw, 1994) or not used (Liebman and Dyck, 1993).

Crop yields in rotation are typically 5 to 15% higher than in monoculture (Crookston, 1984; Heichel and Barnes, 1984; Voss and Shrader, 1984; Roder et al., 1989; Crookston et al., 1991; Bullock, 1992; Karlen et al., 1994), even when nutrients are supplied at high levels or N-supplying legumes are excluded from the crop sequence (Heichel and Barnes, 1984; Voss and Shrader, 1984). Increases in crop yield through rotation may accrue from reductions in pest pressure and improvements in soil conditions, or from other factors that remain incompletely understood (Crookston, 1984; Crookston et al., 1988; Crookston and Kurle, 1989; Johnson et al., 1992).

Crop rotation has traditionally been regarded as an important strategy for weed management (Leighty, 1938; Froud-Williams, 1988). However, in most experiments examining the effects of crop rotation on weed dynamics, crop rotation is confounded with herbicide rotation, making it difficult to separate ecological from chemical effects. Table 1 summarizes results of 27 test crops by rotation combinations for which comparisons with suitable monoculture systems were possible and not confounded by herbicide use. In the 25 cases where weed plant data were reported, weed plant density in rotation was less than in monoculture in 19 cases, higher than in monoculture in 2 cases, and equivalent with monoculture in 4 cases. In the 12 cases where weed seed data were reported, weed seed density in crop rotation was lower in 9 cases and equivalent in 3 cases, compared to monocultures of the component crops. In the 12 cases where crop yield data were reported, yields of test crops were higher in rotation than in monoculture in 9 cases and equivalent in 3 cases. These results suggest that crop rotation can be an important component of strategies to reduce weed density and maintain or increase crop yield. They lend support to Leighty's (1938) contention that "weed problems are likely to be least severe on farms where crop diversification is practiced and most severe on farms devoted for one reason or another to a single crop."

III. Weed Management in Crop Rotation Systems: Basic Theory

A crop and its associated management practices create a matrix of stresses, mortality risks, and resource conditions into which weeds must fit in order to survive and proliferate. Crop characteristics and management practices defining this matrix can act on the full range of weed life stages. For example, weed seed dormancy, survival, germination, and seedling emergence can be strongly affected by tillage operations used to prepare seedbeds (Mohler, 1993), residues of previous crops (Putnam and DeFrank, 1983; Einhellig and Rasmussen, 1989), management of animal manures used as nutrient sources (Bloemhard et al., 1992; Sarapatka et al., 1993), and fertilizer application (Freyman et al., 1989; Dyck and Liebman, 1994). Weed and crop resource use, competitive interactions, and growth and reproductive output may be determined, in part, by planting density and spatial pattern (Teasdale and Frank, 1983; Minotti, 1991), fertilizer regime (Dotzenko et al., 1969; Alkamper, 1976; Tollenaar et al., 1994), and crop species and cultivar identity (Lotz et al., 1991; Callaway, 1992; Callaway and Forcella, 1993). Substantial damage to or death of weed vegetative and reproductive stages may result from the cultivation (Buhler et al., 1992; Rasmussen, 1992; Rydberg, 1993; Mulder and Doll, 1994) and grazing practices (Heard, 1963; Dowling and Wong, 1993; Thomsen et al., 1993) associated with particular crops.

Timing is a critical component of the matrix determining a crop's impact on weeds. For example, Dotzenko et al. (1969) observed that weed seed density was highest when sugar beet (*Beta vulgaris* L.) followed corn, lowest when sugar beet followed dry bean (*Phaseolus vulgaris* L.), and intermediate when it followed barley (*Hordeum vulgare* L.). Much of this effect was attributed to differences among rotation crops in planting and cultivation dates. Because bean had the latest planting date of the three crops grown before sugar beet, it allowed the highest proportion of weeds to emerge and be killed by cultivation prior to planting. Variation in weed mortality was then reflected in the seed bank present during the sugar beet phase of the rotations.

Variations in the timing of cropping practices may also be reflected in differences in weed species composition. In a study of weed dynamics during 20 years of arable cropping, Chancellor (1985) observed that season of planting the crop was one of the most important factors determining weed floristic composition in any year. Spring germinating weed species (e.g., fool's parsley [*Aethusa cynapium* L.]) predominated in spring sown crops (barley and potato), whereas autumn germinating weeds (e.g., annual bluegrass [*Poa annua* L.]) predominated in autumn sown crops (winter barley, wheat, and oat).

Continuous production of a single crop species, in combination with a fixed set of management practices, exposes weeds to a consistent pattern of stresses, mortality risks, resource conditions, and growth and reproduction opportunities. Such a pattern acts to shift the composition of weed populations and communities toward those genotypes and species best adapted to grow and reproduce under conditions similar to those chosen for production of the crop. The resulting weed

Table 1. Weed control in rotation versus monoculture of various test crops

Study	Rotation	Test crop	Weed density in rotation[a]	Test crop yield in rotation
Stewart and Pittman (1931)[b]	Wheat/sugar beet	Wheat	Lower	Higher
	Potato/potato/pea/sugar beet/sugar beet/bean	Wheat	Lower	Higher
	Wheat/red clover/potato/sugar beet/wheat	Wheat	Lower	Higher
	Alfalfa (5 years)/potato/sugar beet/pea/sugar beet/wheat	Wheat	Lower	Higher
	Potato/oat/corn/bean	Oat	Lower	Higher
	Alfalfa (3 years)/oat/sugar beet/sugar beet/oat	Oat	Lower	Higher
Dospekhov (1967)	Fallow/winter rye/potato/oat + clover/clover/flax	Winter rye	Lower	Not available
		Potato	Equal	Not available
		Oat	Lower	Not available
		Clover	Lower	Not available
		Flax	Lower	Not available
MacHoughton (1973)[c]		Corn	(Lower)	Higher
	Corn/corn/soybean	Soybean	(Lower)	Higher
	Corn/soybean/wheat			
Dowler et al. (1974)		Corn	Lower	Equal
	Corn/peanut/cotton/soybean	Peanut	Higher	Equal
		Cotton	Equal	Equal
		Soybean	Lower	Higher

Table 1. continued--

Study	Rotation	Test crop	Weed density in rotation[a]	Test crop yield in rotation
Zawislak (1980)[d]	Sugar beet/faba bean/spring barley/winter rye/ winter rape/winter wheat	Sugar beet	Higher (equal)	Not available
		Faba bean	Lower (lower)	Not available
		Spring barley	Lower (equal)	Not available
		Winter rye	Lower (lower)	Not available
		Winter rape	Lower (equal)	Not available
		Winter wheat	Lower (lower)	Not available
Covarelli and Tei (1988)	Winter wheat/corn	Corn	Lower (lower)	Not available
	Corn/winter wheat/corn	Corn	Equal (lower)	Not available
	Winter wheat/corn/corn	Corn	Equal (lower)	Not available
	Winter wheat/winter wheat/corn	Corn	Lower (lower)	Not available

[a]Where data were available, comparison of weed seed number is given in parentheses.
[b]Actual weed numbers were not reported; comparisons between rotation and monoculture treatments are based on qualitative observations by the study's authors.
[c]For further discussion of data in this study, see Aldrich, 1984 (p. 381-383).
[d]Weed density data represent means of combined herbicide and nonherbicide treatments. However, no interactive effects between cropping system (monoculture versus rotation) and weed control regime (herbicide versus nonherbicide) were reported.
(Adapted from Liebman and Dyck, 1993.)

flora is generally highly competitive with the associated crop and difficult to control.

This principle is illustrated by three cropping system studies involving wheat. In experiments conducted in Utah, Stewart and Pittman (1931) could not control wild oat (*Avena fatua* L.) in continuous wheat, but could control the weed when wheat was grown in various rotation systems (Table 1). The use of cultivation in row crops (sugar beet, potato, dry bean) was particularly important for controlling wild oat in the rotation sequences. Similarly, Martin and Felton (1993), working in New South Wales, observed a massive buildup of wild oat in continuous wheat, whether or not herbicides were used. In contrast, rotation of wheat (a winter crop) with sorghum (*Sorghum bicolor* [L.] Moench; a summer crop) was highly effective for controlling wild oat and caused an exponential decline in the soil weed seed bank. The effectiveness of the rotation in controlling wild oat was attributed to winter cultivation or application of herbicides during fallow periods. In experiments conducted in Alberta, Blackshaw (1994) observed that densities of downy brome (*Bromus tectorum* L.) were an order of magnitude higher in continuous winter wheat than in wheat grown in rotation with spring-sown canola (*Brassica napus* L.). Increases in downy brome density were correlated with reductions in wheat yield. Failure to control the weed in wheat was attributed to a lack of effective herbicides. Successful control of the weed in the wheat-canola rotation was attributed to cultivation (used to prepare a seedbed for canola), and application of herbicides tolerated by canola but effective against downy brome. Use of a wheat-fallow sequence provided control of downy brome comparable to that achieved in the wheat-canola sequence, an effect ascribed to the use of tillage or herbicides for weed suppression during the fallow period.

Compared with monoculture cropping practices, rotation systems confront weeds with more complex patterns of resource availability, soil disturbance, herbivory, and mechanical damage, to create a less stable and more frequently inhospitable environment (Liebman and Janke, 1990; Liebman and Dyck, 1993). The use of crops with different planting dates disrupts the habitat requirements of weeds whose germination and establishment are concentrated at particular times of the year. Cultivation, used to prepare seedbeds or clean row crops, can damage or kill weeds before they reproduce, thus reducing the soil seedbank affecting subsequent crops. Small grain cereals, planted at high density in narrow rows, may compete strongly with weeds for light, water, and nutrients. Forage grasses and legumes sown as hay or pasture crops within a rotation sequence may offer opportunities to suppress weeds through mowing, grazing, crop competition, seed decay, and predation. Tillage used to terminate hay or pasture phases can damage perennial weed species and control the placement of weed seeds within the soil profile. The integration of these approaches is exemplified by the four-course Norfolk rotation system that emerged in England during the 18th century (Grigg, 1974; Karlen et al., 1994). The rotation included a row crop (turnip; *Brassica rapa* L.) that was cultivated for weed control before being grazed by sheep or fed to cattle; a spring-sown cereal (barley) that can be highly effective in suppressing weed growth (Overland, 1966); a forage legume (clover) that was grazed and used for restoration of soil N fertility; and a fall-sown cereal (wheat).

As a result of rotation through a sequence of diverse crops and management practices, and thus through a sequence of radically changing environments, weed communities in rotation systems may be less dense and less skewed toward a single dominant species, compared to monoculture systems. An illustrative example is provided by Covarelli and Tei (1988), who compared weed community characteristics after six years of continuous corn or various wheat-corn rotations (Table 1). Although the different cropping systems contained equal numbers of weed species, they differed in the relative importance of the weed species. In continuous corn, four species comprised 94% of all weed individuals, with barnyardgrass (*Echinochloa crus-galli* [L.] Beauv.) accounting for 71% of the total weed density. In contrast, in wheat-corn rotation sequences, no one species comprised >43% of total weed individuals. Total weed seedling density was 62% lower and total weed seed density was 66% lower in an alternate year wheat-corn rotation than in continuous corn. Similarly, Forcella and Lindstrom (1988) reported lower weed seed and seedling densities in corn rotated with soybean than continuous corn.

The possible effects of increasing crop diversity from two to three crops can be seen in a long-term study conducted in Indiana by Schreiber (1992). Population dynamics of giant foxtail (*Setaria faberi* Herrm.) were compared in continuous corn, corn rotated with soybean, and corn following wheat in a soybean-wheat-corn rotation. Included in the experiment was a no-tillage, minimum weed management treatment that received herbicides at lower concentrations than those recommended for commercial production and thus allowed measurement of crop rotation effects on weeds. In four out of seven years, giant foxtail plant density was greatest in continuous corn, intermediate in corn following soybean, and lowest in corn following soybean-wheat. In two years when the wheat crop failed, no difference was observed in weed density between the soybean-corn and soybean-wheat-corn rotation systems, although both contained less foxtail than the corn monoculture. In one year, when weed densities in all cropping sequences were low, the three year rotation contained the least giant foxtail, but no difference was observed between continuous corn and the soybean-corn sequence. The marked reductions in giant foxtail density measured in the three-year rotation sequences were attributed to allelopathic effects of wheat straw.

Case studies of commercial farms that avoid the use of herbicides and other agrichemicals indicate that rotation sequences are almost always longer than three years and often contain legumes used for hay, pasture, or green manure. Existing rotation sequences include:

- corn-soybean-wheat plus red clover (*Trifolium pratense* L.)-red clover hay, on a mixed crop and livestock farm in Ohio (NRC, 1989, p. 253-265);

- corn-soybean-corn-oat plus grass/clover hay-grass/clover hay, on a mixed crop and livestock farm in Iowa (Exner et al., 1990);

- flax (*Linum usitatissimum* L.) plus sweet clover (*Melilotus officinalis* Lam.)-sweet clover (for green manure)-wheat-lentil (*Lens culinaris* Medik.; for seed)-lentil (for green manure), on a cash grain farm in Saskatchewan (Matheson et al., 1991, p. 46-51);

- wheat-pea (*Pisum sativum* L.; for seed)-pea (for green manure)-wheat-pea (for seed)-alfalfa (*Medicago sativa* L.)/grass (for green manure)-summer fallow, on a cash grain farm in Washington (Reganold, 1988).

The high proportion of legume species in these rotations is a direct consequence of strategies to reduce reliance on synthetic N fertilizer. However, data drawn from a diversity of sources suggest that exploitation of legume residues as sources of N for subsequent crops may also have important beneficial effects on weed management. Here we examine reasons why weed management benefits associated with use of legumes in crop rotations may be linked to temporal patterns of N availability and allelopathy.

IV. Nitrogen, Crop Rotation, and Weed Management

A. Fertilizer Application Can Alter Weed-Crop Interactions

Nitrogen is often the most limiting element for plant growth in agricultural ecosystems and its availability in the soil can have profound effects on weed-crop interactions. A substantial number of studies show that detrimental effects of weeds on crops may be exacerbated by N fertilization. Nitrogen fertilizer can increase the rate and total amount of weed seed germination (Egley, 1986; Taylorson, 1987; Karssen and Hilhorst, 1992; Dyck and Liebman, 1994), and can promote the growth of weeds more than that of crops (Alkamper, 1976; Dyck and Liebman, 1994). These effects are important because weed populations that emerge early in the growing season at high densities and grow quickly are generally more detrimental to crop performance than are less dense weed populations that emerge later and grow more slowly (Cousens et al., 1987; Weaver et al. 1987; Graham et al., 1988; Blackshaw, 1993).

Carlson and Hill (1985) found that application of fertilizer N to wild oat/spring wheat mixtures grown at controlled densities in northern California greatly increased wild oat growth and decreased wheat yields. Nitrogen fertilization resulted in a positive yield response in wheat only when density of wild oat was <1.6% of the total combined densities of the weed and crop. In experiments conducted in Oregon, yield reductions of wheat due to interference from Italian ryegrass (*Lolium multiflorum* Lam.) were also greater under N fertilized than unfertilized conditions (Appleby et al., 1976). Studies with rice (*Oryza sativa* L.; Okafor and De Datta, 1976), sugar beet (Scott and Wilcockson, 1976), cabbage (*Brassica oleracea* var. *capitata* L.; Lawson and Wiseman, 1979), fava bean (*Vicia faba* L.; Patriquin et al., 1988), barley/pea intercrops (Liebman, 1989), and oat

(Sindel and Michael, 1992) indicate that N application can intensify weed interference against crops.

Exacerbation of weed problems through N fertilization may not be confined to a single season. Staniforth (1962) reported that weed growth in soybean and soybean yield reductions were greater when N fertilizer had been applied to a previous year's corn crop than when it had not been applied. Dotzenko et al. (1969) observed that weed seed densities in sugar beet increased with increasing rates of N applied to previous crops of barley, corn, or bean.

Despite the prevalence of studies demonstrating that weed problems may be worsened by N fertilizer, it should be noted that the phenomenon is not universal. Wells (1979) reported that application of N fertilizer to wheat in Australian experiments increased wheat yield and had no effect on the strength of weed interference against the crop. Tollenaar et al. (1994) found that applying N to corn under weed-infested conditions in Ontario resulted in less weed biomass and greater corn yield, a pattern suggesting stronger crop interference against weeds at higher fertility levels. These positive effects of N on crop performance under weed-infested conditions, in contrast to the negative effects cited above, suggest that the impact of N fertilization on weed-crop interactions defies a simple summary. Results may depend on crop cultural practices, such as planting date; the identities of the species and genotypes involved and their intrinsic shoot and root growth responses to N additions; and differential responses to other environmental factors, including ratios of N to other nutrients (Tilman, 1982). A useful approach for studying the variation in outcomes of weed-crop interactions might involve (i) monitoring the availability of nutrients, light, and water in the environment; (ii) determining how the availability of these resources is affected by the presence of competitors and different management practices; (iii) quantifying resource capture by weeds and crops grown in pure stands and mixtures; and (iv) measuring plant physiological performance, growth, biomass partitioning, and yield in pure stand and mixtures (Berkowitz, 1988; Liebman and Dyck, 1993; Liebman and Gallandt, in press).

B. The Timing of Nutrient Additions Can Affect Weed-Crop Interactions

Because many weeds are capable of earlier and more rapid uptake of nutrients than are crops (Alkamper, 1976), some research attention has focused on whether delayed application of nutrients deprives weeds of nutrients early in the growing season and better matches crop uptake ability. Data indicate that temporal differences in nutrient supply can be important in determining the outcome of weed-crop interactions, but results are not consistent for all years, sites, and species.

Alkamper et al. (1979) conducted pot experiments to examine the effects of temporal variation in nutrient supply on corn, wild mustard (*Sinapis arvensis* L. syn. *Brassica kaber* [DC] L.C. Wheeler var. *pinnatifida* (Stokes) L.C. Wheeler), and common lambsquarters (*Chenopodium album* L.) grown in mixture at fixed densities. Treatments consisted of different rates of NPK fertilizer applied in either a split dose (half at planting and half at corn ear emergence) or a full dose at planting. Split application of fertilizer improved crop growth and reduced weed growth, compared with a single early application of the same amount of fertilizer. In field competition experiments conducted in Arkansas with flooded rice and unregulated densities of barnyardgrass (*Echinochloa crus-galli* [L.] Beauv.), Smith and Shaw (1966) reported that application of N fertilizer after barnyardgrass had reached the reproductive stage increased rice yields much more than did earlier application; unfortunately, effects on the weed were not reported. Delayed application of N had inconsistent effects on interactions between winter wheat and downy brome in field experiments conducted in Colorado (Anderson, 1991). Compared to the effects of N applied at planting, N application during the subsequent winter dormancy period reduced weed density and biomass and increased wheat yield in a year of nearly normal precipitation, but increased weed density and biomass and decreased wheat yield in a drier second year. Data generated by the one year of Carlson and Hill's (1985) study that included treatments involving delayed and split N applications indicate that those treatments had no effect on wheat yield loss due to wild oat interference, compared to full application of N before planting.

C. The Timing of Nitrogen Availability Can Differ Between Fertilizer and Green Manure Sources

If delayed availability of N can favor crop growth over that of weeds in competitive situations, then use of legume residues as substitutes for N fertilizer may have important implications for weed management. Studies comparing N availability from legume green manures and fertilizers indicate that for a given amount of total N applied, the availability of N to plants is lower from green manure than from fertilizer, especially early in the growing season.

In a field experiment investigating N dynamics during flooded rice production in California, Westcott and Mikkelsen (1987) observed that soil inorganic N levels were lower for the first 48 days following incorporation of purple vetch, compared to application of ammonium sulfate containing the same amount of N. Total recovery of ^{15}N from the contrasting N sources was equivalent, but a higher percentage of ^{15}N derived from the legume remained in the soil in organic form. Similarly, in a field experiment investigating N dynamics in a Kentucky corn field, Varco et al. (1993) noted that percentage recovery of soil inorganic ^{15}N from fertilizer (ammonium nitrate) exceeded that of soil-incorporated hairy vetch for 30 days after treatments were applied in one year and for 45 days in a second year. A larger fraction of legume N than fertilizer N was present in the soil organic fraction. In a field experiment conducted in Pennsylvania, Dou et al. (1995) studied seasonal soil nitrate dynamics in corn supplied with N from synthetic fertilizer or red clover and hairy vetch green manures. Throughout the season, nitrate levels in the surface 25 cm were lower in the clover and vetch green manure treatments than in a synthetic fertilizer treatment supplied with 200 kg N ha^{-1}. Corn grain yields, however, were comparable in all three treatments.

Data concerning plant uptake of N also indicate that legume residues function as slow release sources of N, compared to synthetic fertilizer. In a greenhouse pot experiment in which corn was grown following application of ^{15}N labelled ammonium sulfate or *Sesbania aculeata* Pers. green manure, Azam et al. (1985) observed that corn had recovered 20% of the labelled N from the fertilizer, but only 5% of the labelled N from the legume residue, after 5 weeks of growth. In a microplot field experiment conducted in a semi-arid area of South Australia, Ladd and Amato (1986) compared the fate of ^{15}N labelled fertilizer (urea, ammonium sulfate, and potassium nitrate) and legume residue (*Medicago littoralis* Rohde ex Lois.) applied to soil planted with wheat. Seventeen percent of the ^{15}N in the legume residue was taken up by wheat in the first year, while 62% of the label remained in the soil organic fraction. In contrast, 47% of the labelled N in the fertilizers was taken up by wheat in the first year, and only 29% remained in the soil organic fraction. Similar results were obtained in field experiments conducted in California by Westcott and Mikkelsen (1987), who compared ^{15}N uptake by rice following addition of labelled ammonium sulfate and purple vetch (*Vicia benghalensis* L.); in western Canada by Janzen et al. (1990), who compared ^{15}N uptake by wheat following addition of labelled ammonium sulfate and lentil and Tangier flatpea (*Lathyrus tinginatus* L.) green manures; and in Pennsylvania by Harris et al. (1994), who compared ^{15}N uptake by corn following addition of labelled ammonium sulfate and red clover. Azam et al. (1985) characterized legume residues as important for long-term maintenance of soil N to ensure adequate N delivery to future crops, but not as immediate N sources.

In all of the comparisons of N release from fertilizer and legume residue cited above, legume materials were incorporated into the soil. Transformation of N from organic into inorganic forms is faster when green manure material is soil-incorporated, compared to when it serves as a surface mulch (Wilson and Hargrove, 1986; Groffman et al., 1987; Sarrantonio and Scott, 1988; Dou et al., 1995). Other factors affecting the rate of N mineralization from legume residues include tissue age and quality (e.g., C:N ratio, lignin and polyphenol contents), and soil temperature and moisture conditions (Fox et al., 1990; Palm and Sanchez, 1991; Ranells and Wagger, 1992; Honeycutt et al., 1991, 1993, 1994). Given the complexity of factors involved, interdisciplinary work by weed scientists, agronomists, soil chemists, and soil microbiologists will be required if weeds and crops are to be better managed through crop rotation practices that regulate temporal patterns of N availability in a manner favorable to crops and detrimental to weeds.

V. Allelopathy, Crop Rotation, and Weed Management

A. Allelopathic Interactions May Be Common

Allelopathy involves the suppression of germination or growth of a target plant species through the release of inhibitory chemical substances by another plant. Because many plant materials contain phytotoxic substances as well as nutrients, the production and subsequent decomposition of large quantities of legume residue to increase soil N levels may also result in allelopathic interactions.

In an early review of the ecological significance of allelopathy, Whittaker and Feeny (1971) noted that (i) there are a variety of routes of allelochemical release, including rainwash from plant surfaces, volatilization, root exudation, and decomposition of plant parts; (ii) allelopathic interactions reducing germination and growth are probably common; (iii) allelopathy may be of widespread significance in determining plant species abundance and distribution; and (iv) known allelopathic compounds can be classified into a few major chemical classes, including phenolic acids, flavonoids, terpenoids, steroids, alkaloids, and organic cyanides. The overall effect of inhibiting germination and/or growth of a target species is a summation of many possible individual processes, including the release of allelochemicals or their precursors from the plant to the soil solution; microbial transformation of allelochemicals; sorption and desorption reactions with soil surfaces; reaction with soil surface constituents; and reaction with inorganic and organic constituents of the soil solution.

One approach to the challenge of reducing reliance on synthetic herbicides for weed control is to adopt crop rotation practices that allow for the beneficial expression of allelopathy (Putnam et al., 1983; Einhellig and Leather, 1988; Worsham, 1989). Understanding mechanisms of allelopathic interactions will aid the development of strategies for manipulating biological, chemical, and physical factors to optimize allelopathic weed control. The following discussion will highlight some of the areas that require further research to develop a mechanistic model of weed-suppressive allelopathic effects of legumes on a field scale.

B. Legume Residues Can Have Allelopathic Effects on Weeds

A considerable amount of research effort has focused on the weed-suppressive allelopathic effects of nonlegume species, such as sunflower (*Helianthus annuus* L.; Leather, 1987), rye (Barnes and Putnam 1983, 1986, 1987; Barnes et al., 1987; Shilling et al., 1985, 1986; Chase et al., 1991; Liebl et al., 1992), oat (Fay and Duke, 1977; Putnam and DeFrank, 1983; Putnam et al., 1983), barley (Overland, 1966; Liu and Lovett, 1993), wheat (Liebl and Worsham, 1983; Shilling et al., 1985), and sorghum (Einhellig and Rasmussen, 1989; Einhellig, 1995; Rice, 1995, p. 30-31). In contrast, very little attention has been focused on weed-suppressive allelopathic effects of commonly grown legume species. Available data indicate, however, that allelopathy may be important in understanding the overall impact of legume-based crop rotations on weed dynamics.

Most reports of allelopathic effects of legumes species focus on inhibition of other crops used as bioassay species. Aqueous extracts of pea (Kimber, 1973; Cochran et al., 1977), lentil (Cochran et al., 1977), broadbean (*Vicia faba* L.; Patrick et al., 1963), vetch (Patrick et al., 1963), sweetclover (McCalla and Duley, 1948; Guenzi and McCalla, 1962), alfalfa (Nielson et al., 1960; Guenzi et al., 1964; Kimber, 1973; Rice, 1995, p. 4-10), and red clover (Rice, 1995, p. 7) have all been shown to inhibit germination or growth of such crops as wheat, oat, lettuce, corn, soybean, and sorghum. Volatile emissions from berseem clover (*Trifolium alexandrinum* L.), hairy vetch, and crimson clover (*Trifolium incarnatum* L.) have been shown to inhibit the germination and growth of onion (*Allium cepa* L.), carrot (*Daucus carota* var. *sativus* L.), and tomato (*Lycopersicon lycopersicon* [L.] Karst. ex Farw.) (Bradow and Connick, 1990).

Fewer reports exist with regard to the allelopathic effects of legumes on weed species. White et al. (1989) examined the effects of aqueous extracts of crimson clover and hairy vetch on pitted morningglory (*Ipomoea lacunosa* L.), wild mustard (*Sinapis arvensis* L.), and Italian ryegrass (*Lolium multiflorum* Lam.). Germination and growth of all three species on filter paper in petri plates were inhibited by clover and vetch extracts, with mustard and ryegrass suffering most. Corn and cotton were also suppressed by the legume extracts. Responses to polyethelene glycol solutions suggested that shifts in osmotic potential were not responsible for reductions in germination and growth. In a greenhouse pot experiment, germination and biomass accumulation of morningglory and cotton declined as increasing rates of clover or vetch residue were mixed into soil. Corn exhibited minimal responses to legume residues; responses of mustard and ryegrass were not tested.

Laboratory and greenhouse studies indicate alfalfa residue and extracts can suppress the germination and growth of a number of weed species. Abdul-Rahman and Habib (1989) reported that decomposed alfalfa roots and their associated soil reduced germination and growth of the grass weed *Imperata cylindrica* (L.) Beauv.; phenolic acids derived from alfalfa were suggested to be responsible for the observed weed suppression. Waller (1989) extracted saponin compounds from alfalfa roots and found that they reduced the growth of barnyardgrass, cheatgrass (*Bromus secalinus* L.), coffeeweed (*Daubentonia punicea* [Cav.] DC.), dandelion (*Taraxacum officinale* L.), and pigweed (*Amaranthus retroflexus* L.). Chung and Miller (1995) observed that alfalfa residue mixed with sand at rates of 1 to 2 g kg^{-1} reduced growth of lambsquarters, pigweed, crabgrass (*Digitaria sanguinalis* [L.] Scop.), and velvetleaf (*Abutilon theophrasti* Medic.). Amending sand with alfalfa residue increased growth of giant foxtail and cheatgrass, however. Aqueous extracts of alfalfa reduced germination and seedling growth of all six of the weed species.

Results of a field study examining effects of crimson clover residue and N fertilizer on germination and growth of common lambsquarters and sweet corn (Table 2; Dyck and Liebman, 1994) strongly suggested selective allelopathic weed suppression by legume residue. Incorporation of clover residue into the soil not only reduced lambsquarters density, but also delayed lambsquarters emergence relative to corn emergence. Application of N increased lambsquarters density and emergence rate, but failed to overcome fully the suppressive effect of clover residue. Clover residue slightly delayed corn emergence and, when N fertilizer was applied, slightly reduced corn density. At the termination of the experiment, 53 days after planting the weed and crop, biomass of the weed was reduced by clover residue and increased by N fertilizer. Sweet corn biomass production was unaffected by clover residue or N application, perhaps because of high background soil fertility or insufficient growth period. (Other experiments at this site [Dyck et al., 1995] indicated that crimson clover had a fertilizer equivalency value of 55 kg

Table 2. Effects of crimson clover residue and ammonium nitrate fertilizer on total emergence and aboveground biomass production of common lambsquarters and sweet corn grown in single-species stands; the field experiment was conducted in 1992 at Stillwater, Maine

	Total emergence[a]				Biomass at 53 days after planting[b]			
	Lambsquarters		Corn		Lambsquarters		Corn	
Fertilizer rate (kg N ha^{-1})	Without residue	With residue[c]	Without residue	With residue[c]	Without residue	With residue[c]	Without residue	With residue[c]
	Plants m^{-2}				g m^{-2}			
	215	150	7.8	7.6	42	30	142	148
	357	287	7.8	6.2	82	61	160	147
	382	260	7.9	7.1	88	75	177	185
	360	257	7.7	6.8	95	60	131	154
Standard error	30		0.3		11		22	
ANOVA								
N fertilizer	$p < 0.001$		$p < 0.10$		$p < 0.001$		Not significant	
Clover residue	$p < 0.001$		$p < 0.001$		$p < 0.05$		Not significant	
Interaction	Not significant		$p < 0.05$		Not significant		Not significant	

[a]Fixed quantities of lambsquarters and corn seed were sown and resultant stands were counted.

[b]Lambsquarters and corn were sown in single species stands and thinned to common densities in all treatments. Lambsquarters density was 144 plants m^{-2}; corn density was 8.8 plants m^{-2}.

[c]Clover produced 5,466 kg ha^{-1} of aboveground biomass.

(Adapted from Dyck, 1994 and Dyck and Liebman, 1994; see those sources for data concerning residue and fertilizer effects on emergence rates of lambsquarters and corn.)

N ha^{-1} for a succeeding crop of sweet corn.) Incorporation of clover residue had no significant effects on soil moisture and temperature, and no disease symptoms were observed during the study.

Dyck and Liebman (1994) concluded that the suppressive effect of crimson clover residue on lambsquarters emergence and growth was not attributable to low N availability. However, given the stimulatory effect of N fertilizer on lambsquarters development, use of crimson clover as a N source would appear to provide weed control benefits by directly suppressing weed emergence and growth and by depriving weeds of the stimulatory effect of N fertilizer.

The differential suppressive effect of crimson clover residue on the weed is a highly desirable feature of this model system. However, there is clearly a need to better understand the mechanisms involved if the principles illustrated in this example are to be extended to other cropping systems. To that end, we now focus our attention on several important methodological issues concerning the study of allelopathic interactions under field conditions.

C. Studying Allelochemicals in Legumes Requires a Focus on Methodology

1. Extraction of Allelochemicals

Sources of allelochemicals or their precursors include canopy-flow, root exudation, and the decomposition of plant residues. Aqueous extraction of shoot and root tissues or their residue is commonly used to extract allelochemicals from source materials (White et al., 1989; Mallik et al., 1994; Ahmed and Wardle, 1994). Soil water will solubilize 25 to 40% of organic matter from plant material within 24 hours (Thurman, 1985), so aqueous extracts are conceptually attractive in mimicking potential release of allelochemicals from plant materials. Unpurified aqueous extracts of plant material contain primary plant substances, such as simple carbohydrates, proteins and organic acids that presumably are not allelopathic. Attempts to elucidate the mechanisms of allelopathy will be assisted by separation of allelochemicals from the complex matrix of aqueous plant extracts. Several investigators have used physical or chemical separation methods to successfully isolate allelochemicals from aqueous extracts.

Mallik et al. (1994) used ultrafiltration through a 1000 molecular weight cutoff filter to isolate active allelochemicals from an aqueous common lambquarters extract. Ultrafiltration involves separation of molecules on a size basis through a microporous membrane under an applied hydrostatic pressure. Bioassays using radish (*Raphanus sativus* L.) as the target species indicated that the filtrate significantly inhibited germination while the retentate did not. Hypocotyl and radicle growth were inhibited more strongly by the filtrate than by the retentate. The authors inferred from these results that the molecular weight of the allelochemicals was <1000. Further separation work using paper chromatography and HPLC identified chlorogenic acid as the principal phytotoxin. Other separation methods based on physical exclusion processes such as dialysis and size exclusion chromatography should also be applicable in isolating active allelochemicals.

Chromatography with non-ionic, macroporus XAD-4 resin has been used to isolate allelochemicals from root exudates (Tang and Young, 1982; Yu and Matsui, 1994). Non-ionic, macroporous sorbents such as the XAD series of resins, particularly XAD-8, have been used extensively to concentrate and purify dissolved organic matter (DOM) from a number of aquatic environments (Aiken, 1985). The principle driving force for sorption of DOM on these resins is nonspecific dipole-dipole interactions between the solute and the resin surface. Generally, samples are acidified before passage through XAD resins. Protonation of the organic matter reduces its charge and increases its affinity for the hydrophobic resin surface. Yu and Matsui (1994) eluted nine organic acids sorbed onto XAD-4 from the root exudates of cucumber (*Cucumis sativa* L.) seedlings that were inhibitory to lettuce root growth.

The use of aqueous extracts for allelopathy studies presents several analytical challenges. Generally, the allelochemical should be separated from both inorganic and organic constituents of the raw extract and concentrated to allow isolation with increased speed and decreased effort. Most importantly, the extraction and purification method must allow for the isolation of allelochemicals without altering their chemical character.

2. Soil Reactions

Phenolic compounds, defined as any molecule containing an aromatic ring with one or more hydroxy groups directly attached to the ring, are common constituents of legumes and other plant species and have been the focus of many studies of allelopathy (for recent reviews, see Siqueira et al. [1991] and Appel [1993]). Low molecular weight phenolic acids based on the substitution of benzoic acid (i.e., *p*-hydroxybenzoic, salicylic and vanillic acids) and cinnamic acid (i.e., ferulic, *p*-coumaric, caffeic acids) have been reported to be allelochemicals (Chou and Patrick, 1976; Blum et al., 1992; Gallet, 1994).

Blum (1995) and Cheng (1995) have stressed the importance of determining the reaction of phenolics in the soil system. Phenolic acids have been shown to rapidly (<2 min) sorb onto the A1 horizon of soils collected from North Carolina, and to show increased sorption with increasing levels of soil organic matter (Dalton et al., 1989). Dalton et al. (1989) speculate that potential mechanisms responsible for phenolic sorption onto soils include ligand exchange, soil surface catalyzed oxidation, and incorporation into soil organic matter. Diffuse-reflectance Fourier-transform infrared spectroscopy interrogation of the bond between *p*-hydroxybenzoate and iron oxide surfaces revealed coordination by an inner-sphere complex through the carboxylate group (Kung and McBride, 1989). Inner-sphere complexes are strong due to a largely covalent bond between the surface metal and the electron donating oxygen group (Stumm, 1992). Results from work by Kung and McBride (1989) and Dalton et al. (1989) clearly indicate that phenolic compounds react with both inorganic and organic components of soil. Reaction of phenolic acid to surfaces through the carboxylate group suggests that sorption will be sensitive to chemical factors that influence the degree of protonation of the carboxylate group.

Rapid sorption kinetics combined with strong inner-sphere bonding result in low levels of phenolic compounds in the soil solution (Whitehead et al., 1982). The reversibility of this sorption has key ecological significance. If the sorption is reversible, this process could serve to buffer the equilibrium concentration over an extended time period despite a single input of these compounds at the time of residue incorporation. Experiments determining the effects of soil pH, organic matter, and texture on the sorption and desorption kinetics of allelochemicals are needed to assess the relative importance of these soil characteristics. Evaluation of management methods to optimize allelopathic efficiency should be aided by understanding how these master soil variables affect the fate of bioassays.

The rapid sorption of phenolic compounds raises questions about the interpretation of traditional bioassays that are conducted with aqueous plant extracts in the absence of soil. The lack of soil which serves as a "sink" for phenolics suggests that bioassays without soil indicate only "potential" allelopathic activity. Similarly, studies that have attempted to investigate the effects of plant decomposition in a soil matrix (Chou and Patrick, 1976; Ahmed and Wardle, 1994; Mallik et al., 1994) are confounded by sorptive properties of the soil used. An alternative soil-less decomposition system is to inoculate the plant material with the supernatant of soil-water extracts (Herman et al., 1977; Reinertsen et al., 1984). This would allow decomposition to occur with similar microorganisms found in soil, yet avoid the confounding caused by sorption onto soil surfaces. Bioassays conducted on extracts of plant residue decomposed without the presence of soil would show the effects of decomposition on allelopathic activity.

3. Soil Solution Reactions

In addition to reaction of allelochemicals with soil surfaces, interaction of these compounds with both inorganic and organic components of soil solution are expected. Carboxyl groups present in phenolic acids form complexes with the inorganic cations present in soil solution (Wolt, 1994). Allelopathic potential of aqueous plant extracts and synthetic phenolic compounds have frequently been conducted using extracts diluted with water (Ahmed and Wardle, 1994), which results in solutions of varying ionic strength. Because the ionic strength of solutions can influence germination and growth responses, care must be taken to control for this factor. The use of Hoagland's solution (Lehman et al., 1994) results in a solution of much higher ionic strength than typical soil solution.

The effects of inorganic soil solution components on allelopathic activity have not been systematically investigated. An approach similar to the use of chemical equilibrium modeling and hydroponic systems for investigating the relationship between chemical speciation and nutrient uptake, transport and function (Parker et al., 1995) should illuminate how allelochemicals may be affected by changes in soil solution chemistry. Chemical speciation attempts to determine the distribution of a chemical element among different complexes that may form in multicomponent solutions (Sposito, 1989). Generally, cations can occur as "free" species where the cation is surrounded only by the solvation shell of water molecules or as "complexed" species where organic or inorganic anions replace one or more water molecules in the solvation sphere (Morel and Hering, 1993). There is growing acceptance that plant responses to elements that are essential to plant growth and those that are phytotoxic are related to the activity of free, uncomplexed ions in solution (Parker et al., 1995). Complexation reactions can be described thermodynamically through the use of stability constants, which form the basic data set used in speciation modeling. If allelopathic activity also follows this approximation, chemical equilibrium modeling of the interaction of allelochemicals with the major inorganic soil ions should be able to predict allelochemical activity under varying soil solution conditions. Conducting bioassays of speciated solutions for allelopathic response could be used to verify the predictions based on speciation of allelochemicals.

Research addressing the extraction, soil reactions, and soil solution reactions of allelochemicals will contribute to an improved mechanistic understanding of the fate of allelochemicals (or their precursors) upon entering the soil system. The challenges of working with organic molecules that offer no clear analytical signals for measurement, in combination with the complex, multicomponent soil environment, are many. However, application of modern chemical methods to probe interactions of allelochemicals with the soil system can be expected to improve mechanistic understanding of this process, and aid the development of weed-suppressive cropping systems that are less reliant on synthetic herbicides.

VI. Field Applications

Results of recent competition experiments conducted in Maine, described below, suggest that integration of legumes into cropping systems holds promise for reducing reliance on herbicides and N fertilizer. These experiments show that the amount of damage inflicted on crops by a given density of weeds may be less when N fertility is gained through green manuring practices than through application of synthetic N fertilizer, and suggest that temporal patterns of N availability and residue-mediated phytotoxicity may be important in determining the outcome of crop-weed interactions. The experiments leave unresolved the possible roles of soil-borne pathogens and other factors whose abundance and influence may be affected by the presence of legume residues. Moreover, because the setting of the experiments has been restricted to one location and relatively few years, additional work is needed in other environments over extended periods of time before effects we have observed can be assumed to occur predictably in other locations.

Dyck et al. (1995) investigated crop-weed interactions in soils amended with crimson clover green manure or ammonium nitrate. In each of two years, plots were (i) planted with crimson clover in early May, tilled in early July, and planted with sweet corn and common lambsquarters, or (ii) maintained in bare fallow until early July, tilled, and planted. Corn was planted alone or in mixture with lambsquarters one day after live clover was incorporated into the soil. Incremental rates of ammonium nitrate were applied to corn and lambsquarters planted in plots that had contained the bare fallow; no N fertilizer was applied to plots that had contained the green manure. Corn and lambsquarters were thinned to common densities in all treatments; weeds other than lambsquarters were removed.

Based on curves relating N accumulation of weed-free corn to varying rates of ammonium nitrate, the calculated fertilizer equivalency value of the green manure was 55 kg N ha^{-1}. Calculation of this value allowed soil and plant responses in the crimson clover treatment to be compared with those in a treatment receiving a similar rate of fertilizer (45 kg N ha^{-1}). One week after planting corn and lambsquarters, soil NO_3^--N levels were > 40% lower in the clover treatment than the 45 kg N ha^{-1} fertilizer treatment. Differences between the two treatments in soil nitrate levels decreased as the season progressed.

Averaged across the two years of the experiment, lambsquarters biomass was 72% lower in the clover treatment than the 45 kg N ha^{-1} treatment at two weeks after emergence, and remained 39% lower in the clover treatment at final harvest (Table 3). In contrast, corn biomass was 31% lower in the clover treatment than the 45 kg N ha^{-1} fertilizer treatment at two weeks after emergence, but recovered to levels attained in the fertilizer treatment by the final harvest date (Table 3). Comparison of data from plots with and without lambsquarters indicated that weed interference reduced corn biomass production by 8% in the clover treatment, but by 28% in the 45 kg N ha^{-1} treatment (Table 3). Thus the weed was smaller and inflicted less damage on the crop when green manure, rather than synthetic fertilizer, was used as the source of N. Results of a companion study, in which lambsquarters was grown alone in soils managed with the same treatments as the sweet corn plus lambsquarters study, indicated that the weed-suppressive effect of crimson clover residue also occurred when corn was not present (Dyck et al., 1995).

Results of a second set of experiments (Dyck and Liebman, 1995) suggest that aging and weathering of legume residues reduce their ability to suppress weed growth. In this study, plots were planted in mid-summer with either oat or crimson clover, which were killed in early winter by cold weather. Winter-killed residues of both species were incorporated into the soil in May of the following year and plots were sown and thinned to a fixed density of silage corn. In half of the plots, the natural background of weeds (which was dominated by lambsquarters) was allowed to grow in a 30 cm band surrounding the crop; in the remaining plots, corn was grown under weed-free conditions. As in the sweet corn experiments described above, incremental rates of ammonium nitrate were applied to corn and lambsquarters planted in plots that had contained the oat cover crop; no N fertilizer was applied to plots that had contained crimson clover. The experiment was conducted twice.

The fertilizer equivalency value of crimson clover, calculated on the basis of N accumulation of weed-free corn in response to ammonium nitrate, was 52 kg N ha^{-1}, which was similar to the value observed in the doublecropping experiments conducted with sweet corn. Consequently, crop and weed performance were compared in the unfertilized crimson clover treatment and the treatment that had been cover-cropped with oats and amended with ammonium nitrate at 45 kg N ha^{-1}. In 1990, corn, lambsquarters, and total weed biomass did not differ between clover

Table 3. Effects of crimson clover residue, ammonium nitrate fertilizer, and interspecific competition on aboveground biomass production by sweet corn and common lambsquarters; the field experiments were conducted in 1989 and 1990 at Stillwater, Maine

		1989[a]			1990[a]		
Crimson clover residue[b]	N fertilizer rate (kg N ha^{-1})	Weed-free corn[c]	Corn with lambsquarters[c]	Lambsquarters with corn[d]	Weed-free corn[c]	Corn with lambsquarters[c]	Lambsquarters with corn[d]
		g m^{-2}					
No	0	239	145	294	609	469	152
No	45	305	195	325	669	529	214
No	90	259	186	343	676	530	301
No	135	295	206	327	807	547	259
No	180	307	220	290	768	581	269
Yes	0	274	235	203	646	635	124
$LSD_{0.05}$		44		76	117		77

[a]Biomass measurements made at 77 and 86 days after planting in 1989 and 1990, respectively.
[b]Clover produced 5,136 and 4,768 kg ha^{-1} of aboveground biomass in 1989 and 1990, respectively; calculated N fertilizer equivalency values of the clover residue were 52 and 58 kg N ha^{-1} in 1989 and 1990, respectively.
[c]Corn densities were 6.8 and 7.3 plants m^{-2} in 1989 and 1990, respectively.
[d]Lambsquarters densities were 35 and 81 plants m^{-2} in 1989 and 1990, respectively.
(Adapted from Dyck, 1994 and Dyck et al., 1995.)

and fertilizer treatments. The reduction in corn biomass due to weed interference was 19 and 14% in the clover and fertilizer treatments, respectively. In contrast, in 1991, lambsquarters and total weed biomass were 65 and 58% lower, respectively, in the clover treatment than in the 45 kg N ha^{-1} fertilizer treatment. Weed interference reduced corn biomass by 62 and 80% in the clover and fertilizer treatments, respectively. The less detrimental effect of weeds on corn in 1990 compared to 1991 was ascribed to drier conditions following planting, which delayed weed emergence relative to crop emergence.

Dyck and Liebman (1995) suggested that the less consistent and less effective weed suppression obtained from crimson clover in the silage corn experiment, compared to that obtained in the above-described sweet corn experiment, may have resulted from using winter-killed, partially decomposed clover residue rather than immature, fresh material. Phytotoxicity can be greater with green residues than mature crop residues, and can decrease as decomposition occurs (Kimber, 1973). Dyck and Liebman (1995) also suggested that use of oat in the nonlegume plots in the silage corn experiment, compared to bare fallow in the sweet corn experiment, may have reduced differences in weed suppression between chemically fertilized and green manure treatments. A lack of weed suppression by soil-incorporated crimson clover residue was observed by Groffman et al. (1987), who compared a system in which sorghum followed rye and received N fertilizer against a system in which sorghum followed crimson clover and received no N fertilizer. In one year, no difference in weed biomass was observed between the two treatments; in a second year, weed biomass was higher following clover. No cause for the difference between years was evident.

The weed-suppressive effect of fresh red clover residue was observed in a recent study examining impacts of green manure, synthetic N fertilizer, and interspecific competition on growth and yield of dry bean and wild mustard (Liebman and Gallandt, unpublished data). A two-year spring wheat-bean rotation was established, with half of the wheat plots undersown with red clover during the first year. Following harvest of the wheat grain, wheat stubble or wheat stubble plus red clover occupied the plots until May of the following year. Wheat stubble and spring regrowth of clover were then incorporated into the soil, and the field was sown with dry bean and wild mustard in June. Both the crop and weed were grown at fixed densities in single-species stands and in mixture. Because the yield of short-season dry bean varieties grown in Maine can be increased by N fertilizer (Liebman et al., 1995), half of the plots received ammonium nitrate at 84 kg N ha^{-1}. The experimental design was thus a full factorial of two rotation sequences (with and without red clover), three species compositions (bean alone, mustard alone, bean/mustard mixtures), and two fertilizer regimes (with and without N). The comparison of major interest, however, was between two contrasting soil management regimes: (+) clover/(-) N fertilizer versus (-) clover/(+) N fertilizer.

Immediately after planting, inorganic N concentrations in the surface 0 to 10 cm of soil were lower in the (+) clover/(-) N fertilizer treatment than the (-) clover/(+) N fertilizer treatment (Figure 1A). Differences in soil inorganic N levels between these two treatments declined as the season progressed. Throughout the growing season, soil inorganic N levels were highest in the (+) clover/(+) N fertilizer treatment and lowest in the (-) clover/(-) N fertilizer treatment, in both the 0 to 10 cm (Figure 1A) and 10 to 20 cm (Figure 1B) layers.

As shown in Table 4, mustard biomass production was reduced by the presence of clover residue and increased by application of N fertilizer. Competition from bean reduced mustard biomass production equally under all soil management regimes (i.e., there were no interactions with rotation sequence or N fertilizer). In contrast, competition from mustard caused greater reductions of bean seed yield when clover residue was not present and when N fertilizer was added. A comparison of data from bean alone and bean plus mustard plots indicates that mustard competition reduced bean yield by 7% under the (+) clover/(-) N fertilizer regime, but by 40% under the (-) clover/(+) N fertilizer regime. Thus, in competitive situations, mustard fared worst and bean fared best when fresh legume material rather than synthetic fertilizer was used as a source of N.

Application of N fertilizer (which resulted in an increase in soil inorganic N concentration; Figure 1) was unable to overcome the growth suppression of mustard effected by the presence of clover residue, suggesting that phytotoxicity rather than N deficiency was involved in weed suppression. Phytotoxicity of clover residue, at least under unfertilized conditions, is also suggested by results of a bioassay conducted with aqueous extracts derived from soils amended with red clover and ammonium nitrate (Figure 2). When treated with extracts from plots not receiving N fertilizer, radicle elongation of mustard was significantly lower in plots amended with clover residue than those without clover residue. No difference was observed in radicle elongation between (+) clover and (-) clover plots when N fertilizer was applied. Identification of the inhibitory agent(s), the mechanisms for their production, the temporal pattern of their abundance, and their interaction with soil fertility conditions will require further experimentation.

It should be noted that weed control in the corn and dry bean studies described herein was the result only of crop competition and soil management. No attempt was made to study the interaction of these factors with other tactics that might be desirable in integrated weed management strategies, such as cultivation and low rates of herbicides. Because pest populations might be best regulated by the combination of multiple, small stress factors (Bottrell and Weil, 1995; Liebman and Gallandt, in press), the integration of legume-based crop rotations with other weed management tactics should be pursued in future studies. Furthermore, because the damage inflicted upon crops by a given density of weeds often differed between crop rotation and fertilization treatments, weed density thresholds for initiating control measures may need to be defined separately for contrasting crop and soil management systems. Research attention should be directed toward this issue.

VII. Conclusions and Research Needs

While crop rotation is not a panacea for weed management problems, it is clearly a valuable pest management strategy that includes many opportunities to stress and kill weeds with reduced reliance on herbicides. In our opinion, effective integration of crop rotation into farming systems will require attention to the following topics:

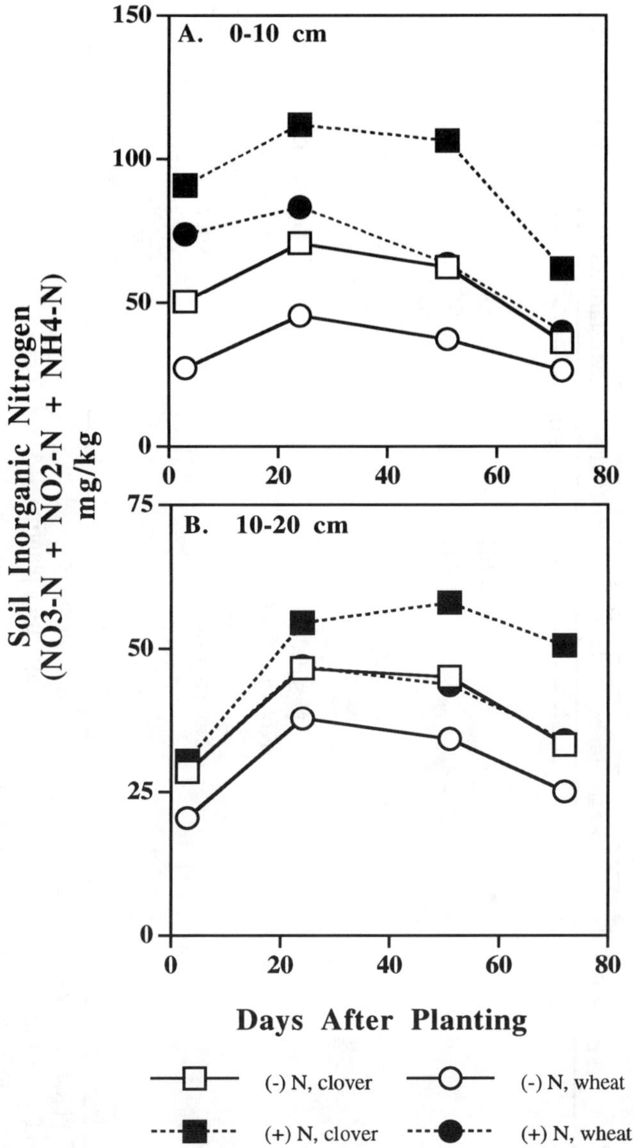

Figure 1. Effects of previous rotation crop and N fertilizer application on soil inorganic N concentrations at 0-10 cm (A) and 10-20 cm (B) depths during the 1994 growing season. Values are averaged over bean alone, wild mustard alone, and bean plus mustard treatments. Significant ($p < 0.05$) effects at 0-10 cm depth: rotation crop at 24, 51, and 72 days after planting (DAP), and N fertilizer at 3, 24, 51, and 72 DAP. Significant effects at 10-20 cm depth: rotation crop and N fertilizer at 24, 51, and 72 DAP. There were no significant rotation crop x N fertilizer interactions. (From M. Liebman and E. Gallandt, unpublished data.)

Table 4. Effects of red clover residue, ammonium nitrate fertilizer, and interspecific competition on bean seed yield (oven-dry basis) and wild mustard (above- and belowground) biomass production in a field experiment conducted in 1994 at Stillwater, Maine

Previous rotation crop	N fertilizer rate (kg N ha^{-1})	Bean seed yield[a]		Mustard biomass[b]	
		Bean alone	Bean with mustard	Mustard alone	Mustard with bean
		$g\ m^{-2}$			
Wheat + red clover[c]	0	297	279	566	94
Wheat + red clover[c]	84	316	262	588	119
Wheat	0	268	242	553	150
Wheat	84	329	199	780	263
Standard error		13		60	
ANOVA					
Rotation crop (R)		$p < 0.01$		$p < 0.05$	
N fertilizer (N)		Not significant		$p < 0.05$	
Competition (C)		$p < 0.001$		$p < 0.001$	
R x N		Not significant		Not significant	
R x C		$p < 0.05$		Not significant	
N x C		$p < 0.01$		Not significant	
R x N x C		$p < 0.10$		Not significant	

[a]Bean density was 14.3 plants m^{-2}; seed was harvested 96 days after planting.
[b]Mustard density was 6.0 plants m^{-2}; biomass was harvested 96 days after planting.
[c]Clover produced 2,388 kg ha^{-1} of aboveground biomass.
(From M. Liebman and E. Gallandt, unpublished data.)

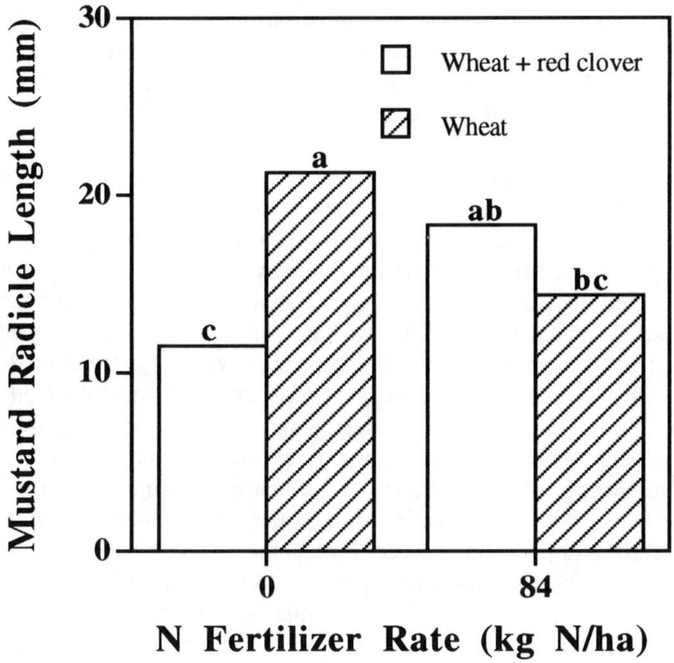

Figure 2. Effects of red clover residue and ammonium nitrate fertilizer on 72-hour radicle elongation of wild mustard. Mustard seeds were treated with 1:1 soil:water extracts. Soil samples were drawn from field plots at Stillwater, Maine on 25 July 1995, which was 66 days after live red clover had been incorporated into the soil, and 46 days after ammonium nitrate had been applied. The clover x N fertilizer interaction was significant ($p < 0.01$); lower case letters indicate mean separations at $p < 0.05$. Source: T. Ohno and K. Doolan, unpublished data.

1) Crop rotation effects on weed life history parameters and population dynamics. Increasing crop diversity through rotation can reduce weed seed and plant density (Table 1; Covarelli and Tei, 1988; Forcella and Lindstrom, 1988; Schreiber, 1992; Blackshaw, 1993; Martin and Felton, 1993) and prevent major shifts in weed community composition toward particularly noxious species (Covarelli and Tei, 1988). These effects appear to be due to the creation of complex and unstable patterns of resource availability, soil disturbance, herbivory, and mechanical damage that reduce weed survival, growth, and reproduction. Rotation characteristics that may strongly influence weed performance include: (i) the sequence and relative proportions of different crop types, e.g., species planted in the spring versus fall, species sown in narrow versus wide rows, annuals versus perennials; (ii) the timing and magnitude of variations in soil conditions resulting from applications of synthetic fertilizers and organic amendments, and management of crop residues; (iii) the timing and types of tillage and cultivation operations; and (iv) the timing, intensity, and duration of grazing and mowing practices. How do these and other factors, alone and in combination, affect weed seed survival, weed seedling emergence and survival, weed growth and competitive ability, and weed seed production? How do these and other factors affect the life history characteristics and population dynamics of perennial weed species?

2) Weed suppression by legume residues. The ability of legume species to supply N and reduce fertilizer requirements for succeeding crops is well documented. Legumes may also contribute to weed management by delaying weed emergence (Dyck and Liebman, 1994), reducing weed density and growth (Tables 2, 3, and 4), and decreasing the damage inflicted on a crop by a given density of weeds (Tables 3 and 4). In this review we have suggested that legume residues may contribute to weed suppression through delays in N availability (relative to patterns observed with application of synthetic N fertilizer) and release of allelopathic substances. To test these hypotheses adequately, intensive measurements of soil and plant nutrient status are needed for weeds and crops grown under a range of soil and residue management conditions. Refinements of methods for studying interactions between allelochemicals and the soil system are also necessary. To test whether selective suppression of weeds by legume residues is a general phenomenon, more legume species and genotypes should be screened for their effects on different weed and crop species. To maximize the beneficial effects of legume residues on weed suppression and crop yield, different residue management strategies (e.g., choice of tillage implements, time of residue incorporation) should be compared. Additional research attention should be directed toward soil-borne pathogens and other organisms whose abundance may be affected by addition of legume residues, and whose activities may regulate weed or crop performance.

3) Long-term experiments. Useful experiments with tillage systems are long-term, since crop yields, weed populations, and soil characteristics generally do not stabilize until management regimes have been established for 4 to 10 years (Swanton et al., 1993). By analogy, it seems reasonable to believe that understanding the effects of crop rotation on weeds also requires long-term experiments. In a review of design considerations for rotation experiments, Cady (1991) emphasized that rotation studies are not only long-term, but also large-scale enterprises: each experimental treatment must be replicated, each entry point of each rotation system should be present each year, and large-scale plots that minimize border effects should be employed. Thus, a critical issue facing weed scientists and others seeking useful data about crop rotation effects on weeds is how to secure adequate research budgets and consistent long-term infrastructural support for field studies.

4) Modelling. One way to increase the value of data collected in field experiments is to use them in modelling efforts. Models can be used to simulate weed population trajectories in different crop rotation sequences and to suggest particularly important control points in weed life cycles (Jordan et al., 1995). They can also be used to evaluate the economics of controlling varying densities of weeds in different rotation systems (Mortensen et al., 1993). While modelling is not a substitute for field work, it is an important means of asking and trying to answer questions about the effects of cropping practices on weeds (Cousens and Mortimer, 1995). Research work combining field experiments and modelling should be encouraged.

5) Hypothesis generation and testing on commercial farms. Observations by farmers and on-farm research provide additional means of generating and testing hypotheses concerning the effects of crop rotation practices on weeds. For example, in a recent survey conducted by Entz et al. (1995), 83% of western Canadian farmers who grew forages in their rotations indicated that they encountered fewer weed problems in crops following forages than following cereals. What ecological and management factors were responsible for this effect? To answer this question, manipulative experiments could be conducted with relevant crops and equipment on nearby research stations. Alternatively, the question could be addressed with statistically rigorous experiments conducted on commercial farms (Thompson and Thompson, 1990).

6) Government policy. For many American farmers, federal farm programs are both a major source of income and a primary factor determining which crops they will plant. In its 1989 report *Alternative Agriculture*, the National Research Council noted that monoculture and rotation sequences of limited diversity are encouraged by government policies that restrict eligibility for subsidy payments to only certain crops and link payments to the number of acres enrolled in support programs. The report clearly describes the financial rewards of not diversifying. For example, a hypothetical farm with 1000 acres of corn eligible for federal subsidy payments would lose $90,000 per year in government support if it changed from continuous corn to a corn-oat-meadow-meadow rotation. In a second example, the impact of government policies on the profitability of continuous corn and four rotation systems in Iowa was evaluated. Without government payments, continuous corn was found to be less profitable than rotations of corn with oat, soybean, and forage crops; with government payments, continuous corn was substantially more profitable than all of the rotations. Thus, it seems appropriate to conclude an analysis of crop rotation effects on weed management by observing that U.S. farm policies can strongly discourage farmers from diversifying, particularly with forage legumes and green manures. Before crop rotation practices and legumes might be better integrated into the weed management systems of most commercial farms, conflicts between economic, agronomic, and environmental policies and objectives will need to be resolved.

Acknowledgments

This is contribution 1948 of the Maine Agricultural and Forest Experiment Station. We thank K. Doolan, E. Dyck, E. Gallandt, and N. Jordan for generously sharing their data and insights.

References

Abdul-Rahman, A.A. and S.A. Habib. 1989. Allelopathic effect of alfalfa (*Medicago sativa*) on bladygrass (*Imperata cylindrica*). *J. Chem. Ecol.* 15:2289-2300.

Ahmed, M. and D.A. Wardle. 1994. Allelopathic potential of vegetative and flowering ragwort (*Senecio jacobaea* L.) plants against associated pasture species. *Plant Soil* 164:61-68.

Aiken, G.R. 1985. Isolation and concentration techniques for aquatic humic substances. p. 363-385. In: G.R. Aiken, D.M. McKnight, R.L. Wershaw, and P. MacCarthy (eds.), *Humic Substances in Soil, Sediment, and Water: Geochemistry, Isolation and Characterization.* J. Wiley & Sons, NY.

Aldrich, R.J. 1984. *Weed-Crop Ecology: Principles in Weed Management.* Breton Publishers, North Scituate, MA.

Alkamper, J. 1976. Influence of weed infestation on effect of fertilizer dressings. *Pflanzenschutz-Nachrichten Bayer* 29:191-235.

Alkamper, J., E. Pessios, and D.V. Long. 1979. Einfluss der duengung auf die entwicklung und naehrstoffaufnahme verschiedener unkrauter in mais. *Proc. European Weed Res. Soc.*:181-192.

Anderson, R.L. 1991. Timing of nitrogen application affects downy brome (*Bromus tectorum*) growth in winter wheat. *Weed Tech.* 5:582-585.

Appel, H.M. 1993. Phenolics in ecological interaction: the importance of oxidation. *J. Chem. Ecol.* 19:1521-1553.

Appleby, A.P., P.D. Olson, and D.R. Colbert. 1976. Winter wheat yield reduction from interference by Italian ryegrass. *Agron. J.* 68:463-466.

Ashby, J.A. and D. Pachico. 1987. Agricultural ecology of the mid-hills of Nepal. p. 195-222. In: B.L. Turner and S.B. Brush (eds.), *Comparative Farming Systems*. Guilford Press, NY.

Aspelin, A.L. 1994. *Pesticides Industry Sales and Usage, 1992 and 1993 market estimates*. U.S. Environmental Protection Agency, Washington, D.C.

Azam, F., K.A. Malik, and M.I. Sajjad. 1985. Transformations in soil and availability to plants of ^{15}N applied as inorganic fertilizer and legume residues. *Plant Soil* 86:3-13.

Barnes, J.P. and A.R. Putnam. 1983. Rye residues contribute weed suppression in no-tillage cropping systems. *J. Chem. Ecol.* 9:1045-1057.

Barnes, J.P. and A.R. Putnam. 1986. Evidence for allelopathy by residues and aqueous extracts of rye (*Secale cereale*). *Weed Sci.* 34:384-390.

Barnes, J.P. and A.R. Putnam. 1987. Role of benzoxazinones in allelopathy by rye (*Secale cereale* L.). *J. Chem. Ecol.* 13:889-906.

Barnes, J.P., A.R. Putnam, B.A. Burke, and A.A. Aasen. 1987. Isolation and characterization of allelochemicals in rye herbage. *Phytochem.* 26:1385-1390.

Beets, W.C. 1982. *Multiple Cropping and Tropical Farming Systems.* Westview Press, Boulder, CO.

Berkowitz, A.R. 1988. Competition for resources in weed-crop mixtures. p. 89-119. In: M.A. Altieri and M. Liebman (eds.), *Weed Management in Agroecosystems: Ecological Approaches*. CRC Press, Boca Raton, FL.

Blackshaw, R.E. 1993. Downy brome (*Bromus tectorum*) density and relative time of emergence affect interference in winter wheat (*Triticum aestivum*). *Weed Sci.* 41:551-556.

Blackshaw, R.E. 1994. Rotation affects downy brome (*Bromus tectorum*) in winter wheat (*Triticum aestivum*). *Weed Tech.* 8:728-732.

Bloemhard, C.M.J., M.W.M.F. Arts, P.C. Scheepens, and A.G. Elema. 1992. Thermal inactivation of weed seeds and tubers during drying of pig manure. *Neth. J. Agric. Sci.* 40:11-19.

Blum, U. 1995. The value of model plant-microbe-soil systems for understanding processes associated with allelopathic interaction. p. 127-131. In: Inderjit [sic], K.M.M. Dakshini, and F.A. Einhellig (Eds.), *Allelopathy: Organisms, Processes, and Applications.* Am. Chem. Soc., Washington, D.C.

Blum, U., T.M. Gerig, A.D. Worsham, L.D. Holappa, and L.D. King. 1992. Allelopathic activity in wheat-conventional and wheat-no-till soils: development of soil extract bioassays. *J. Chem. Ecol.* 18:2191-2221.

Bottrell, D.R. and R.R. Weil. 1995. Protecting crops and the environment: striving for durability. p. 55-74. In: A.S.R. Juo and R.D. Freed (eds.), *Agriculture and Environment: Bridging Food Production and Environmental Protection in Developing Countries.* Amer. Soc. Agron., Madison, WI.

Bradow, J.M. and W.J. Connick. 1990. Volatile seed germination inhibitors from plant residues. *J. Chem. Ecol.* 16:645-666.

Brust, G.E. and L.R. King. 1994. Effects of crop rotation and reduced chemical inputs on pests and predators in maize agroecosystems. *Agric., Ecosyst. Envir.* 48:77-89.

Buhler, D.D., J.L. Gunsolus, and D.F. Ralston. 1992. Integrated weed management techniques to reduce herbicide inputs in soybean. *Agron. J.* 84:973-978.

Bullock, D.G. 1992. Crop rotation. *Crit. Rev. Plant Sci.* 11:309-326.

Cady, F.B. 1991. Experimental design and data management of rotation experiments. *Agron. J.* 83:50-56.

Callaway, M.B. 1992. A compendium of crop varietal tolerance to weeds. *Amer. J. Alter. Agric.* 7:169-180.

Callaway, M.B. and F. Forcella. 1993. Crop tolerance to weeds. p. 100-131. In: M.B. Callaway and C.A. Francis (eds.), *Crop Improvement for Sustainable Agriculture*. Univ. Nebraska Press, Lincoln, NE.

Carlson, H.L. and J.E. Hill. 1985. Wild oat (*Avena fatua*) competition with spring wheat: effects of nitrogen fertilization. *Weed Sci.* 34:29-33.

Chancellor, R.J. 1985. Changes in the weed flora of an arable field cultivated for 20 years. *J. Appl. Ecol.* 22:491-501.

Chase, W.R., M.G. Nair, and A.R. Putnam. 1991. 2,2'-oxo-1,1'-azobenzene: selective toxicity of rye (*Secale cereale* L.) allelochemicals to weed and crop species: II. *J. Chem. Ecol.* 17:9-19.

Cheng, H.H. 1995. Characterization of the mechanisms of allelopathy. p. 132-141. In: Inderjit [sic], K.M.M. Dakshini, and F.A. Einhellig (eds.), *Allelopathy: Organisms, Processes, and Applications.* Am. Chem. Soc., Washington, D.C.

Chou, C.-H. and Z.A. Patrick. 1976. Identification and phytotoxic activity of compounds produced during decomposition of corn and rye residues in soil. *J. Chem. Ecol.* 2:369-387.

Chung, I.-M. and D.A. Miller. 1995. Natural herbicide potential of alfalfa residue on selected weed species. *Agron. J.* 87:920-925.

Cochran, V.L., L.F. Elliot, and R.I. Papendick. 1977. The production of phytotoxins from surface crop residues. *Soil Sci. Soc. Am. J.* 41:903-908.

Cousens, R. and M. Mortimer. 1995. *Dynamics of Weed Populations.* Cambridge Univ. Press, Cambridge, U.K.

Cousens, R., P. Brain, J.T. O'Donovan, and P.A. O'Sullivan. 1987. The use of biologically realistic equations to describe the effects of weed density and relative time of emergence on crop yield. *Weed Sci.* 35:720-725.

Covarelli, G. and F. Tei. 1988. Effet de la rotation culturale sur la flore adventice du mais. p. 477-484. In: *VIIIeme colloque international sur la biologie, l'ecologie et la systematique des mauvaises herbes, Vol. 2.* Comite Francais de Lutte Contre les Mauvaises Herbes, Paris, France and European Weed Research Society, Leverkusen, Germany.

Crookston, R.K. 1984. The rotation effect: what causes it to boost yields? *Crops Soils* 36:12-14.

Crookston, R.K. and J.E. Kurle. 1989. Corn residue effect on the yield of corn and soybean grown in rotation. *Agron. J.* 82:229-232.

Crookston, R.K., J.E. Kurle, and W.E. Lueschen. 1988. Relative ability of soybean, fallow, and tricontanol to alleviate yield reductions associated with growing corn continuously. *Crop Sci.* 28:145-147.

Crookston, R.K., J.E. Kurle, P.J. Copeland, J.H. Ford, and W.E. Lueschen. 1991. Rotation cropping sequence affects yield of corn and soybean. *Agron. J.* 8-3:108-113.

Dabney, S.M., E.C. McGawley, D.J. Boethel, and D.A. Berger. 1988. Short-term crop rotation systems for soybean production. *Agron. J.* 80:197-204.

Dalton, B.R., U. Blum, and S.B. Weed. 1989. Differential sorption of exogenously applied ferulic, p-coumaric, p-hydroxybenzoic, and vanillic acids in soil. *Soil Sci. Soc. Am. J.* 53:757-762.

Dospekhov, B.A. 1967. The effect of long term fertilizer application and crop rotation on the infestation of fields by weeds. *Weed Abstr.* 17:132.

Dotzenko, A.D., M. Ozkan, and K.R. Storer. 1969. Influence of crop sequence, nitrogen fertilizer, and herbicides on weed seed populations in sugar beet fields. *Agron. J.* 61:34-37.

Dou, Z., R.H. Fox, and J.D. Toth. 1995. Seasonal soil nitrate dynamics in corn as affected by tillage and nitrogen source. *Soil Sci. Soc. Am. J.* 59:858-864.

Dowler, C.C., E.W. Hauser, and A.W. Johnson. 1974. Crop-herbicide sequences on a southeastern coastal plain soil. *Weed Sci.* 22:500-505.

Dowling, P.M. and P.T.W. Wong. 1993. Influence of preseason weed management and in-crop treatments on two successive wheat crops. 1. Weed seedling numbers and wheat grain yield. *Aust. J. Exper. Agr.* 33:167-172.

Dyck, E. 1994. *Toward integration of soil fertility management and weed control: experiments with crimson clover (Trifolium incarnatum L.), corn (Zea mays L.), and lambsquarters (Chenopodium album L.).* Ph.D. dissertation, Univ. of Maine, Orono.

Dyck, E. and M. Liebman. 1994. Soil fertility management as a factor in weed control: the effect of crimson clover residue, synthetic nitrogen fertilizer, and their interaction on emergence and early growth of lambsquarters and sweet corn. *Plant Soil* 167:227-237.

Dyck, E. and M. Liebman. 1995. Crop-weed interference as influenced by a leguminous or synthetic fertilizer nitrogen source: II. Rotation experiments with crimson clover, field corn, and lambsquarters. *Agric., Ecosyst. Environ.* 56:109-120.

Dyck, E., M. Liebman, and M.S. Erich. 1995. Crop-weed interference as influenced by a leguminous or synthetic fertilizer nitrogen source: I. Doublecropping experiments with crimson clover, sweet corn, and lambsquarters. *Agric., Ecosyst. Environ.* 56:93-108.

Economic Research Service (ERS). 1992. *Agricultural resources: inputs situation and outlook report.* Publ. AR-28. Economic Research Service, U.S. Dept. of Agriculture, Washington, D.C.

Egley, G.H. 1986. Stimulation of weed seed germination in soil. *Rev. Weed Sci.* 2:67-89.

Einhellig, F.A. 1995. Mechanism of action of allelochemicals in allelopathy. p. 96-116. In: Inderjit [sic], K.M.M. Dakshini, and F.A. Einhellig (eds.), *Allelopathy: Organisms, Processes, and Applications.* Amer. Chem. Soc., Washington, D.C.

Einhellig, F.A. and G.R. Leather. 1988. Potentials for exploiting allelopathy to enhance crop protection. *J. Chem. Ecol.* 14:1829-1844.

Einhellig, F.A. and J.A. Rasmussen. 1989. Prior cropping with grain sorghum inhibits weeds. *J. Chem. Ecol.* 15:951-960.

Emmond, G.S. and R.J. Ledingham. 1972. Effects of crop rotation on some soil-borne pathogens of potato. *Can. J. Plant. Sci.* 52:605-611.

Entz, M.H., W.J. Bullied, and F. Katepa-Mupondwa. 1995. Rotational benefits of forage crops in Canadian prairie cropping systems. *J. Prod. Agric.* 8:521-529.

Exner, D., R. Thompson, and S. Thompson. 1990. Case study: a resource-efficient farm with livestock. p. 263-280. In: C.A. Francis, C.B. Flora, and L.D. King (eds.), *Sustainable Agriculture in Temperate Zones.* John Wiley, NY.

Fay, P.K. and W.B. Duke. 1977. An assessment of allelopathic potential in *Avena* germ plasm. *Weed Sci.* 25:224-228.

Forcella, F. and M.J. Lindstrom. 1988. Movement and germination of weed seeds in ridge-tillage crop production systems. *Weed Sci.* 36:56-59.

Fox, R.H. and W.P. Pielielek. 1988. Fertilizer N equivalence of alfalfa, birdsfoot trefoil, and red clover for succeeding corn crops. *J. Prod. Agric.* 1:313-317.

Fox, R.H., R.J.K. Myers, and I. Vallis. 1990. The nitrogen mineralization rate of legume residues in soil as influenced by their polyphenol, lignin, and nitrogen contents. *Plant Soil* 129:251-259.

Freyman, S., C.G. Kowalenko, and J.W. Hall. 1989. Effect of nitrogen, phosphorus and potassium on weed emergence and subsequent weed communities in south coastal British Columbia. *Can. J. Plant. Sci.* 69:1001-1010.

Froud-Williams, R.J. 1988. Changes in weed flora with different tillage and agronomic practices. p. 213-236. In: M.A. Altieri and M. Liebman (eds.), *Weed Management in Agroecosystems: Ecological Approaches.* CRC Press, Boca Raton, FL.

Frye, W.W., R.L. Blevins, M.S. Smith, S.J. Corak, and J.J. Varco. 1988. Role of annual legume cover crops in efficient use of water and nitrogen. p. 129-154. In: W.L. Hargrove (ed.), *Cropping Strategies for Efficient Use of Water and Nitrogen.* Amer. Soc. Agron., Madison, WI.

Gallet, C. 1994. Allelopathic potential in bilberry-spruce forests: influence of phenolic compounds on spruce seedlings. *J. Chem. Ecol.* 20:1009-1024.

Gantzer, C.J., S.H. Anderson, A.L. Thompson, and J.R. Brown. 1991. Evaluation of soil loss after 100 years of soil and crop management. *Agron. J.* 83:74-77.

Goolsby, D.A., W.A. Battaglin, and E.M. Thurman. 1993. *Occurrence and transport of agricultural chemicals in the Mississippi River basin, July through August 1993.* U.S. Geological Survey Circular 1120-C. U.S. Govt. Printing Office, Washington, D.C.

Goolsby, D.A., R.C. Coupe, and D.J. Markovchick. 1991. *Distribution of selected herbicides and nitrate in the Mississippi and its major tributaries, April through June 1991.* Water Resources Investigation Report 91-4163, U.S. Geological Survey, Denver, CO.

Graham, P.L., J.L. Steiner, and A.F. Wiese. 1988. Light absorption and competition in mixed sorghum-pigweed communities. *Agron. J.* 80:415-418.

Grigg, D.B. 1974. *The Agricultural Systems of the World: an Evolutionary Approach.* Cambridge Univ. Press, Cambridge, U.K.

Groffman, P.M., P. Hendrix, and D.A. Crossley. 1987. Nitrogen dynamics in conventional and no-tillage agroecosystems with inorganic fertilizer or legume nitrogen inputs. *Plant Soil* 97:315-332.

Guenzi, W.D. and T.M. McCalla. 1962. Inhibition of germination and seedling development by crop residues. *Soil Sci. Soc. Am. Proc.* 26:456-458.

Guenzi, W.D., W.R. Kehr, and T.M. McCalla. 1964. Water soluble phytotoxic substances in alfalfa forage: variation with variety, cutting, year, and stage of growth. *Agron. J.* 59:163-169.

Hallberg, G.R. 1989. Pesticide pollution of groundwater in the humid United States. *Agric., Ecosyst. Envir.* 26:299-367.

Harris, G.H., O.B. Hesterman, E.A. Paul, S.E. Peters, and R.R. Janke. 1994. Fate of legume and fertilizer nitrogen-15 in a long-term cropping systems experiment. *Agron. J.* 86:910-915.

Heard, A.J. 1963. Weed populations on arable land after four-course rotations and after short leys. *Ann. Appl. Biol.* 52:177-184.

Heichel, G.H. and D.K. Barnes. 1984. Opportunities for meeting crop nitrogen needs from symbiotic nitrogen fixation. p. 49-60. In: D.F. Bezdicek, J.F. Power, D.R. Keeney, and M.J. Wright (eds.), *Organic Farming: Current Technology and its Role in a Sustainable Agriculture.* Amer. Soc. Agron., Madison, WI.

Hepler, P.R., L.H. Long, K.J. LaFlamme, and J.H. Wenderoth. 1984. Field appraisal of resource management systems: crop yield and quality relationships with soil erosion-1981. Maine Agric. Expt. Stn. Bull. 805, Univ. of Maine, Orono, ME.

Hepler, P.R., L.H. Long, K.J. LaFlamme, and J.H. Wenderoth. 1985. Field appraisal of resource management systems: crop yield and quality relationships with soil erosion-1982. Maine Agric. Expt. Stn. Bull. 811, Univ. of Maine, Orono, ME.

Herman, W.A., W.B. McGill, and J.F. Dormaar. 1977. Effects of initial chemical composition on decomposition of roots of three grass species. *Can. J. Soil Sci.* 57:205-215.

Hesterman, O.B. 1988. Exploiting forage legumes for nitrogen contribution in cropping systems. p. 155-166. In: W.L. Hargrove (ed.), *Cropping Strategies for Efficient Use of Water and Nitrogen.* Amer. Soc. Agron., Madison, WI.

Honeycutt, C.W., W.M. Clapham, and S.S. Leach. 1994. A functional approach to efficient nitrogen use in crop production. *Ecol. Modelling* 73:51-61.

Honeycutt, C.W., L.J. Potaro, and W.A. Halteman. 1991. Predicting nitrate formation from soil, fertilizer, crop residue, and sludge with thermal units. *J. Environ. Qual.* 20:850-856.

Honeycutt, C.W., L.J. Potaro, K.L. Avila, and W.A. Halteman. 1993. Residue quality, loading rate and soil temperature relations with hairy vetch (*Vicia villosa* Roth) residue carbon, nitrogen and phosphorus mineralization. *Biol. Agric. Hort.* 9:181-199.

Jackson, L.E., L.J. Wyland, and L.J. Stivers. 1993. Winter cover crops to minimize nitrate losses in intensive lettuce production. *J. Agric. Sci. Cambridge.* 121:55-62.

Janzen, H.H., J.B. Bole, V.O. Biederbeck, and A.E. Slinkard. 1990. Fate of N applied as green manure or ammonium sulphate fertilizer to soil subsequently cropped with spring wheat in three sites in western Canada. *Can. J. Soil Sci.* 70:313-323.

Johnson, N.C., P.J. Copeland, R.K. Crookston, and F.L. Pfleger. 1992. Mycorrhizae: possible explanation for yield decline with continuous corn. *Agron. J.* 84:387-390.

Jordan, N., D.A. Mortensen, D.M. Prenzlow, and K. Curtis-Cox. 1995. Simulation analysis of crop rotation effects on weed seedbanks. *Amer. J. Bot.* 82:390-398.

Karlen, D.L., G.E. Varvel, D.G. Bullock, and R.M. Cruse. 1994. Crop rotations for the 21st century. *Adv. Agron.* 53:1-45.

Karssen, C.M. and H.W.M. Hilhorst. 1992. Effect of chemical environment on seed germination. p. 327-348. In: M. Fenner (ed.), *Seeds: The Ecology of Regeneration in Plant Communities*. C.A.B. International, Wallingford, U.K.

Kass, D.C.L., C. Folletti, L.T. Szott, R. Landaverde, and R. Nolasco. 1993. Traditional fallow systems of the Americas. *Agrofor. Syst.* 23:207-218.

Kimber, R.W.L. 1973. Phytotoxicity from plant residues. II. The effect of time of rotting of straw from some grasses and legumes on the growth of wheat seedlings. *Plant Soil* 38:347-361.

Kung, K.-H. and M.B. McBride. 1989. Coordination complexes of p-hydroxybenzoate on Fe oxides. *Clays Clay Minerals* 37:333-340.

Ladd, J.N. and M. Amato. 1986. The fate of nitrogen from legume and fertilizer sources in soils successively cropped with wheat under field conditions. *Soil Biol. Biochem.* 18:417-425.

Lashomb, J.H. and Y.S. Ng. 1984. Colonization by Colorado potato beetles, *Leptinotarsa decemlineata* (Say) (Coleoptera: Chrysomelidae), in rotated and nonrotated fields. *Environ. Entomol.* 13:1352-1356.

Lawson, H.M. and J.S. Wiseman. 1979. Competition between weeds and transplanted spring cabbage: effects of nitrogen top-dressing. *Hort. Res.* 19:25-34.

Leather, G.R. 1987. Weed control using allelopathic sunflowers and herbicide. *Plant Soil* 98:17-23.

Lehman, M.E., U. Blum, and T.M. Gerig. 1994. Simultaneous effects of ferulic and p-coumaric acids on cucumber leaf expansion in split-root experiments. *J. Chem. Ecol.* 20:1773-1782.

Leighty, C.E. 1938. Crop rotation. p. 406-430. In: U.S. Dept. of Agriculture, *Soils and Men, Yearbook of Agriculture 1938*. USDA, Washington, D.C.

Lewis, W.M. and J.A. Phillips. 1976. Double cropping in the eastern United States. p. 41-50. In: R.I. Papendick, P.A. Sanchez, and G.B. Triplett (eds.), *Multiple Cropping*. Amer. Soc. Agron., Madison, WI.

Liebl, R.A. and A.D. Worsham. 1983. Inhibition of pitted morning glory (*Ipomoea lacunosa* L.) and certain other weed species by phytotoxic components of wheat (*Triticum aestivum* L.) straw. *J. Chem. Ecol.* 9:1027-1043.

Liebl, R.A., F.W. Simmons, L.M. Wax, and E.W. Stoller. 1992. Effect of rye (*Secale cereale*) mulch on weed control and soil moisture in soybean (*Glycine max*). *Weed Tech.* 6:838-846.

Liebman, M. 1989. Effects of nitrogen fertilizer, irrigation, and crop genotype on canopy relations and yields on an intercrop/weed mixture. *Field Crops Res.* 22:83-100.

Liebman, M. 1992. Research and extension efforts for improving agricultural sustainability in the north central and northeastern United States. *Agric., Ecosyst., Envir.* 39:101-122.

Liebman, M. and E. Dyck. 1993. Crop rotation and intercropping strategies for weed management. *Ecol. Applic.* 3:92-122.

Liebman, M. and E. Gallandt. In press. Many little hammers: ecological approaches for managing crop-weed interactions. In: L.E. Jackson (ed.), *Agricultural ecology*. Academic Press, San Diego, CA.

Liebman, M. and R.R. Janke. 1990. Sustainable weed management practices. p. 111-143. In: C.A. Francis, C.B. Flora, and L.D. King (eds.), *Sustainable Agriculture in Temperate Zones*. John Wiley, New York.

Liebman, M., S. Corson, R.J. Rowe, and W.A. Halteman. 1995. Dry bean responses to nitrogen fertilizer in two tillage and residue management systems. *Agron. J.* 87:538-546.

Liu, D.L. and J.V. Lovett. 1993. Biologically active secondary metabolites of barley. I. Developing techniques and assessing allelopathy in barley. *J. Chem. Ecol.* 19:2217-2230.

Lockeretz, W., G. Shearer, and D.H. Kohl. 1981. Organic farming in the corn belt. *Science* 211:540-547.

Lotz, L.A.P., R.M.W. Groeneveld, B. Habekotte, and H. van Oene. 1991. Reduction of growth and reproduction of *Cyperus esculentus* by specific crops. *Weed Res.* 31:153-160.

MacHoughton, J. 1973. Ecological changes in weed populations as a result of crop rotations and herbicides. *Diss. Abstr.* 34:4160B-4161B.

MacRae, R.J. and G.R. Mehuys. 1985. The effect of green manuring on the physical properties of temperate-area soils. *Adv. Soil Sci.* 3:71-93.

Mallik, M.A.B., R. Puchala, and F.A. Grosz. 1994. A growth inhibitory factor from lambsquarters (*Chenopodium album*). *J. Chem. Ecol.* 20:957-967.

Maloy, O.C. and W.H. Burkholder. 1959. Some effects of crop rotation on the fusarium root rot of bean. *Phytopath.* 49:583-587.

Martin, R.J. and W.L. Felton. 1993. Effect of crop rotation, tillage practice, and herbicides on the population dynamics of wild oats in wheat. *Aust. J. Exp. Agric.* 33:159-165.

Matheson, N., B. Rusmore, J.R. Sims, M. Spengler, and E.L. Michalson. 1991. *Cereal-legume cropping systems: nine farm case studies in the dryland northern plains, Canadian prairies and intermountain northwest.* Alternative Energy Resources Organization, Helena, MT.

McCalla, T.M. and F.L. Duley. 1948. Stubble mulch studies: effect of sweetclover extract on corn germination. *Science* 108:163.

McEwen, J., R.J. Darby, M.V. Hewitt, and D.P. Yeoman. 1989. Effects of field beans, fallow, lupins, oats, oilseed rape, peas, ryegrass, sunflowers and wheat on nitrogen residues in the soil and on the growth of a subsequent wheat crop. *J. Agric. Sci., Cambridge* 115:209-219.

Minotti, P.L. 1991. Role of crop interference in limiting losses from weeds. p. 359-368. In: D. Pimentel (ed.), *CRC Handbook of Pest Management in Agriculture. Vol. 2.* CRC Press, Boca Raton, FL.

Mohler, C.L. 1993. A model of the effects of tillage on emergence of weed seedlings. *Ecol. Applic.* 3:53-73.

Morel, F.M.M. and J.G. Hering. 1993. *Principles and Applications of Aquatic Chemistry*. J. Wiley & Sons, New York.

Mortensen, D.A., A.R. Martin, T.E. Harvill, and T.A. Bauer. 1993. The influence of rotational diversity on economic optimum thresholds in soybean. In: *Quantitative approaches in weed and herbicide research and their practical application.* Proc. European Weed Res. Soc. 8:815-823.

Mulder, T.A. and J.D. Doll. 1994. Reduced input corn weed control: the effects of planting date, early season weed control, and row crop cultivator selection. *J. Prod. Agric.* 7:256-260.

National Research Council (NRC). 1989. *Alternative Agriculture.* National Academy Press, Washington, D.C.

National Research Council (NRC). 1991. *Sustainable Agriculture Research and Education in the Field: a Proceedings.* National Academy Press, Washington, D.C.

Nelson, H. and R.D. Jones. 1994. Potential regulatory problems associated with atrazine, cyanazine, and alachlor in surface water source drinking water. *Weed Tech.* 8:852-861.

Nielson, K.F., T.F. Cuddy, and W.B. Woods. 1960. The influence of the extracts of some crop and soil residues on germination and growth. *Can. J. Plant. Sci.* 40:188-197.

Okafor, L.I. and S.K. De Datta. 1976. Competition between upland rice and purple nutsedge for nitrogen, moisture, and light. *Weed Sci.* 24:43-46.

Overland, L. 1966. The role of allelopathic substances in the "smother crop" barley. *Amer. J. Bot.* 53:423-432.

Palm, C.A. and P.A. Sanchez. 1991. Nitrogen release from the leaves of some tropical legumes as affected by their lignin and polyphenolic contents. *Soil Biol. Biochem.* 23:83-88.

Parker, D.R., W.A. Norvell, and R.L. Chaney. 1995. GEOCHEM-PC: a chemical speciation program for IBM and compatible personal computers. p. 253-269. In: R.H. Loeppert, A.P. Schwab, and S. Goldberg (eds.), *Chemical equilibrium and reaction models.* Soil Sci. Soc. Am., Madison, WI.

Patrick, Z.A., T.A. Toussoun, and W.C. Snyder. 1963. Phytotoxic substances in arable soils associated with decomposition of plant residues. *Phytopath.* 53:152-161.

Patriquin, D.G., D. Baines, J. Lewis, and A. MacDougall. 1988. Aphid infestation of faba beans on an organic farm in relation to weeds, intercrops, and added nitrogen. *Agric., Ecosyst. Environ.* 20:279-288.

Power, J.F. 1990. Legumes and crop rotation. p. 178-204. In: C.A. Francis, C.B. Flora, and L.D. King (eds.), *Sustainable agriculture in temperate zones*. John Wiley, NY.

Power, J.F. and R.F. Follett. 1987. Monoculture. *Sci. Amer.* 256(3):78-86.

Putnam, A.R. and J. DeFrank. 1983. Use of phytotoxic plant residues for selective weed control. *Crop. Prot.* 2:173-181.

Putnam, A.R., J. DeFrank, and J.P. Barnes. 1983. Exploitation of allelopathy for weed control in annual and perennial cropping systems. *J. Chem. Ecol.* 9:1001-1010.

Ranells, N.N. and M.G. Wagger. 1992. Nitrogen release from crimson clover in relation to plant growth stage and composition. *Agron. J.* 84:424-430.

Rasmussen, J. 1992. Testing harrows for mechanical control of annual weeds in agricultural crops. *Weed Res.* 32:267-274.

Reganold, J.P. 1988. Comparison of soil properties as influenced by organic and conventional farming systems. *Amer. J. Alter. Agric.* 3:144-155.

Reganold, J.P., R.I. Papendick, and J.F. Parr. 1990. Sustainable agriculture. *Sci. Amer.* 262(6):112-120.

Reinertsen, S.A., L.F. Elliott, V.L. Cochran, and G.S. Campbell. 1984. Role of available carbon and nitrogen in determining the rate of wheat straw decomposition. *Soil Biol. Biochem.* 16:459-464.

Rice, E.L. 1995. *Biological Control of Weeds and Plant Diseases: Advances in Applied Allelopathy*. Univ. of Oklahoma Press, Norman, OK.

Roder, W., S.C. Mason, M.D. Clegg, and K.R. Kniep. 1989. Yield-soil water relationships in sorghum-soybean cropping systems with different fertilizer regimes. *Agron. J.* 81:470-475.

Rydberg, T. 1993. Weed harrowing--driving speed at different stages of development. *Swed. J. Agric. Res.* 23:107-113.

Sarapatka, B., M. Holub, and M. Lhotska. 1993. The effect of farmyard manure anaerobic treatment on weed seed viability. *Biol. Agric. Hort.* 10:1-8.

Sarrantonio, M. and T.W. Scott. 1988. Tillage effects on availability of nitrogen to corn following a winter green manure crop. *Soil Sci. Soc. Amer. J.* 52:1661-1668.

Schreiber, M.M. 1992. Influence of tillage, crop rotation, and weed management on giant foxtail (*Setaria faberii*) population dynamics and corn yield. *Weed Sci.* 40:645-653.

Scott, R.K. and S.J. Wilcockson. 1976. Weed biology and growth of sugar beet. *Ann. Appl. Biol.* 83:331-335.

Shilling, D.G., R.A. Liebl, and A.D. Worsham. 1985. Rye (*Secale cereale* L.) and wheat (*Triticum aestivum* L.) mulch: the suppression of certain broadleaved weeds and the isolation and identification of phytotoxins. p. 243-271. In: A.C. Thompson (ed.), *The Chemistry of Allelopathy: Biochemical Interactions among Plants.* Amer. Chem. Soc., Washington, D.C.

Shilling, D.G., L.A. Jones, A.D. Worsham, C.E. Parker, and R.F. Wilson. 1986. Isolation and identification of some phytotoxic compounds from aqueous extracts of rye (*Secale cereale* L.). *J. Agric. Food Chem.* 34:633-638.

Shipley, P.R., J.J. Meisinger, and A.M. Decker. 1992. Conserving residual corn fertilizer nitrogen with winter cover crops. *Agron. J.* 84:869-876.

Sindel, B.M. and P.W. Michael. 1992. Growth and competitiveness of *Senecio madagascariensis* Poir. (fireweed) in relation to fertilizer use and increases in soil fertility. *Weed Res.* 32:399-406.

Siqueira, J.O., M.G. Nair, R. Hammerschmidt, and G.R. Safir. 1991. Significance of phenolic compounds in plant-soil-microbial systems. *Crit. Rev. Plant Sci.* 10:63-121.

Smith, R.J. and W.C. Shaw. 1966. *Weeds and their control in rice production.* Agric. Handbook No. 292. U.S.D.A., Washington, D.C.

Smith, M.S., W.W. Frye, and J.J. Varco. 1987. Legume winter cover crops. *Adv. Soil Sci.* 7:95-139.

Sposito, G. 1989. *Chemistry of Soils*. Oxford University Press, New York.

Staniforth, D.W. 1962. Responses of soybean varieties to weed competition. *Agron. J.* 54:11-13.

Stewart, G. and D.W. Pittman. 1931. Twenty years of rotation and manuring experiments at Logan, Utah. *Utah Agric. Exp. Stn. Bull. 228*, Utah St. Univ., Logan, UT.

Stinner, B.R. and G.J. House. 1989. The search for sustainable agricultural systems. *J. Soil Water Cons.* 44:111-116.

Stumm, W. 1992. *Chemistry of the Solid-Water Interface: Processes at the Mineral-Water and Particle-Water Interface in Natural Systems*. J. Wiley & Sons, New York.

Sturz, A.V. and C.C. Bernier. 1987. Survival of cereal root pathogens in the stubble and soil of cereal versus noncereal crops. *Can. J. Plant Path.* 9:205-213.

Sumner, D.R. 1982. Crop rotation and plant productivity. p. 273-313. In: M. Rechcigl (ed.), *CRC Handbook of Agricultural Productivity. Vol. 1. Plant productivity.* CRC Press, Boca Raton, FL.

Swanton, C.J., D.R. Clements, and D.A. Derksen. 1993. Weed succession under conservation tillage: a hierarchical framework for research and management. *Weed Tech.* 7:286-297.

Tang, C.S. and C.C. Young. 1982. Collection and identification of allelopathic compounds from the undisturbed root system of bigatta limpograss (*Hemathria altissima*). *Plant Physiol.* 69:155-160.

Taylor, D.C., T.L. Dobbs, and J.D. Smolik. 1992. Beliefs and practices of sustainable farmers in South Dakota. *J. Prod. Agric.* 5:545-550.

Taylorson, R.B. 1987. Environmental and chemical manipulation of weed seed dormancy. *Rev. Weed Sci.* 3:135-154.

Teasdale, J.R. and J.R. Frank. 1983. Effect of row spacing on weed competition with snap beans (*Phaseolus vulgaris*). *Weed Sci.* 31:81-85.

Thompson, D. and S. Thompson. 1990. The on-farm research program of Practical Farmers of Iowa. *Amer. J. Alter. Agric.* 5:163-167.

Thomsen, C.D., W.A. Williams, M. Vayssieres, F.L. Bell, and M.R. George. 1993. Controlled grazing on annual grassland decreases yellow starthistle. *Calif. Agric.* 47(6):36-40.

Thurman, E.M. 1985. *Organic Geochemistry of Natural Waters.* Martinus Nijhoff/Dr. W. Junk, Cordrecht, The Netherlands.

Thurman, E.M., D.A. Goolsby, M.T. Meyer, and D.W. Kolpin. 1991. Herbicides in surface waters of the midwestern United States: the effect of the spring flush. *Envir. Sci. Tech.* 25:1794-1796.

Tilman, D. 1982. *Resource Competition and Community Structure.* Princeton Univ. Press, Princeton, NJ.

Tollenaar, M., S.P. Nissanka, A. Aguilera, S.F. Weise, and C.J. Swanton. 1994. Effect of weed interference and soil nitrogen on four maize hybrids. *Agron. J.* 86:596-601.

U.S. Dept. of Agriculture (USDA). 1973. *Monoculture in agriculture: extent, causes, and problems. Report of the task force on spatial heterogeneity in agricultural landscapes and enterprises.* USDA, Washington, D.C.

U.S. Environmental Protection Agency (USEPA). 1992. *Another look: national survey of pesticides in drinking water wells. Phase 2 report.* USEPA, Washington, D.C.

Varco, J.J., W.W. Frye, M.S. Smith, and C.T. MacKown. 1993. Tillage effects on legume decomposition and transformation of legume and fertilizer nitrogen-15. *Soil Sci. Soc. Amer. J.* 57:750-756.

Voss, R.D. and W.D. Shrader. 1984. Rotation effects and legume sources of nitrogen for corn. p. 61-68. In: D.F. Bezdicek, J.F. Power, D.R. Keeney, and M.J. Wright (eds.), *Organic Farming: Current Technology and its Role in a Sustainable Agriculture.* Amer. Soc. Agron., Madison, WI.

Waller, G.R. 1989. Allelochemical action of some natural products. p. 129-154. In: C.K. Chou and G.R. Waller (ed.), *Phytochemical Ecology: Allelochemicals, Mycotoxins, and Insect Pheromones and Allomones.* Monograph Series No. 9. Inst. Botany, Acad. Sinica, Taipei, Taiwan.

Weaver, S.E., N. Smits, and C.S. Tan. 1987. Estimating yield losses of tomato (*Lycopersicon esculentum*) caused by nightshade (*Solanum* spp.) interference. *Weed Sci.* 35:163-168.

Wells, G.J. 1979. Annual weed competition in wheat crops: the effect of weed density and applied nitrogen. *Weed Res.* 19:185-191.

Westcott, M.P. and D.S. Mikkelsen. 1987. Comparison of organic and inorganic nitrogen sources for rice. *Agron. J.* 79:937-943.

White, R.H., A.D. Worsham, and U. Blum. 1989. Allelopathic potential of legume debris and aqueous extracts. *Weed Sci.* 37:674-679.

Whitehead, D.C., H. Dibb, and R.D. Hartley. 1982. Phenolic compounds in soil as influenced by the growth of different plant species. *J. Appl. Ecol.* 19:579-588.

Whittaker, R.H. and P.P. Feeny. 1971. Allelochemics: chemical interactions between species. *Science* 171:757-770.

Wilson, D.O. and W.L. Hargrove. 1986. Release of nitrogen from crimson clover residue under two tillage systems. *Soil Sci. Soc. Amer. J.* 50:1251-1254.

Wolt, J.D. 1994. *Soil Solution Chemistry: Applications to Environmental Science and Agriculture.* J. Wiley & Sons, New York.

Worsham, A.D. 1989. Current and potential techniques using allelopathy as an aid in weed management. p. 275-291. In: C.H. Chou and G.R. Waller (eds.), *Phytochemical Ecology: Allelochemicals, Mycotoxins and Insect Pheromones and Allomones*. Monograph Series No. 9, Inst. Botany, Acad. Sinica, Taipei, China.

Wright, R.J. 1984. Evaluation of crop rotation for control of Colorado potato beetles (Coleoptera: Chrysomelidae) in commercial potato fields on Long Island. *J. Econ. Entomol.* 77:1254-1259.

Yu, J.Q. and Y. Matsui. 1994. Phytotoxic substances in root exudates of cucumber (*Cucumis sativus.* L.). *J. Chem. Ecol.* 20:21-31.

Zawislak, K. 1980. Stopien specjalizacji zmianowan a aktualne i potencjalne zachwaszczenie stanowisk. *Zesz. Nauk. Akad. Roln. Tech. Olszt. Rolnictwo.* 29:99-118.

Tillage Systems and Weed Population Dynamics and Management

Douglas D. Buhler

I. Introduction

Soil tillage has been a major input of crop production for centuries. While an integral component of many cropping systems, tillage has negative side-effects, including increased soil erosion and high labor and fuel requirements (Mannering et al., 1987). Methods for reducing tillage were proposed as early as the 1920s (Graber, 1928). For row-crop production, no-tillage methods were widely discussed beginning in the late 1950s and 1960s as herbicides became available to replace tillage for control of unwanted vegetation (Sprague, 1986).

Conservation tillage has been used as a general term for tillage systems that leave at least 30% of the soil surface covered with plant residue after crop harvest and maintain that cover following planting of the succeeding crop (Anonymous, 1982). Conventional tillage refers to systems that leave the soil surface with little plant residue and often include moldboard plowing followed by secondary tillage with a disk, field cultivator, and/or harrow. Reduced tillage systems are those that include primary tillage (i.e., chisel plow or disk), yet leave at least 30% of the soil surface covered with plant residue. Ridge-tillage is characterized by building a ridge of soil by cultivation the previous growing season. During the planting operation, the ridges are truncated and the truncated soil moved into the interrow area. Ridge building and truncation is repeated annually. No-tillage is defined as a system with no preplant tillage with soil disturbance limited to that created by the planting operation.

The advantages and disadvantages of conservation tillage systems have been extensively debated over the past two decades. In the central United States, the most important advantage of conservation tillage is control of soil erosion. The magnitude of soil erosion reduction varies among production systems, soil types, and location. In a review of soil erosion research, Triplett and VanDoren (1977) found that no-tillage reduced soil erosion by as much as 50-fold compared with moldboard plowing on highly-erodible land. Fuel and labor requirements can be reduced 60 to 75% by reducing or eliminating preplant tillage (Frye, 1984). Reducing or eliminating preplant tillage may also improve soil water retention, reduce evaporation, and improve soil tilth (Doran and Linn, 1994). In a recent review of the impact of conservation tillage on the environment, Sims et al. (1994) concluded that the available literature overwhelmingly supports the success of conservation tillage practices in achieving the original goal: reduction of soil erosion and runoff. It was also apparent that off-site movement of nutrients and pesticides was reduced due to reduced runoff volume and sediment transport. Unfortunately, it is necessary to rely more heavily on pesticides, especially herbicides, when tillage is reduced or eliminated. Thus, even though a smaller fraction of applied material may leave the field, more kinds of pesticides and possibly higher rates are used. In a summary of the literature on herbicide runoff, Fawcett et al. (1994) concluded that all conservation tillage practices reduced herbicide movement from the treated field on highly-erodible land when compared to fields tilled with a moldboard plow.

The potential benefits of conservation tillage are substantial; however, significant problems remain. Soil temperatures early in the growing season are often reduced by residue on the soil surface, slowing crop emergence and growth (Johnson and Lowery, 1985). The lack of tillage may cause stratification of nutrients and pH levels (Thomas, 1986). The environment created in reduced tillage systems may increase incidence of certain insect, rodent, and disease pests, especially during the first 3 to 5 years after the reduction or elimination of tillage (Forcella et al., 1994). One of the greatest obstacles to conservation tillage continues to be effective and economical weed management (Buhler, 1991; Gebhardt et al., 1985; Koskinen and McWhorter, 1986; Wicks et al., 1994). Because of the importance of herbicides in conservation tillage systems, a high level of herbicide efficacy is required, often increasing herbicide use and weed control costs (Buhler, 1992; Erbach and Lovely, 1975; Griffith et al., 1973; Kells et al., 1980).

Weed management is often difficult, herbicide-intensive, and economically costly in conservation tillage because of our poor understanding of the relationships between tillage systems and weed population dynamics (Buhler, 1991; Liebman and Dyck, 1993a; Wyse, 1992). Within a cropping season, there are two major options for controlling weeds growing with the crop: herbicides and mechanical removal. The challenge is to control weeds with minimal herbicide use and tillage to protect the environment and maintain crop residue on the soil surface for erosion control and soil water management. One of the first steps in effectively managing weeds in conservation tillage is understanding weed populations and their responses to tillage systems.

As the impacts of tillage systems on weed population dynamics are examined, we must apply basic ecological principles. Weeds are successful because of their genetic diversity, and ability to adapt to and take advantage of conditions created by crop production systems. Therefore, modifying crop management inputs results in an altered competitive environment in which the morphological and physiological traits that confer success are altered (Tilman, 1990). Understanding weed population shifts will identify vulnerable stages in weed life cycles that can be exploited in management systems. Understanding population shifts will also identify species that are favored as management practices are changed, so that weed management systems that target the appropriate weed species can be developed.

II. Impacts of Tillage on Weed Management

A. General Effects

Weeds present in crop production fields reflect management practices (Froud-Williams et al., 1983; Liebman and Dyck, 1993b; Pollard et al., 1982). Crop production practices exert selection pressure on weed communities and create niches that favor or disfavor various species. Since tillage has been an integral part of many cropping systems, weeds that are present in fields have been greatly influenced by tillage for seed bed preparation and weed control.

Tillage for seed bed preparation can greatly reduce densities of annual weed populations, especially if planting is delayed to allow weed seed germination prior to the final tillage operation (Gunsolus, 1990; Robinson and Dunham, 1956; Stoller and Wax, 1973). Tillage just prior to crop planting also kills germinated seedlings that have yet to emerge from the soil. In cropping systems that include perennial crops, tillage has been used to destroy the perennial crop prior to the seeding of an annual crop (Triplett, 1985). When tillage is eliminated, vegetation management in these systems is especially challenging because of the difficulty in controlling a diverse population of established perennial crop and weed species with herbicides (Buhler and Mercurio, 1988; Griffith et al., 1984; Triplett, 1985).

Tillage buries crop residue and alters the characteristics of the surface soil, greatly influencing the germination environment of seeds by reducing soil surface cover (Aase and Tanaka, 1987; Andraski et al., 1985), altering soil temperature and moisture patterns (Al-Darby and Lowery, 1987; Johnson et al., 1984; Johnson and Lowery, 1985), and altering weed seed distribution in the soil (Buhler and Mester, 1991; Roberts and Feast, 1973; Yenish et al., 1992). Harper et al. (1965) used the term "safe site" to describe the complex conditions that are required for successful seed germination and seedling establishment. Several studies have pointed out that "safe site" conditions vary greatly among plant species (Dao, 1987; Evans and Young, 1972; Harper et al., 1965; Mester and Buhler, 1991; Sheldon, 1974). In addition, weed species vary greatly in temperature requirements for germination (Horowitz and Taylorson, 1984; Wiese and Binning, 1987). These factors, along with the physical disruption caused by tillage (Pareja and Staniforth, 1985), combine to regulate weed population dynamics in crop production systems.

B. Control Options and Efficacy

Tillage systems impact weed management options available to producers. Certain herbicides may not be used in conservation tillage systems because of the need for mechanical incorporation into the soil after application (Doersch and Buhler, 1988). With less tillage and more plant residue on the soil surface, mechanical weed control operations may also become less effective (Springman et al., 1989). Rotary hoeing is especially difficult in untilled soil covered with residue of the previous crop. Interrow cultivation remains as an option in some conservation tillage systems. Interrow cultivation is an integral component of ridge-tillage and contributes to weed control (Buhler and Daniel, 1988; Buhler, 1992; Forcella and Lindstrom, 1988b). Combining interrow cultivation with reduced herbicide use provided weed control similar to full-rate herbicide treatments in conservation tillage systems and reduced problems associated with herbicide dependence (Buhler et al., 1994a; Buhler et al., 1995; Mulder and Doll, 1993). However, the impact of interrow cultivation on the long-term impacts of no-tillage to soil properties and soil erosion reduction have not been documented.

Plant residue on the soil surface may alter the behavior of soil-applied herbicides and resulting weed control (Banks and Robinson, 1982; Bauman and Ross, 1983; Ghadiri et al., 1984; Johnson et al., 1989). Depending on percentage surface cover, residue type, and herbicide formulation, up to 60% of the herbicide applied may be intercepted by residue. However, much of this herbicide may be washed off by subsequent rainfall and irrigation.

In continuous corn production (*Zea mays* L.), giant foxtail (*Setaria faberi* Herrm.) became more difficult to control with soil-applied herbicides as tillage was reduced, while velvetleaf (*Abutilon theophrasti* Medik.) control increased (Figure 1) (Buhler and Daniel, 1988). A similar weed control response with soil-applied herbicides was observed in solid-seeded soybean [*Glycine max* (L.) Merr.] (Buhler and Oplinger, 1990; Freed et al., 1987). The effects of tillage systems on the control of small-seeded broadleaf weed species, such as common lambsquarters (*Chenopodium album* L.) and redroot pigweed (*Amaranthus retroflexus* L.), varied by species and study (Buhler, 1991; Buhler, 1992; Buhler and Oplinger, 1990; Johnson et al., 1989; Triplett and Lytle, 1972; Wrucke and Arnold, 1985).

Foliar-applied, selective herbicides are becoming increasingly important in weed management, especially in conservation tillage systems. Herbicides applied following weed emergence may be influenced less by tillage systems because they are applied directly to the target weeds and will not interact with residue on the soil surface. For this reason, postemergence herbicides have been suggested to be preferable to soil-applied herbicides for weed management in conservation tillage systems (Lewis, 1985).

Figure 1. Effect of tillage systems on control of giant foxtail and velvetleaf in corn averaged over several soil-applied herbicide combinations at Janesville, WI, averaged over two years. Vertical bars are the standard error of the mean. (Adapted from Buhler and Daniel, 1988.)

Velvetleaf control in solid-seeded soybean with foliar-applied bentazon [3-(1-methylethyl)-(1H)-2,1,3-benzothiadiazin-4(3H)-one 2,2-dioxide] or acifluorfen {5-[2-chloro-4-(trifluoromethyl)phenoxy]-2-nitrobenzoic acid} was not influenced by tillage systems (Freed et al., 1987). In similar research (Buhler et al., 1990), velvetleaf control with acifluorfen, bentazon, or a combination of the two, was less in moldboard plow than chisel plow or no-tillage. It is possible that the extended period of velvetleaf emergence and greater densities with moldboard plow compared to other tillage systems (Buhler and Daniel, 1988; Mester, 1988) reduced control.

Foliar-applied sethoxydim {2-[1-(ethoxyimino)butyl]-5-[2-(ethylthio)propyl]-3-hydroxy-2-cyclohexen-1-one} provided similar giant foxtail control in moldboard plow and chisel plow at both early (V2 soybean) and late (V4 soybean) application times (Buhler et al., 1990). Early postemergence application controlled less giant foxtail in no-tillage than in moldboard plow or chisel plow, but delaying sethoxydim application overcame differences in control. The longer period of giant foxtail emergence in no-tillage compared with the other tillage systems (Buhler and Daniel, 1988; Buhler, 1992) allowed seedlings to escape early postemergence application.

Based on the literature reviewed, the effect of tillage systems on herbicide efficacy was due to the combination of surface residue interception of herbicides and changes in weed population dynamics. The effect of tillage systems on control varied by species and herbicide. For example, control of annual grass species was often reduced as tillage was reduced, while velvetleaf control increased. These variable responses indicate that choosing the appropriate herbicide and application timing is critical in conservation tillage systems. With the reduced number of herbicide options and the reduced effectiveness of others, knowledge of weed biology and ecology is essential to managing weeds in conservation tillage without increasing herbicide use to levels that are environmentally and economically unacceptable.

III. Biology and Ecology of Weed Responses to Tillage Systems

Herbicides are not an inherent component of crop production systems, but are used in an effort to minimize crop yield losses due to weeds. Since weed populations are greatly influenced by production practices such as tillage, it is essential that these changes be understood and anticipated so that appropriate weed management systems can be devised.

A. Classification of Weed Species

Weeds can be classified in many ways, and understanding weed life cycles is important in predicting population shifts and designing management strategies (Burnside, 1993; Wicks et al., 1994; Wyse, 1992). Weed species with different life cycles have different requirements for seed germination, seedling establishment, growth, and reproduction. Because tillage greatly alters the environment where weed species survive, altering tillage systems favors certain life cycles over others. Wicks et al. (1994) concluded that in reduced tillage systems in the Great Plains of the United States, major weeds include winter and summer annual plus bulbous and creeping perennial species. In no-tillage systems, they concluded that there will be a marked drop in the variety of annual species. Concomitantly, there will be an increase in winter annual, biennial, and all types of perennial weed species.

Weeds of agronomic crops are a unique group of plant species because of their ability to infest and often thrive in intensively disturbed habitats despite extensive efforts to eliminate them. In addition to classification by life cycle, a classification appropriate to understanding weed population dynamics under different tillage systems (Wicks et al., 1994) follows:

1. Arable response weed species that require periodic disturbance of the soil for survival.
2. Inverse response weed species that require nondisturbed soil for survival.
3. Intermediate response weed species that survive in both disturbed and nondisturbed soil.

Combining characteristics of life cycles and tillage response of weed species is useful in predicting weed population shifts. For example, weed populations in reduced tillage systems with summer annual crops are dominated by arable and intermediate response summer annual species because the soil is disturbed annually and management practices favor summer annual species. In no-tillage systems, weed populations shift towards inverse and intermediate response species. Winter annual, biennial, and perennial species are well adapted to no-tillage because there is little or no tillage to interrupt the life cycle. In addition, summer annual species that do not require burial for establishment are well-adapted to no-tillage systems. However, since soil disturbance is required to plant seeds and incorporate fertilizers in annual cropping systems, arable response species may continue to exist in no-tillage fields. Seed buried in the soil from previous tillage may also persist for several years and continue to be a source of arable response species in no-tillage systems (Lueschen and Andersen, 1980; Yenish et al., 1992).

B. Summer Annual Species

Extensive inputs of tillage and herbicide are used to control arable response, summer annual weed species in summer annual crop production systems (Gianessi and Puffer, 1991). These weed species have become problems in production systems dominated by summer annual crops because they are well-adapted to the environment created by the production systems (i.e., high fertility, annual disturbance, production in rows, and herbicide use). Changes that occur in the dynamics of annual weed communities following changes in tillage practices are regulated by species present in the field, biology of the species, and efficacy of weed management practices.

The most commonly reported change in annual weed populations is an increase in summer annual grass species, including giant foxtail (Buhler and Daniel, 1988; Buhler and Oplinger, 1990; Johnson et al., 1989), green [*Setaria viridis* (L.) Beauv.] and yellow foxtail [*Setaria glauca* (L.) Beauv.] (Buhler, 1992; Forcella and Lindstrom, 1988b; Wrucke and Arnold, 1985), fall panicum (*Panicum dichotomiflorum* Michx.) (Williams and Wicks, 1978), and field sandbur (*Cenchrus incertus* M.A. Curtis) (Williams and Wicks, 1978; Wrucke and Arnold, 1985). The magnitude of changes in summer annual grass densities has varied among experiments, but the potential for rapid increases in the density and changes in emergence patterns of these weed species clearly exists.

In no-tillage, giant foxtail emergence continued through 77 days after corn planting and total density reached 1400 plants m^{-2} without herbicide treatment (Buhler and Daniel, 1988). No emergence was observed after 56 days after planting

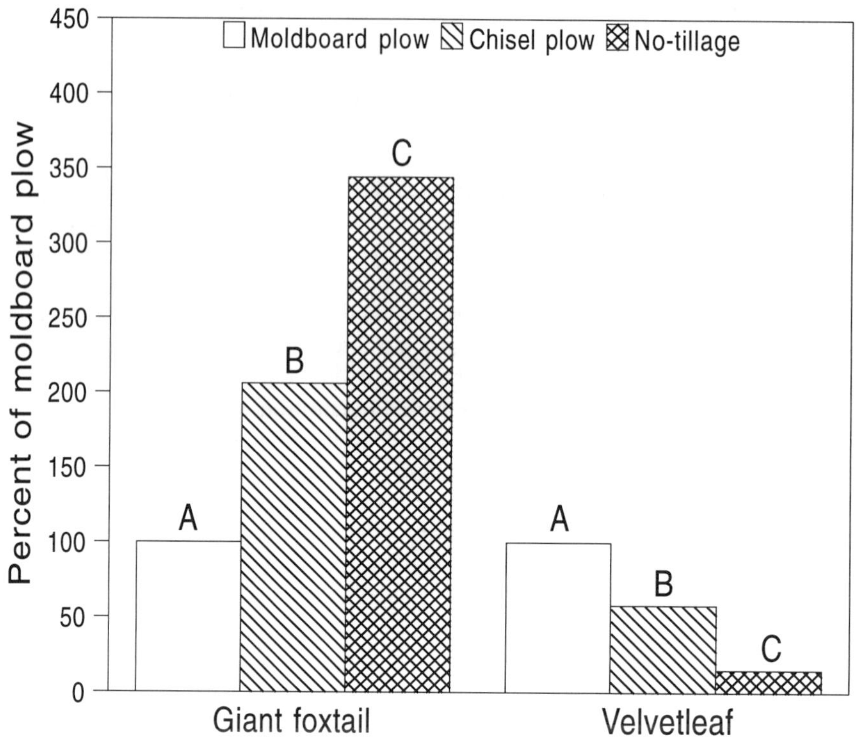

Figure 2. Effect of tillage systems on giant foxtail and velvetleaf densities in soybean relative to moldboard plowing without herbicide treatment at Janesville, WI, averaged over two years. Bars with the same letter within a species do not differ using LSD (0.05). (Adapted from Buhler and Oplinger, 1990.)

in chisel plow and the total density was less than no-tillage. Most of the giant foxtail with moldboard plowing emerged by 42 days after planting and total density was less than 200 plants m^{-2}. Giant foxtail density in ridge-tillage was similar to moldboard plowing throughout the growing season. In soybean grown in rotation with corn (Buhler and Oplinger, 1990), giant foxtail densities in chisel plow were twice that of moldboard plow and in no-tillage, densities were almost twice that of chisel plow (Figure 2). Johnson et al. (1989) found that chisel plow corn and soybean had greater giant foxtail densities than moldboard plow and ridge-tillage. No-tillage was similar to chisel plow and moldboard plow, but greater than ridge-tillage. Species shifts have also been observed within the foxtail complex (Buhler and Mester, 1991). Different tillage systems were established in a field with a history of moldboard plowing and crop rotation with a mixed population of giant and green foxtail. After four years of continuous corn production, green foxtail was the dominant species in moldboard plow, while giant foxtail dominated in chisel plow and no-tillage.

Changes in the population dynamics of summer annual broadleaf species have been less consistent. The most common response among annual broadleaf species has been reduced densities of large-seeded species such as velvetleaf (Buhler and Daniel, 1988; Buhler and Oplinger, 1990), common cocklebur (*Xanthium strumarium* L.) (Wrucke and Arnold, 1985), and sicklepod (*Cassia obtusifolia* L.) (Banks et al., 1986) as tillage is reduced. For example, velvetleaf density with moldboard plowing was 130 plants m^{-2} compared with 60, 42, and 25 plants m^{-2} in chisel plow, ridge-tillage, and no-tillage, respectively (Buhler and Daniel, 1988). In soybean grown in rotation with corn (Buhler and Oplinger, 1990), velvetleaf continued to emerge through 60 days after soybean planting with moldboard plowing, while few velvetleaf emerged after 40 days after planting in chisel plow and no-tillage. Total velvetleaf densities were greater in moldboard plow than chisel plow and no-tillage (Figure 2).

The response of small-seeded summer annual broadleaf weed species to changes in tillage practices has been variable among weed species and experiments. This variable response is typified by common lambsquarters. In research conducted in soybean grown in rotation with corn (Buhler and Oplinger, 1990), common lambsquarters densities were affected by tillage at one location, but not at another location with similar edaphic and cropping histories. At the location where no-tillage had greater common lambsquarters densities than chisel plow and moldboard plow, the planting date was 3 to 5 days earlier than at the other location. In another study, common lambsquarters densities in corn were several times greater in chisel plow than no-tillage, ridge-tillage, or moldboard plow (Buhler, 1992). Johnson et al. (1989) observed no differences in common lambsquarters densities among tillage systems. Conversely, Putnam et al. (1983) observed common lambsquarters to decrease under no-tillage. Fawcett (1987) pointed out that species that germinate under cool soil conditions, such as common lambsquarters, may be prevalent in no-tillage fields prior to crop planting. This may allow them to suppress later emerging weeds or a significant portion of the population may be destroyed by tillage or herbicide prior to crop planting. Therefore, the effects of planting date and spring soil temperatures on common lambsquarters germination prior to crop planting probably caused the inconsistent responses among studies.

Redroot pigweed is another small-seeded summer annual broadleaf species that has responded inconsistently to changes in tillage systems. No-tillage and chisel plow systems had greater redroot pigweed densities than moldboard plowing at one location of an experiment conducted at two locations (Figure 3). At the other location, chisel plow had more redroot pigweed than the moldboard plow system and no-tillage was similar to both moldboard plow and chisel plow (Buhler and Oplinger, 1990). In another study, redroot pigweed densities in corn were similar in chisel plow and no-tillage, but both were several times greater than moldboard plow and ridge-tillage (Buhler, 1992). Forcella et al. (1994) concluded that redroot pigweed may be best suited to systems with moderate tillage because its seed are well adapted to shallow burial.

Soil disturbance patterns in ridge-tillage are different from other tillage systems and represent a unique environment for weed population development. After 7 to 8 years of standard chemical and mechanical weed control, 1500 to 3000 weed seed

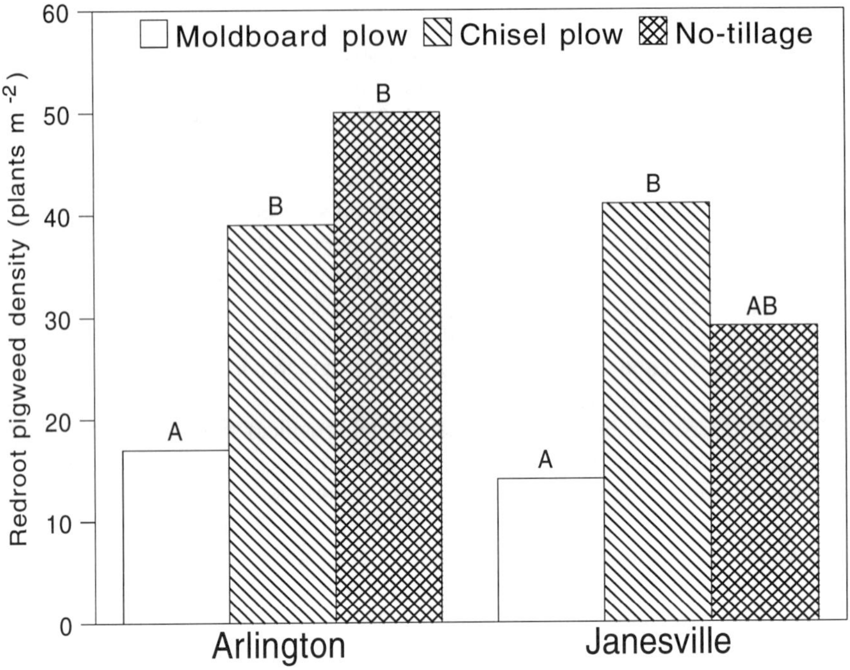

Figure 3. Effect of tillage systems on redroot pigweed densities 60 days after soybean planting at Arlington and Janesville, WI, averaged over two years. Bars with the same letter within a location do not differ using LSD (0.05). (Adapted from Buhler and Oplinger, 1990.)

m^{-2} were found in the top 10 cm of soil in continuous corn with ridge tillage whereas about two-thirds fewer seeds were found in conventional tillage (Forcella and Lindstrom, 1988b). Annual loss of weed seed from the soil through germination was from 3 to 12% in ridge-tillage and 11 to 43% in conventional tillage. Additions to the seed bank were supplied by late-emerging weeds whose germination was stimulated by cultivation, with up to 10 times more emergence and 140 times more seed production in ridge than in conventional tillage. Withholding herbicides for 1 year reduced yields of continuous corn by 10 to 27% in ridge-tillage and only 2 to 4% in conventional tillage. It was concluded that reducing seed production from late emerging weeds in ridge-tillage may aid in weed control.

In a related experiment in ridge-tillage (Forcella and Lindstrom, 1988a), soils harbored at least twice as many weed seed under continuous corn production than a corn/soybean rotation. Truncation of the ridges at the time of crop planting removed 31 to 37% of the weed seed from the ridges of continuous corn, and 80 to 100% of the seed from ridges of the corn/soybean rotation. From 14 to 36% of weed seed originally in ridges were lost through germination either on the ridge or in the furrow. Ridging the soil just before canopy closure stimulated germination of weed seed. The resulting weed population produced up to 1000 seed m^{-2} in continuous corn and about 100 seed m^{-2} in the corn/soybean rotation.

It is often difficult to determine whether changes in the population dynamics are due to biological properties of the weed species or differences in control efficacy. To address this issue, research has been conducted to separate various aspects of tillage/weed species/weed control efficacy interactions to elucidate the mechanisms of weed population shifts.

C. Weed Seed Bank and Seedling Biology

The primary source of arable response, summer annual weed species is the weed seed bank in the soil, with the largest source of seed being that produced in previous years in the same field (Cavers, 1983). Changes in the weed seed bank occur with time and weed management programs (Chauvel et al., 1989). Tillage is the primary cause of vertical seed movement in arable soils (Cousens and Moss, 1990; Roberts, 1963; Staricka et al., 1990). Since tillage systems affect weed management, weed seed production, and soil disturbance, changing tillage systems will change the distribution and density of weed seed in agricultural soils.

The effects of chisel and moldboard plowing on the vertical distribution of weed seed in the soil were evaluated using ceramic spheres (Staricka et al., 1990). Spheres, with similar size and density of weed seed, were found to depths of 12 cm following chisel plowing and 32 cm following moldboard plowing. In chisel plowed plots, 48% of the spheres were found within 4 cm of the soil surface compared to 4% in moldboard plow plots.

Moldboard plow plots had fewer weed seed in the upper 20 cm of soil than chisel plow or no-tillage after five years (Yenish et al., 1992). Moldboard plowing resulted in the most uniform distribution of seed over soil depths. In the no-tillage system, over 60% of all weed seed were found in the upper 1 cm of soil and few seed were found below 10 cm. The concentration of weed seed in no-tillage decreased logarithmically with increasing depth. In the chisel plow system, over 30% of the weed seed were in the upper 1 cm of soil, and seed concentration decreased linearly with depth. Pareja et al. (1985) found 85% of all weed seed in the upper 5 cm of soil in a reduced tillage system, but only 28% in the same zone in the moldboard plow system.

Seed and tracer distribution data (Pareja et al., 1985; Staricka et al., 1990; Yenish et al., 1992) substantiate differences in emergence depths of giant and green foxtail in different tillage systems in the field (Buhler and Mester, 1991). Mean seedling emergence depths in no-tillage were the least, followed by chisel plow and moldboard plow systems in two soil types over three years (Figure 4). At least 40% of the giant and green foxtail plants emerged from the upper 1 cm of soil in no-tillage compared with about 25% in chisel plow and less than 15% in moldboard plow plots. As many as 25% of the plants emerged from greater than 4 cm in moldboard plow compared to about 10% in chisel plow and less than 5% in no-tillage.

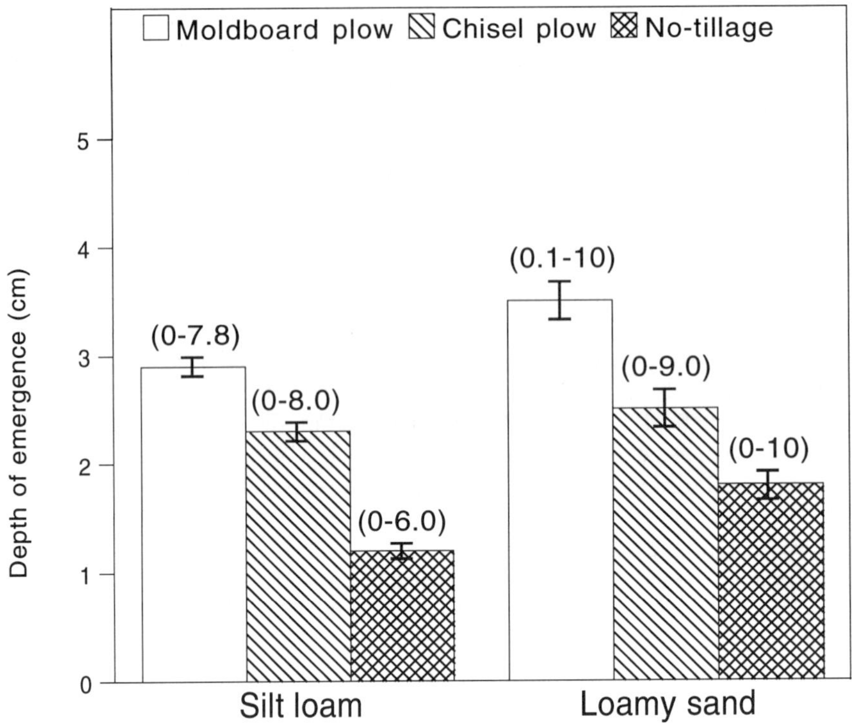

Figure 4. Effect of tillage systems on the depth of emergence of foxtail species in two soil types averaged over three years. Vertical bars are the standard error of the mean. Numbers in parentheses are the minimum and maximum emergence depths for each soil type by tillage system combination. (Adapted from Buhler and Mester, 1991.)

Forcella et al. (1994) proposed the use of a tillage index to assess the degree of soil mixing and surface residue incorporation with different tillage practices. A uniformly mixed soil surface layer (0 to 20 cm) has a tillage index of one. Tillage systems leaving a disproportionate number of seed on the soil surface have a tillage index less than one, whereas tillage systems that bury most of the seed in the lower half of the surface layer have a tillage index greater than one. Using the tillage index to assess the effect of several tillage systems on velvetleaf populations, it was concluded that velvetleaf density was not so much a function of tillage type, but rather a linear function of the degree of soil mixing associated with tillage. It was concluded that a tillage index allows more quantitative comparisons for weed population responses than commonly used tillage system designations.

Changes in seed depth in the soil and the corresponding differences in emergence depth may contribute to shifts among weed species under different tillage systems. In greenhouse research (Buhler, 1991), velvetleaf establishment from seed germinating on the soil surface was only 18% of seed planted 6 cm deep (Figure 5).

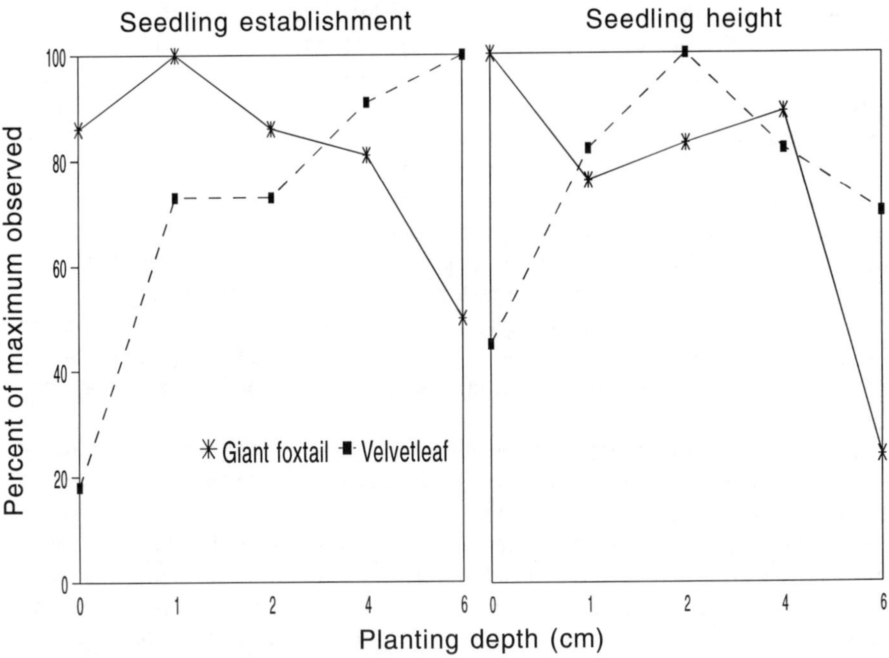

Figure 5. Effect of planting depth on the establishment and growth of velvetleaf and giant foxtail after 28 days in the greenhouse. (From Buhler, 1991.)

Seedlings from surface-placed seed were less vigorous and reached only 45% of the height of seedlings emerging from 2 cm deep. Giant foxtail seed germinating on the soil surface had an establishment percentage similar to seed planted 1 to 4 cm deep. Giant foxtail establishment was reduced by 50% when the seed were planted 6 cm deep. The seedlings established from surface-placed seed were also among the most vigorous.

Giant foxtail and velvetleaf seed germinated on a soil surface kept moist by mulch or frequent watering (Mester and Buhler, 1990; Mester and Buhler, 1991). All giant foxtail seedlings survived following germination on the soil surface; however, velvetleaf often failed to establish seedlings. Giant foxtail did not germinate at 5°C and failed to emerge from 6 cm deep within 21 days at 10°C. Increasing soil temperature above 10°C increased giant foxtail germination and emergence. Velvetleaf germinated at 5°C, but did not emerge within 21 days. Velvetleaf emerged within 21 days from soil depths of 2 to 6 cm at temperatures of 10, 15, and 20°C. The differential response to planting depth exhibited by velvetleaf and giant foxtail may be a major reason for the shifts observed in these species in the field when tillage practices are changed. However, soil temperature changes associated with conservation tillage systems did not appear to contribute to shifts in giant foxtail and velvetleaf populations.

Surface topography influenced the emergence location of velvetleaf in a field that had not been tilled the previous two growing seasons (Mester, 1988). No velvetleaf seedlings emerged from bare soil and the majority of the velvetleaf seedlings emerged from cracks in the soil. This emergence pattern was probably due to seedlings originating from seed buried by moldboard plowing prior to the conversion to no-tillage, or by seed that were washed into soil cracks and established sufficient seed-to-soil contact for germination. Another significant portion of the velvetleaf emerged from under surface residue. Surface residue increased soil moisture and reduced soil crusting compared with bare soil, thus providing adequate moisture conditions for germination for a longer period of time. These data suggest that the "safe-site" concept (Harper et al., 1965) should be extended below the soil surface to the depth where soil surface conditions influence germination conditions and that soil cracks may be a factor in weed emergence patterns.

Mester (1988) attempted to separate the effects of tillage and surface residue by establishing tilled and untilled plots with various levels of corn residue. Velvetleaf densities were greatest in tilled plots without residue. Velvetleaf densities decreased as residue levels were increased in both tilled and untilled plots. Redroot pigweed densities were greatest when the residue was removed from untilled plots, and the base level of residue reduced densities by up to 70% compared with no residue. Giant foxtail densities were several times greater in plots that were not tilled when averaged over residue levels. Base residue reduced giant foxtail densities by about 50% compared to no residue. These data indicate that soil disturbance plays a more important role in weed population shifts among tillage systems than surface residue. It was surprising that base levels of corn residue reduced weed densities, especially with untilled soil. Several studies have suggested that surface residue increased germination and establishment of seed at or near the soil surface (Harper, 1977; Mester and Buhler, 1990; Mester and Buhler, 1991). The effect of surface residue on weed populations appears to be complex and controlled by several interacting factors including type and quantity of residue, allelopathy, weed species, seed depth in the soil, soil type, and climatic conditions (Crutchfield et al., 1986; Forcella et al., 1994; Wicks et al., 1994).

The effect of conservation tillage systems on the population dynamics of summer annual weed species is complex and involves several factors. However, seed depth in the soil appears to be the most important factor. Weed species that have the ability to germinate and become established when the seed are at or near the soil surface have the greatest potential to increase under conservation tillage systems. These species tend to be small-seeded and are represented by small-seeded annual broadleaf and most annual grass species. Deep burial of seed of small-seeded species by moldboard plowing reduces germination and emergence. Conversely, seed of large-seeded species remain near the soil surface in conservation tillage systems, inhibiting establishment and preventing them from contributing to reinfestation following subsequent tillage operations (Lueschen and Andersen, 1980; Warnes and Andersen, 1984).

D. Winter Annual and Biennial Species

Weeds not previously present in fields have rapidly appeared following elimination of preplant tillage. Species most rapidly and commonly observed are winter annual and biennial species that have an inverse tillage response (Wicks et al., 1994). These weed species are unable to complete their life cycles in association with summer annual crops if the soil is disturbed prior to planting.

Changes in soil surface characteristics and disturbance patterns in many conservation tillage systems create an environment where winter annual and biennial species may become established and complete their life cycle (Bazzaz, 1990). Winter annual species germinate in the late summer and fall, survive the winter, and flower the following spring. Because of their rosette growth habit and ability to adjust photosynthetic responses to temperature, many winter annual species can grow even during the winter (Bazzaz, 1990; Regehr and Bazzaz, 1979). Therefore, some of the rosettes reach considerable size before they bolt in the spring. The winter annual species preempt space, suppress summer annual species, and can be very competitive with summer annual crops. Rapid growth early in the spring can also make winter annual species difficult to control with herbicides (Kapusta, 1979; Triplett, 1985; Wilson et al., 1985).

When the soil is tilled in the spring, rosettes of winter annual and biennial species recruited during the previous growing season are destroyed, allowing summer annual species to predominate (Bazzaz, 1990). However, new seedlings of winter annual and biennial species may emerge under the canopy of summer annual crop and weeds. Some of these seedlings survive and grow the following fall and winter and become established during the second year if tillage is not conducted. Thus, these species can rapidly infest fields (often during the first year) when tillage is eliminated. The rapid appearance of winter annual species does not appear to result from differences in innate competitive abilities relative to summer annual species, but from the timing of disturbance in the fields, which changes competitive hierarchies (Bazzaz, 1990). Either group of species could suppress the other depending on the timing of disturbance. Thus, changing tillage patterns can rapidly change the species balance of fields. However, in agronomic situations, management inputs such as herbicides and large seed banks often override these ecological processes.

Horseweed [*Conyza canadensis* (L.) Cronq.] is one of the most common and troublesome winter annual weeds in no-tillage systems (Bruce and Kells, 1990; Triplett, 1985) and an excellent example of an inverse tillage response winter annual species that infests summer annual crop production fields when preplant tillage is eliminated (Brown and Whitwell, 1988; Kapusta, 1979). Horseweed can infest no-tillage fields very rapidly because it is commonly found in much of the central and eastern United States in field edges and roadsides (Anonymous, 1981), produces many wind-disseminated seed (Bhowmik and Bekech, 1993; Fernald, 1950), is tolerant to many commonly used herbicides (Bruce and Kells, 1990), can become established under a wide-range of soil and climatic conditions (Fernald, 1950), and is well adapted to the no-tillage soil environment (Bhowmik and Bekech, 1993; Brown and Whitwell, 1988). Other winter annual species, such as downy brome (*Bromus tectorum* L.) and shepherd's purse [*Capsella bursa-pastoris* (L.) Medik], have also been observed to infest no-tillage fields (Triplett, 1985; Wicks et al., 1994).

E. Perennial Species

Changes in soil disturbance patterns associated with conservation tillage systems may be especially important for perennial weed species. Triplett (1985) suggested that perennial weed populations increase as tillage is reduced because the root system is not disturbed and most herbicides used to control annual weeds are not effective on established perennial plants. Wicks et al. (1994) also concluded that perennial weed species will become more important as tillage decreases. Froud-Williams et al. (1981) reviewed changes in weed floras associated with reduced tillage systems and cited several studies where perennial monocot and dicot species increased in the absence of tillage. They concluded that perennial monocots might be the greatest threat to the adoption of reduced tillage systems. They also suggested that reduced tillage would favor rhizome and stolon bearing perennials over annual species. While most research has indicated that reducing tillage favors perennial species, Derkson et al. (1993) found that an increase of perennial species did not generally occur with reduced tillage in several fields in Saskatchewan.

Triplett and Lytle (1972) found that hemp dogbane (*Apocynum cannabinum* L.) became a significant problem in some no-tillage plots after several years of continuous corn, while little or no hemp dogbane occurred in tilled areas. After establishment, hemp dogbane was not affected by herbicides used to control annual weeds, and populations expanded in the absence of competition from other weed species. In these studies, Canada thistle [*Cirsium arvense* (L.) Scop.] and dandelion (*Taraxacum officinale* Weber in Wiggers) also increased in untilled compared to tilled plots. Pollard and Cussans (1976) also noted that perennial species such as dandelion, field bindweed (*Convolvulus arvensis* L.), and quackgrass [*Elytrigia repens* (L.) Nevski] became more prevalent in no-tillage compared to systems that included preplant tillage. Foxtail barley (*Hordeum jubatum* L.) has also been observed to increase after several years of no-tillage production (Donald, 1990).

Impacts of tillage and cropping patterns on perennial weed populations in corn and soybean production were evaluated over a 14-year period (Buhler et al., 1994b). By the end of the experiment, continuous corn with no-tillage had greater hemp dogbane densities than other crop rotation by tillage system combinations (Figure 6). Field bindweed developed primarily in the corn/soybean rotations with the greatest densities occurring in no-tillage. Greater and more diverse populations of perennial weeds developed in conservation tillage systems than in the moldboard plow system. However, practices used to control annual weeds, climatic conditions, and other management practices interacted with tillage to regulate perennial weed populations.

In most studies, more diverse and dense populations of perennial weeds developed in conservation tillage systems than where moldboard plowing was conducted. Increases in perennial weed populations varied among studies because

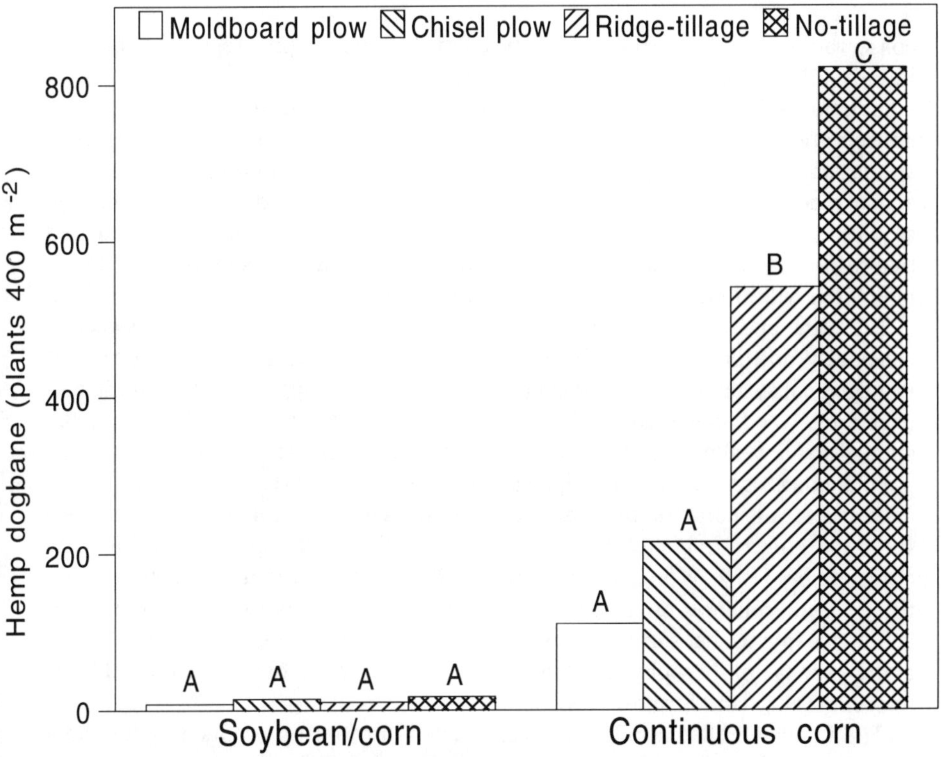

Figure 6. Effect of tillage systems and crop rotation on hemp dogbane densities after 14 growing seasons at Nashua, IA. Bars with the same letter within a rotation do not differ using LSD (0.05). (From Buhler et al., 1994b.)

tillage interacted with weed management practices, environment, and initial perennial weed populations. Therefore, the general statement that perennial weeds increase as tillage decreases must be made with caution. Management practices such as selective application of glyphosate [N-(phosphonomethyl)glycine] to tall-growing perennial weeds during soybean production (Figure 6) and interrow cultivation helped manage perennial species regardless of tillage system (Buhler et al., 1994b). Conservation tillage systems often increase densities and species diversity of perennial weeds, but management practices such as crop rotation, selective application of glyphosate or other herbicides, and interrow cultivation can prevent perennial weed species from increasing to levels that reduce crop yields or increase production costs.

IV. Summary, Conclusions, and Research Needs

Although general trends in weed population dynamics in conservation tillage have been observed, it is important to note that location, environment, the type of conservation tillage system used, and weed management inputs regulate responses in individual fields. For example, Derkson et al. (1993) found that an increased association of perennial and annual grass species did not generally occur with conservation tillage in several fields in Saskatchewan. However, wind-dispersed species and volunteer crops were associated with conservation tillage and summer annual dicots with conventional tillage, but exceptions occurred. More importantly, species responded differently among sites or within a site over time. This indicated that different weed management systems and cultural practices used on the fields interacted with tillage to regulate weed population shifts.

Effective and economical weed management in conservation tillage systems, especially no-tillage, is a challenge. Due to the changes in the biological and management systems, it is important to take an organized approach to planning and executing weed management in conservation tillage systems. Assessing the strengths and weaknesses of the available control options relative to expected weed populations is essential when designing weed management systems. Heavy reliance on herbicides is common in conservation tillage systems. Reliance on herbicides for weed control may result in establishment of resistant species and biotypes (Chancellor, 1979; Worsham and Lewis, 1985). Long-term use of conservation tillage with reliance on herbicides for weed control favors establishment of geophytes (plants that survive unfavorable climatic conditions as subsurface buds) at the expense of therophytes (plants that survive unfavorable climatic conditions as seeds) (Haas and Streibig, 1982).

The intensive tillage traditionally conducted in row crop production has had a profound effect on weeds and weed management. Effective, economical, and environmentally sound weed management in conservation tillage systems over the long-term will require integration of new information with established principles of weed management. Crop and tillage rotations, judicious use of tillage, and innovative application methods for herbicides will aid weed management in conservation tillage systems. New technologies are also needed to deal with the altered ecosystems created by conservation tillage production systems. New control options, such as biological agents and allelopathic cover crops, would provide useful new management tools. Many weed species and weed control tactics behave differently as tillage is reduced or eliminated. These changes must be taken into consideration to develop economically and environmentally sound weed management systems. It is essential that crop producers and pest managers consider the idiosyncrasies of the biological interactions among weeds, crops, residues, and environment as they develop weed management plans for conservation tillage systems.

Acknowledgments

The author thanks the researchers whose work contributed to this review, especially former graduate students and technicians. The assistance of Keith A. Kohler in preparing and reviewing this manuscript is appreciated.

References

Aase, J.K. and D.L. Tanaka. 1987. Soil water evaporation comparisons among tillage practices in the northern great plains. *Soil Sci. Soc. Am. J.* 51:436-440.

Al-Darby, A.M. and B. Lowery. 1987. Seed zone soil temperature and early corn growth with three conservation tillage systems. *Soil Sci. Soc. Am. J.* 51:768-774.

Andraski, B.J., D.H. Mueller, and T.C. Daniel. 1985. Effects of tillage and rainfall simulation date on water and soil losses. *Soil Sci. Soc. Am. J.* 49:1512-1517.

Anonymous. 1981. p. 244. In: *Weeds of the North Central States*. North Cent. Reg. Res. Public. No. 281., Univ. of Illinois, Champaign. 303 pp.

Anonymous. 1982. p. 10-12. In: *Resource Conservation Glossary*. Soil Conserv. Soc. Am., Ankeny, IA. 65 pp.

Banks, P.A. and E.L. Robinson. 1982. The influence of straw mulch on the soil reception and persistence of metribuzin. *Weed Sci.* 30:164-168.

Banks, P.A., T.N Tripp, J.W. Wells, and J.E. Hammel. 1986. Effects of tillage on sicklepod (*Cassia obtusifolia*) interference with soybeans (*Glycine max*) and soil water use. *Weed Sci.* 34:143-149.

Bauman, T.T. and M.A. Ross. 1983. Effect of three tillage systems on the persistence of atrazine. *Weed Sci.* 31:423-426.

Bazzaz, F.A. 1990. Plant-plant interactions in successional environments. p. 239-263. In: J.B. Grace and D. Tilman (eds.), *Perspectives on Plant Competition*. Academic Press, Inc., San Diego, CA.

Bhowmik, P.C. and M.M. Bekech. 1993. Horseweed (*Conyza canadensis*) seed production, emergence, and distribution in no-tillage and conventional-tillage corn (*Zea mays*). *Agron. (Trends in Agric. Sci.)* 1:67-71.

Brown, S.M. and T. Whitwell. 1988. Influence of tillage on horseweed, *Conyza canadensis*. *Weed Technol.* 2:269-270.

Bruce, J.A. and J.J. Kells. 1990. Horseweed (*Conyza canadensis*) control in no-tillage soybean with preplant and preemergence herbicides. *Weed Technol.* 4:642-647.

Buhler, D.D. 1991. Influence of tillage systems on weed population dynamics and control in the northern corn belt of the United States. *Advances in Agron. (Council of Sci. Integ., Kaithamukki, Trivandrum, India)* 1:51-60.

Buhler, D.D. 1992. Population dynamics and control of annual weeds in corn (*Zea mays*) as influenced by tillage systems. *Weed Sci.* 40:241-248.

Buhler, D.D. and T.C. Daniel. 1988. Influence of tillage systems on giant foxtail (*Setaria faberi*) and velvetleaf (*Abutilon theophrasti*) population and control in corn (*Zea mays*). *Weed Sci.* 36:642-647.

Buhler, D.D. and J.C. Mercurio. 1988. Vegetation management and corn growth and yield in untilled mixed-species perennial sod. *Agron. J.* 80:454-462.

Buhler, D.D. and T.C. Mester. 1991. Effect of tillage systems on the emergence depth of giant foxtail (*Setaria faberi*) and green foxtail (*Setaria viridis*). *Weed Sci.* 39:200-203.

Buhler, D.D. and E.S. Oplinger. 1990. Influence of tillage systems on annual weed densities and control in solid-seeded soybean (*Glycine max*). *Weed Sci.* 38:158-164.

Buhler, D.D., B.D. Philbrook, and E.S. Oplinger. 1990. Velvetleaf and giant foxtail control for solid-seeded soybean in three tillage intensities. *J. Prod. Agric.* 3:302-308.

Buhler, D.D., J.D. Doll, R.T. Proost, and M.R. Visocky. 1994a. Interrow cultivation to reduce herbicide use in corn following alfalfa without tillage. *Agron. J.* 86:66-72.

Buhler, D.D., D.E. Stoltenberg, R.L. Becker, and J.L. Gunsolus. 1994b. Perennial weed populations after 14 years of variable tillage and cropping practices. *Weed Sci.* 42:205-209.

Buhler, D.D., J.D. Doll, R.T. Proost, and M.R. Visocky. 1995. Integrating mechanical weeding with reduced herbicide use in conservation tillage corn production systems. *Agron. J.* 87:507-512.

Burnside, O.C. 1993. Weed science-the step child. *Weed Technol.* 7:515-518.

Cavers, P.B. 1983. Seed demography. *Can. J. Bot.* 61:3678-3590.

Chancellor, R.J. 1979. The long-term effects of herbicides on weed populations. *Ann. Appl. Biol.* 91:141-144.

Chauvel, B., J. Gasquez, and H. Darmency. 1989. Changes of weed seed bank parameters according to species, time and environment. *Weed Res.* 29:213-219.

Cousens, R. and S.R. Moss. 1990. A model of the effects of cultivation on the vertical distribution of weed seeds within the soil. *Weed Res.* 30:61-70.

Crutchfield, D.A., G.A. Wicks, and O.C. Burnside. 1986. Effect of winter wheat (*Triticum aestivum*) straw mulch level on weed control. *Weed Sci.* 34:110-114.

Dao, T.H. 1987. Crop residue and management of annual grass weeds in continuous no-till wheat (*Triticum aestivum*). *Weed Sci.* 35:395-400.

Derkson, D.A., G.P. Lafond, A.G. Thomas, H.A. Loeppky, and C.J. Swanton. 1993. Impact of agronomic practices on weed communities: tillage systems. *Weed Sci.* 41:409-417.

Doersch, R.E. and D.D. Buhler. 1988. Weed management in conservation tillage corn production. *Univ. Wis. Coop. Ext. Serv. Bull.* A3425.

Donald, W.W. 1990. Primary tillage for foxtail barley (*Hordeum jubatum*) control. *Weed Technol.* 4:318-321.

Doran, J.W. and D.M. Linn. 1994. Microbial ecology of conservation management systems. p. 1-28. In: J.L. Hatfield and B.A. Stewart (eds.), *Soil Biology: Effects on Soil Quality, Adv. Soil Sci.*, Lewis Publishers, Boca Raton, FL.

Erbach, D.C. and W.G. Lovely. 1975. Effect of plant residue on herbicide performance in no-tillage corn. *Weed Sci.* 23:512-515.

Evans, R.A. and J.A. Young. 1972. Microsite requirements for establishment of annual rangeland weeds. *Weed Sci.* 20:350-356.

Fawcett, R.S. 1987. Overview of pest management for conservation tillage systems. p. 19-38. In: T.J. Logan, J.M. Davidson, J.L. Baker, and M.R. Overcash (eds.), *Effects of Conservation Tillage on Groundwater Quality*. Lewis Publishers, Chelsea, MI.

Fawcett, R.S., B.R. Christensen, and D.P. Tierney. 1994. The impact of conservation tillage on pesticide runoff into surface water: A review and analysis. *J. Soil Water Conserv.* 49:126-135.

Fernald, M.L. 1950. p. 1447. In: *Gray's Manual of Botany, 8th ed.*, American Book Co., New York.

Forcella, F. and M.J. Lindstrom. 1988a. Movement and germination of weeds in ridge-till crop production systems. *Weed Sci.* 36:56-59.

Forcella, F. and M.J. Lindstrom. 1988b. Weed seed populations in ridge and conventional tillage. *Weed Sci.* 36:500-502.

Forcella, F., D.D. Buhler, and M.E. McGiffen. 1994. Pest management and crop residues. p. 173-189. In: J.L. Hatfield and B.A. Stewart (Eds.), *Crop Residue Management, Adv. Soil Sci.*, Lewis Publishers, Boca Raton, FL.

Freed, B.E., E.S. Oplinger, and D.D. Buhler. 1987. Velvetleaf control for solid-seeded soybeans in three corn residue management systems. *Agron. J.* 79:119-123.

Froud-Williams, R.J., R.J. Chancellor, and D.S.H. Drennen. 1981. Potential changes in weed flora associated with reduced-cultivation systems in cereal production in temperate regions. *Weed Res.* 21:99-109.

Froud-Williams, R.J., R.J. Chancellor, and D.S.H. Drennen. 1983. Influence of cultivation regime upon buried weed seeds in arable cropping systems. *J. Applied Ecol.* 20:199-206.

Frye, W.W. 1984. Energy requirements in no-tillage. p. 127-151. In: R.E. Phillips and S.H. Phillips (eds.), *No-Tillage Agriculture: Principles and Practices*. Van Nostrand Reinhold, New York.

Gebhardt, M.R., T.C. Daniel, E.E. Schweizer, and R.R. Allmaras. 1985. Conservation tillage. *Science* 230:625-630.

Ghadiri, H., P.J. Shea, and G.A. Wicks. 1984. Interception and retention of atrazine by wheat (*Triticum aestivum*) stubble. *Weed Sci.* 32:24-27.

Gianessi, L.P. and C. Puffer. 1991. Herbicide use in the United States: National summary report, *Resources for the Future*, Washington, D.C. 127 pp.

Graber, L.F. 1928. Evidence and observations on establishing sweet clovers in permanent bluegrass pastures. *J. Am. Soc. Agron.* 20:1197-1205.

Griffith, D.R., J.V. Mannering, H.M. Galloway, S.D. Parsons, and C.B. Rickey. 1973. Effect of eight tillage-planting systems on soil temperature, percent stand, plant growth, and yield of corn on five Indiana soils. *Agron. J.* 65:321-326.

Griffith, D.R., D.B. Mengel, S.D. Parsons, J.L. Williams Jr., D.H. Scott, F.T. Turpin, and D.H. Doster. 1984. A guide to no-till sod planting in Indiana. *Purdue Univ. Ext. Serv. Bull.* ID-133.

Gunsolus, J.L. 1990. Mechanical and cultural weed control in corn and soybeans. *Am. J. Alter. Agric.* 5:114-119.

Haas, H. and J.C. Streibig. 1982. Changing patterns of weed distribution as a result of herbicide use and other agronomic factors. p. 57-79. In: H.M. LeBaron and J. Gressel (eds.), *Herbicide Resistance in Plants*. John Wiley and Sons, New York.

Harper, J.L. 1977. Chapter 5 In: *Population Biology of Plants*. Academic Press, New York.

Harper, J.L., J.T. Williams, and G.R. Sager. 1965. The behavior of seeds in soil. I. The heterogeneity of soil surfaces and its role in determining the establishment of plants from seed. *J. Ecol.* 53:273-286.

Horowitz, M. and R.B. Taylorson. 1984. Hardseededness and germinability of velvetleaf (*Abutilon theophrasti*) as affected by temperature and moisture. *Weed Sci.* 32:111-115.

Johnson, M.D. and B. Lowery. 1985. Effect of three conservation tillage practices on soil temperature and thermal properties. *Soil Sci. Soc. Am. J.* 49:1547-1552.

Johnson, M.D., B. Lowery, and T.C. Daniel. 1984. Soil moisture regimes of three conservation tillage systems. *Trans. Am. Soc. Agric. Engin.* 27:1385-1390.

Johnson, M.D., D.L. Wyse, and W.E. Lueschen. 1989. The influence of herbicide formulation on weed control in four tillage systems. *Weed Sci.* 37:239-249.

Kapusta, G. 1979. Seedbed tillage and herbicide influence on soybean (*Glycine max*) weed control and yield. *Weed Sci.* 27:520-526.

Kells, J.J., C.E. Rieck, R.L. Blevins, and W.M. Muir. 1980. Atrazine dissipation as affected by surface pH and tillage. *Weed Sci.* 28:101-104.

Koskinen, W.C. and C.G. McWhorter. 1986. Weed control in conservation tillage. *J. Soil and Water Conserv.* 41:365-370.

Lewis, W.M. 1985. Weed control in reduced-tillage soybean production. p. 41-50. In: A.F. Wiese (ed.), *Weed Control in Limited Tillage Systems*. Weed Sci. Soc. Am., Champaign, IL.

Liebman, M. and E. Dyck. 1993a. Weed management-A need to develop ecological approaches. *Ecol. Applic.* 3:39-41.

Liebman, M. and E. Dyck. 1993b. Crop rotation and intercropping strategies for weed management. *Ecol. Applic.* 3:92-122.

Lueschen, W.E. and R.N. Andersen. 1980. Longevity of velvetleaf (*Abutilon theophrasti*) seed in soil under agricultural practices. *Weed Sci.* 28:341-346.

Mannering, J.V., D.L. Schertz, and B.A. Julian. 1987. Overview of conservation tillage. p. 3-17. In: T.J. Logan, J.M. Davidson, J.L. Baker, and M.R. Overcash (eds.), *Effects of Conservation Tillage on Groundwater Quality: Nitrates and Pesticides*. Lewis Publishers, Inc., Chelsea, MI.

Mester, T.C. 1988. Effects of tillage systems on weed ecology and control in corn production. *Ph.D. Dissertation*. Univ. Wisconsin, Madison. 253 pp.

Mester, T.C. and D.D. Buhler. 1990. Effects of seed depth on velvetleaf (*Abutilon theophrasti*) seedling development and response to cyanazine. *Weed Sci.* 38:34-38.

Mester, T.C. and D.D. Buhler. 1991. Effects of soil temperature, seed depth, and cyanazine on giant foxtail (*Setaria faberi*) and velvetleaf (*Abutilon theophrasti*) seedling development. *Weed Sci.* 39:204-209.

Mulder, T.A. and J.D. Doll. 1993. Integrating reduced herbicide use with mechanical weeding in corn (*Zea mays*). *Weed Technol.* 7:382-389.

Pareja, M.R. and D.W. Staniforth. 1985. Soil-seed microsite characteristics in relation to seed germination. *Weed Sci.* 33:190-193.

Pareja, M.R., D.W. Staniforth, and G.P Pareja. 1985. Distribution of weed seed among soil structural units. *Weed Sci.* 33:182-189.

Pollard, F. and G.W. Cussans. 1976. The influence of tillage on the weed flora of four sites sown to successive crops of spring barley. *Proc. 13th Brit. Weed Control Conf.* 13:1019-1028.

Pollard, F., S.R. Moss, G.W. Cussans, and R.J. Froud-Williams. 1982. The influence of tillage on the weed flora in a succession of winter wheat crops on a clay loam soil and a silt loam soil. *Weed Res.* 22:129-136.

Putnam, A.R., J. DeFrank, and J.P. Barnes. 1983. Exploitation of allelopathy for weed control in annual and perennial cropping systems. *J. Chem. Ecol.* 9:1001-1010.

Regehr, D.L. and F.A. Bazzaz. 1979. The population dynamics of *Eriergon canadensis*, a successional winter annual. *J. Ecol.* 67:923-933.

Roberts, H.A. 1963. Studies on the weeds of vegetable crops. III. Effect of different primary cultivations on the weed seeds in the soil. *J. Ecol.* 51:83-95.

Roberts, H.A. and P.M. Feast. 1973. Emergence and longevity of seeds of annual weeds in cultivated and undisturbed soil. *J. Applied Ecol.* 10:133-143.

Robinson, R.G. and R.S. Dunham. 1956. Pre-planting tillage for weed control in soybeans. *Agron. J.* 48:493-495.

Sheldon, J.C. 1974. The behavior of seeds in soil. III. The influence of seed morphology and the behavior of seedlings on the establishment of surface-lying seeds. *J. Ecol.* 62:47-65.

Sims, G.K., D.D. Buhler, and R.F. Turco. 1994. Residue management impact on the environment. p. 77-98. In: P.W. Unger (ed.), *Managing Agricultural Residues*. Lewis Publishers, Boca Raton, FL.

Sprague, M.A. 1986. Overview. p. 1-18. In: M.A. Sprague and G.B. Triplett (eds.), *No-Tillage and Surface Tillage Agriculture*. John Wiley, New York.

Springman, R., D. Buhler, R. Schuler, D. Mueller, and J. Doll. 1989. Row crop cultivators for conservation tillage systems. *Univ. of WI Extension Pub.* A3483.

Staricka, J.A., P.M. Burford, R.R. Allmaras, and W.W. Nelson. 1990. Tracing the vertical distribution of simulated shattered seeds as related to tillage. *Agron. J.* 82:1131-1134.

Stoller, E.W. and L.M. Wax. 1973. Periodicity of germination and emergence of some annual weeds. *Weed Sci.* 21:574-580.

Thomas, G.W. 1986. Mineral nutrition and fertilizer placement. p. 93-116. In: M.A. Sprague and G.B. Triplett (eds.), *No-Tillage and Surface Tillage Agriculture*. John Wiley, New York.

Tilman, D. 1990. Mechanisms of plant competition for nutrients: the elements of a predictive theory of competition. p. 146-173. In: J. Grace and D. Tilman (eds.), *Perspectives on Plant Competition*. Academic Press, Inc., San Diego, CA.

Triplett, G.B., Jr. 1985. Principles of weed control for reduced-tillage corn production. p. 26-40. In: A.F. Wiese (ed.), *Weed Control in Limited Tillage Systems*. Weed Sci. Soc. Am., Champaign, IL.

Triplett, G.B., Jr. and G.D. Lytle. 1972. Control and ecology of weeds in continuous corn without tillage. *Weed Sci.* 20:453-457.

Triplett, G.B., Jr. and D.M. VanDoren, Jr. 1977. Agriculture without tillage. *Sci. Am.* 236:28-33.

Warnes, D.D. and R.N. Andersen. 1984. Decline of wild mustard (*Brassica kaber*) seeds in soil under various cultural and chemical practices. *Weed Sci.* 32:214-217.

Wicks, G.A., O.C. Burnside, and W.L. Felton. 1994. Weed control in conservation tillage systems. p. 211-244. In: P.W. Unger (ed.), *Managing Agricultural Residues*. Lewis Publishers, Boca Raton, FL.

Wiese, A.M. and L.K. Binning. 1987. Calculating the threshold temperature of development for weeds. *Weed Sci.* 35:177-179.

Williams, J.L. and G.A. Wicks. 1978. Weed control problems associated with crop residue systems. p. 165-172. In: W.R. Oshwald (ed.), *Crop Residue Management Systems*. ASA Special Publication 31, Madison, WI.

Wilson, H.P., T.E. Hines, R.R. Bellinder, and J.A. Grande. 1985. Comparisons of HOE-39866, SC-0224, paraquat, and glyphosate in no-till corn (*Zea mays*). *Weed Sci.* 33:531-536.

Worsham, A.D. and W.M. Lewis. 1985. Weed management: Key to no-tillage crop production. p. 177-188. In: W.L. Hargrove, F.C. Boswell, and G.W. Langdall (eds.), *Proc. South. Region No-Till Conf.*, Southern Weed Sci. Soc., Dallas, TX.

Wrucke, M.A. and W.E. Arnold. 1985. Weed species distribution as influenced by tillage and herbicides. *Weed Sci.* 33:853-856.

Wyse, D.L. 1992. Future of weed science research. *Weed Technol.* 6:162-165.

Yenish, J.P., J.D. Doll, and D.D. Buhler. 1992. Effects of tillage on vertical distribution and viability of weed seed in soil. *Weed Sci.* 40:429-433.

Cover Crops, Smother Plants, and Weed Management

John R. Teasdale

I. Introduction

Cover crops are included in cropping systems because of long- and short-term improvements in soil fertility and crop performance. Long-term improvements may result from erosion control, improved soil tilth, and increased soil organic matter (Smith et al., 1987). Short-term influences affecting radiation balance, soil temperature, soil moisture, nutrient availability, runoff/infiltration, crop establishment, and weed/pest populations may influence current crop performance. This paper will review influences on weed populations and discuss cover crop applications to weed management. No-tillage systems will be emphasized because of greater opportunity for cover crops to interact with weed germination and establishment at the surface of soils.

The goal of using cover crops for weed control is replacing an unmanageable weed population with a manageable cover crop species. This is accomplished by adjusting the phenology of the cover crop to preempt niches occupied by weed populations. Two major types of cover crops will be addressed in this review: off-season cover crops (usually winter annuals in temperate climates) and smother crops (a plant grown during part or all of the cropping season). The objective of using an off-season winter annual cover crop for weed control is production of sufficient residue to create an unfavorable environment for weed germination and establishment. The objective of using a smother crop is displacement of weeds by competition from an established cover crop.

II. Off-Season Winter Annual Cover Crops

Winter annual cover crops are planted in late summer or fall, become established before winter, and produce most biomass during early spring before planting a summer crop. Typically, the cover crop is killed or senesces when the summer crop is planted leaving a layer of residue on the soil surface. There are many legume or cereal species that can be used in coastal or southern areas (Holderbaum et al., 1990; Sarrantonio, 1994). Hairy vetch (*Vicia villosa* Roth) is the most winter hardy legume and can be grown in most areas of the U.S. except the northern-most states. Rye (*Secale cereale* L.) is the most winter hardy cereal species and will overwinter anywhere in the U.S. Winter annual cover crops will be most useful in areas where there is a sufficient establishment period in the fall and soil moisture is not limiting in the spring.

A. Weed Control by Cover Crop Residue

There are numerous reports in the literature of cover crops suppressing or controlling weeds. There are fewer reports that demonstrate the effect of cover crops on weeds in field situations and in the absence of herbicides or other confounding treatments. This review will focus on the capacity of cover crop residue without herbicides to reduce weed density and biomass in agroecosystems. In addition, there is informative literature on the influence of plant litter on plant communities in natural systems (Facelli and Pickett, 1991a).

1. Weed Density and Biomass

Early research was done in Michigan and North Carolina with small grain cover crops desiccated by contact herbicides. Rye, wheat (*Triticum aestivum* L. emend. Thell.), and barley (*Hordeum vulgare* L.) residue reduced total weed density and biomass and were superior to various frost-killed summer annual grasses (Putnam et al., 1983). Desiccated rye residue reduced redroot pigweed (*Amaranthus retroflexus* L.) and common lambsquarters (*Chenopodium album* L.) density compared to a no-tillage control (Shilling et al., 1985).

Natural levels of residue biomass found at the time of planting are used for most field research with cover crops. These rates of residue can vary widely among experiments or even within the same experiment making it difficult to compare weed suppression by different cover crops. Experiments with a range of residue rates established by the investigator can provide a quantitative relationship between weed density and residue biomass. Crutchfield et al. (1985) showed that greater than 3400 to 5100 kg/ha of wheat residue was necessary to reduce weed density and biomass by more than 50% in herbicide-free plots. Mohler and Teasdale (1993) showed an exponential decline in total weed density with an increase in rye or hairy vetch residue. A weed density decline of 50% was achieved at residue rates of approximately 1500 to 3000 kg/ha depending on location. At equivalent residue biomass levels, hairy vetch and rye suppressed weeds equivalently (Mohler and Teasdale, 1993). Occasional increases in weed density or biomass were observed at low residue rates in both of these studies.

Response to residue tends to vary according to weed species (Figure 1). Velvetleaf (*Abutilon theophrasti* Medicus), a relatively large-seeded, light-insensitive weed species, was influenced less than small-seeded, light-sensitive species such as common lambsquarters, redroot pigweed, and various annual grasses by rye and hairy vetch residue (Mohler and Teasdale, 1993). Establishment of small-seeded vegetable crops was inhibited more than that of large-seeded crops by residues of various grass species (Putnam and DeFrank, 1983). Litter in old-field communities reduced the density of annual more than perennial species (Monk and Gabrielson, 1985; Carson and Peterson, 1990; Facelli and Pickett, 1991b). Generally, annual species that are small-seeded and have a light requirement for germination are most sensitive to surface residue.

Cover crop residue often influences weed biomass less than weed density, probably because weeds emerging through residue can compensate for reduced density by producing greater biomass per plant. Rye, wheat, and barley residue reduced weed density by an average of 90% relative to a no-tillage control whereas weed biomass was reduced only by 39% (Putnam et al., 1983). Facelli and Pickett (1991b) demonstrated that giant foxtail (*Seteria faberi* Herrm.) density was reduced by oak (*Quercus alba* L.) leaf litter but biomass at the end of the year was unaffected. Despite reductions in emergence by residue, the number of emerged weed seedlings usually will be sufficient to produce a competitive level of weed biomass. In general, cover crop residue can be expected to provide early-season weed suppression but not full-season weed control.

2. Weed Emergence Models

Bioeconomic models are being developed that predict weed emergence and weed control requirements based on knowledge of the weed seedbank (Forcella et al., 1992). Various approaches have been taken for predicting weed emergence. Models can predict accurately the timing and extent of velvetleaf (Forcella, 1993) and common lambsquarters (Harvey and Forcella, 1993) emergence based on soil temperature and moisture data. Alm et al. (1993) reported a model that added seedling elongation functions to account for emergence rates from different depths. Mohler (1993) developed a model for predicting emergence for successive seasons as a function of seedling emergence rates by depth, seed survival by depth, and seed movement by tillage implements.

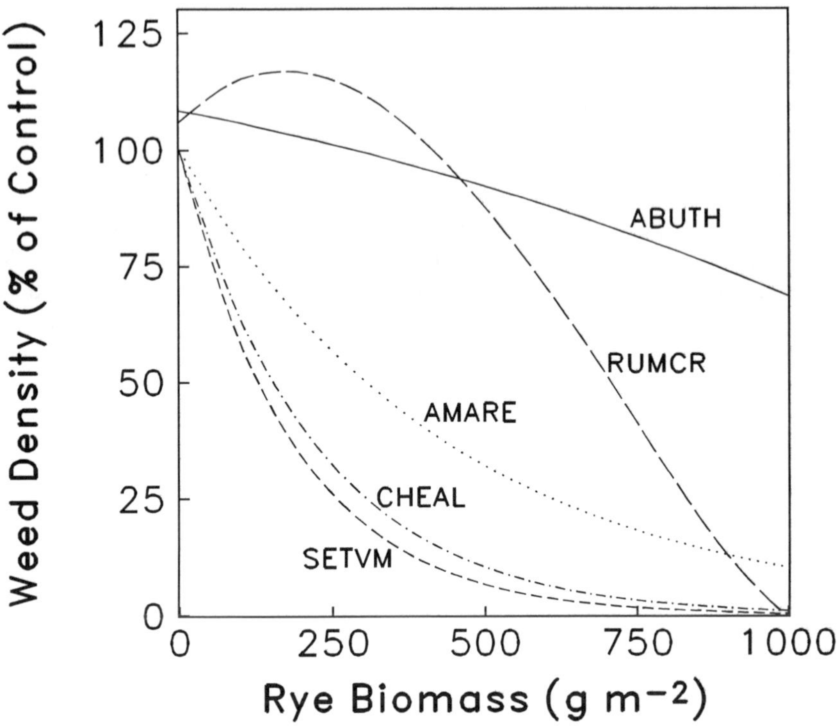

Figure 1. Density of selected weed species 4 to 5 weeks after establishing various rates of rye mulch. ABUTH = velvetleaf (solid line), AMARE = redroot pigweed (dots), CHEAL = common lambsquarters (dash-dot), RUMCR = curly dock (long dash), and SETVM = giant green foxtail (short dash). (Adapted from Mohler and Teasdale, 1993.)

These models assume a homogeneous soil medium without surface residue. There is a need to extend these models to account for residue layers on and/or within the soil. Research is needed to determine 1) what properties of residues (e.g., biomass, thickness, density, surface area index) define weed suppressiveness, 2) whether critical properties defining suppressiveness vary with cover crop species or weed species, 3) how residue and seed depth interact to determine emergence, and 4) how residue influences dormancy and survival of weed seeds.

Most research documenting the influence of cover crops on weed populations has been conducted with natural weed populations. These populations usually are spatially variable as well as variable in relative abundance of species. This confounds attempts to quantitate response to cover crops as well as differences among species. To develop accurate parameters for emergence models, seedbanks should be determined in advance or known densities of weed seed should be established at known depths with known germination requirements.

3. Weed Population Dynamics

Most research has focused on influences of cover crop residue on density and biomass of established weed populations. Little research has been conducted on other processes in the dynamics of weed populations such as seed production, seed dispersal, and seed survival in soils. Research with natural systems has demonstrated that plant litter can influence the community structure of various ecosystems (Facelli and Pickett, 1991a). Monk and Gabrielson (1985) found that pioneer species were affected by litter more than successional species. The timing of litter removal or addition influenced both species richness and abundance (Carson and Peterson, 1990). Facelli and Pickett (1991b) showed that litter reduced seed production by giant foxtail while Facelli and Facelli (1993) found that timing of litter removal controlled the dominance of giant foxtail in successive generations.

These results suggest that manipulation of cover crops can have long-term impacts on weed population dynamics. More long-term research is needed to understand the consequences of addition and degradation of cover crops in multi-year crop rotations. Attention should be focused on defining the impact of cover crops on important rate-defining steps in the life-cycle of weeds. This knowledge will help characterize how to use cover crops most effectively to disrupt the succession of important weed species.

B. Factors Controlling Weed Germination and Establishment

Cover crop residue influences many factors that control weed germination and establishment. Most of these factors relate to modifications of the radiation environment by an organic mulch. Interception and reflection of short-wave radiation by mulch elements reduce the quantity of light available to activate phytochrome-mediated germination and promote seedling growth, reduce heat absorption by soils during the day, and reduce evaporative soil water loss.

1. Light

Transmittance of photosynthetic photon flux density (PPFD) declined according to an exponential decay function of residue biomass (Facelli and Pickett, 1991b; Teasdale and Mohler, 1993). Similar decay functions described both weed density and light transmittance (Teasdale and Mohler, 1992) suggesting that light transmittance may be a good indicator of the capacity of mulches to suppress weed emergence. This would be expected since most weed species require light to activate a phytochrome-mediated germination process (Cone and Kendrick, 1986).

Table 1. Light extinction by various plant residues

Residue	Extinction coefficient[a]	Biomass (g m^{-2}) required for 90% extinction
Oak leaves[b]	0.0237	97
Giant foxtail[b]	0.0059	390
Rye[c]	0.0046	501
Hairy vetch[c]	0.0042	548
Goldenrod[b]	0.0018	1279

[a]From exponential decay function $y = e^{-bx}$ where Y = fraction transmittance, x = residue biomass, and b = extinction coefficient.
[b]From Facelli and Pickett, 1991b.
[c]From Teasdale and Mohler, 1993.

Other factors discussed below are confounded with light transmittance and may contribute to inhibition of weed seed germination.

Mulch element structure influences light transmittance through residue (Table 1). Oak leaves had a greater and goldenrod (*Solidago* spp.) had a lower extinction coefficient than that observed for hairy vetch or rye residue. Hairy vetch and rye residue had very similar light extinction coefficients. Light extinction appears to be greatest for residue consisting of large leafy material, intermediate for residue consisting of fine stems such as hairy vetch or rye, and lowest for residue consisting of larger stems such as goldenrod. If light transmittance is a good indicator of weed suppression by residue, then cover crops that have a greater proportion of large leafy elements and less stems should be best adapted for controlling weeds. Support for this hypothesis can be found in Facelli and Pickett (1991b) who showed that oak leaves reduced plant emergence and growth more than other litter types despite being present at lower biomass levels. More research is needed to determine the optimum structural properties of mulches for controlling weeds.

Light transmittance through cover crop residue was highly heterogeneous despite the appearance of uniform soil coverage (Teasdale and Mohler, 1993). Natural rates of hairy vetch or rye residue had 17 to 25% of sites transmitting greater than 50% of incoming PPFD and twice the natural residue rate permitted 16 to 25% of sites to transmit greater than 10% of incoming PPFD. The heterogeneity of residue may explain why complete weed control is not obtained.

Light quality is influenced minimally by desiccated cover crop residue. Transmittance increased slightly as wavelength increased from 0.4 to 1.1 μm (Teasdale and Mohler, 1993). The red to far-red ratio was not altered sufficiently to influence phytochrome-mediated germination (Teasdale and Daughtry, 1993).

2. Soil Temperature

Residue on the soil surface reduced maximum soil temperature and slightly increased minimum soil temperature compared to bare soil (Bristow, 1988;

Teasdale and Mohler, 1993). Natural residue levels can reduce maximum soil temperatures by 2 to 5° C and raise minimum soil temperatures by 1° C in temperate climates although this will vary according to radiation intensity, soil moisture, and soil type. Mulch architecture also can influence radiation exchange with soils. Bristow (1988) showed that maximum surface temperature under a vertical mulch could be significantly higher than that under a horizontal mulch. Likewise, Fortin and Hamill (1994) found that maximum soil temperatures were higher under standing than flat rye residue and this difference accounted for faster corn (*Zea mays* L.) development in the standing residue.

Most weed seeds will germinate over a wide range of temperatures and, therefore, the degree of reduction in maximum soil temperature by residue would not be sufficient to prevent germination of weed seeds. However, reduced maximum soil temperatures could account for the delay of germination observed for many species (Mohler and Teasdale, 1993). Because of the decrease in maximum and increase in minimum soil temperature, soil temperature amplitude is reduced considerably by residue. Teasdale and Mohler (1993) reported soil temperature amplitudes of 9.6 to 10.9° C for bare soil versus 6.2 to 6.7° C under natural levels of hairy vetch residue in two experiments. Temperature amplitudes of 10° C are often required to break the dormancy of selected weed species (Taylorson, 1987). Therefore, the reduction in soil temperature amplitude by cover crop residue would be sufficient to prevent germination of species with this requirement.

3. Soil Moisture

Residue on the soil surface increased soil moisture by increasing infiltration (McVay et al., 1989) and by decreasing evaporative moisture loss (Bristow, 1988). However, under selected circumstances, mulches with a high leaf area index can intercept and retain more rainfall and result in lower soil moisture than mulches with a lower leaf area index (Bussière and Cellier, 1994).

Higher soil moisture under cover crop residue could either benefit or retard weed germination and emergence. Under saturated soil conditions, residue could slow evaporation and lead to reduced germination of species inhibited by excess soil moisture. Under droughty conditions, retention of soil moisture could enhance weed germination and seedling survival. This could explain an increase in density of some species observed at low residue levels compared to bare soil (Mohler and Teasdale, 1993). Soil moisture retention also could indirectly influence weed emergence and seedling survival by enhancing pathogen or predatory activity.

4. Soil Organisms

Residues have been shown to influence populations and activity of various soilborne pests. Reduction of cotton (*Gossypium hirsutum* L.) emergence and survival by legume mulches was associated with an increase in *Rhizoctonia* infestation (Rickerl et al., 1992). Hairy vetch was found to support higher populations of various bacteria and fungi including *Rhizoctonia* and *Pythium* species than other cover crops (Rothrock et al., 1995). Hairy vetch also enhanced injury by southern corn rootworm (*Diabrotica unidecimpunctata howardi* Barber)(Buntin et al., 1994) and nematode populations (Johnson et al., 1992) compared to other cover crops.

These results suggest that various soil pathogens and pests could contribute to suppression of weed emergence by cover crop residue. Residue could reduce weed seedling survival by both creating an environment favorable to pathogen/pest activity and by inducing seedlings to become etiolated and weakened during growth through residue layers. Cover crops could be used in combination with selective biocontrol organisms to synergize weed control activity. Residue would have potential to provide a substrate and suitable environmental conditions for growth of biocontrol organisms. Stroo et al. (1988) found that populations of toxin-producing bacteria could be established on wheat straw and were highest in no-tillage plots. Cover crops may offer an opportunity for delivery of novel biocontrol organisms.

5. Allelopathy

Extracts of numerous plant species are toxic to germination and growth of plants (Putnam, 1985). Allelopathic compounds have been isolated and identified (Barnes et al., 1987; Shilling et al., 1986). Toxins released from cover crop residue have been presumed to inhibit weed germination and growth in natural environments (Putnam and DeFrank, 1983; Shilling et al., 1985) but this is difficult to demonstrate *in situ*. Allelopathic influences of residues are confounded with the physical influences described above. In addition, cover crop residues can alter the nutrient status of soils (Smith et al., 1987) resulting in either stimulation or inhibition of plant growth. At minimum, demonstration of allelopathy requires that active toxins must remain in the soil environment for sufficiently long at sufficient concentration to inhibit weed germination and growth. Separating the degree of allelopathic influence from that of other factors is a formidable undertaking.

Various approaches have been proposed to demonstrate the presence of allelopathy. Putnam and DeFrank (1983) suggested that superior weed suppression by rye residue compared to a chemically inert wood shavings control demonstrated that more than physical effects were involved. Fortin and Pierce (1991) compared an oat (*Avena sativa* L.) with a wood shavings mulch and showed, using simulated corn response to thermal time, that corn development was influenced by more than changes in soil temperature. Creamer (1994) compared weed suppression by leached and unleached residue to separate physical and allelopathic effects. Dyck and Liebman (1994) showed that failure of supplemental nitrogen treatments to reverse suppression of common lambsquarters by crimson clover (*Trifolium incarnatum* L.) suggested that allelopathy may have been involved. Blum et al. (1992) provided a comprehensive analysis of soil extraction procedures and bioassays that demonstrated the release of allelopathic compounds from wheat residue in no-tillage soils. Yenish et al. (1995) showed that disappearance of allelochemicals from rye residue corresponded more closely with the duration of weed suppression by rye than physical disappearance of rye residue.

Enhancement of allelopathic compounds in cover crops could enhance their capacity to suppress weeds. More research is needed to identify selective toxins that could be enhanced in cover crops and would persist sufficiently to control weeds.

6. Physical Impedance

Surface residue probably inhibits the emergence of germinated weed seeds by forming a mechanical barrier to the upward growth of seedlings. The dense mat of leaf litter in deciduous forests likely limits the distribution of many species in forest habitats (Facelli and Pickett, 1991a). More research is needed to distinguish between the influence of cover crop residue on weed seed germination and emergence from soil and the influence on subsequent seedling growth through layers of residue. Study of this factor will be subject to the same confounding with environmental and allelopathic factors as confronts study of any single factor in isolation of the others.

C. Role of Cover Crop Residue in Weed Management

Research indicates that cover crops can contribute to weed control in reduced-tillage systems but herbicides usually are required for achieving optimum weed control and crop yield. Brown and Whitwell (1985) found that various herbicide programs provided optimum weed control in cotton but cover crop residue improved control of selected weed species. Wallace and Bellinder (1989) showed that a rye mulch reduced early emergence of weeds without herbicides but optimum potato (*Solanum tuberosum* L.) yields were obtained with herbicides. Johnson et al. (1993) found that herbicides were required for optimum weed control by cover crop mulches. Teasdale (1993) showed that herbicides at registered rates were required with a hairy vetch cover crop to achieve consistently good weed control and high corn yields. Wicks et al. (1994) performed a comprehensive analysis of the interaction between wheat residue and herbicides (Table 2) and found better weed control and corn yields with a combination of wheat straw and herbicide compared to either alone. The report by Liebl et al. (1992) that good weed control and soybean [*Glycine max* (L.) Merr.] yields could be achieved by a rye mulch without herbicides is an exception to this body of literature.

These reports provide evidence that complete elimination of herbicides is not a realistic objective when using cover crops. However, cover crops could permit a reduction of herbicide inputs and a shift toward total postemergence herbicide programs. Early weed suppression provided by cover crop residue should permit crops to become established before weeds. Postemergence herbicides then could control later-emerging weeds until the crop is past the critical period for weed control. This approach could contribute to sustainable farming systems by

Table 2. The effect of wheat straw mulch and metolachlor [2-chloro-*N*- (2-ethyl-6-methylphenyl)-*N*- (2-methoxy-1-methylethyl) acetamide] on weed biomass in no-tillage corn grown in Nebraska[a]

Wheat straw	Weed biomass (kg ha^{-1}) at various metolachlor rates[b]			
(Mg ha^{-1})	0	0.5X	1.0X	1.5X
0	1242	478	294	174
1.7	1664	206	148	124
3.4	740	88	96	60
5.1	444	156	68	36
6.8	412	92	46	24

[a]Data are averaged over five location/years.
[b]X = recommended rate for the soil type.
(Adapted from Wicks et al., 1994.)

replacing preemergence herbicides that are frequently detected in ground and surface waters (National Research Council, 1989) with postemergence herbicides that are used at lower rates and are less persistent. Wallace and Bellinder (1992) proposed that cover crops combined with an as-needed postemergence herbicide program could enhance sustainable vegetable production. A postemergence approach has been used successfully with corn grown in hairy vetch (Figure 2).

If elimination of herbicides is a primary objective, new approaches will be needed to develop effective cover crop-based weed management systems. New cover crops will have to be identified or developed that can interfere more effectively with the requirements for weed germination and establishment as described in the previous section. Present methods of mowing cover crops to create a mulch have not been effective (Hoffman et al., 1993; Teasdale, 1993; Curran et al., 1994) but creative approaches to rolling, chopping, or undercutting may be possible (Hoffman et al., 1993; Creamer, 1994). Delayed planting may permit greater cover crop biomass to accumulate and bypass the optimum period for emergence of many weed species (Hoffman et al., 1993). Integration with other nonchemical methods of weed control such as denser crop populations (Teasdale, 1994) or interrow sod cultivation (Buhler et al., 1994) may be advantageous. The combination of early weed suppression by cover crop residue and a shortened critical period for weed control by dense crop populations may be effective. Alternatively, development of high-residue cultivators that eliminate weeds mechanically with minimal disruption of the cover crop mulch may extend weed control by cover crops throughout the critical period for weed control. Use of seedbank information and weed management models could contribute to determining the potential efficacy and advisability of herbicide-free weed management systems.

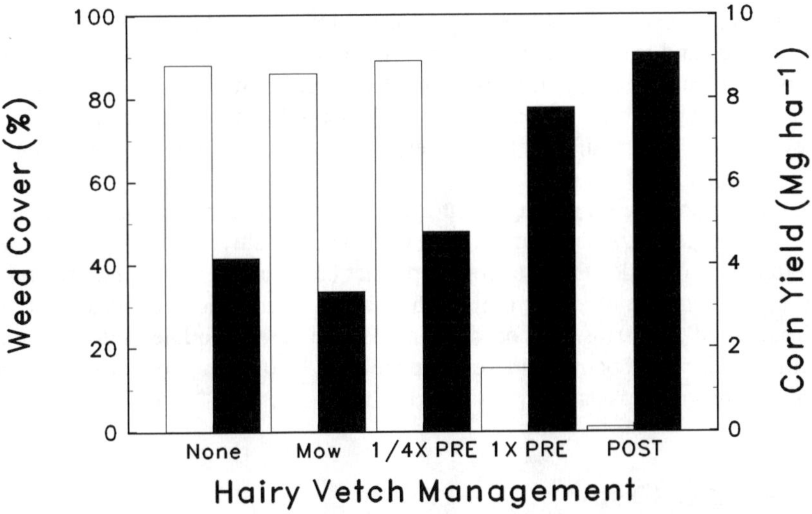

Figure 2. Response of weed cover (open bars) and corn grain yield (solid bars) to various hairy vetch management programs. Hairy vetch was either unmanaged (none), mowed after planting (mow), treated with one-fourth of the labelled rate of preemergence herbicides (1/4X PRE), treated with the labelled rate of preemergence herbicides (1X PRE), or treated with postemergence herbicides only (POST). (Adapted from Teasdale, 1993.)

D. Integrated Management Systems

Cover crops influence all aspects of cropping systems including nutrient availability, soil tilth, soil temperature, soil moisture, erosion, leaching, crop establishment, and weed control. This makes component research difficult because the influence of one factor is confounded with that of other factors. On the other hand, cover crops offer the opportunity to manipulate many facets of cropping systems through management of a single component. The manifold influences of cover crops require an integrated approach for managing and assessing cover crop performance in cropping systems.

Increasing the biomass of cover crop residue to increase weed control provides an illustrative example. Greater cover crop biomass could be achieved by growing species with superior biomass production, by permitting cover crops to grow longer before termination, or by growing cover crop combinations that produce more biomass than each cover crop alone. Any or all of these approaches should lead to enhanced weed control because of greater cover crop residue. However, this approach could have several negative consequences. Crop establishment often is difficult through dense layers of residue (Brown and Whitwell, 1985; Kumwenda et al., 1993; Liebl et al., 1992; Teasdale, 1993). High residue biomass may be achieved at the expense of a high C:N ratio leading to nitrogen immobilization (Holderbaum et al., 1990). Greater or longer cover crop growth in spring could deplete soil moisture (Liebl et al., 1992). Dense residue could lower soil temperatures and reduce early crop growth (Fortin and Hamill, 1994). These effects could nullify the positive weed control benefits of high residue levels and result in a net yield loss. Wicks et al. (1994) found that maximum corn yield was achieved at 4.4 Mg ha^{-1} of wheat straw despite greater weed suppression by higher straw rates (Table 2).

Design of cover crop management systems needs to account for antagonistic influences and identify the optimum balance between positive and negative effects. Development of models that accurately predict cover crop growth and influences would aid in simulating cover crop performance under various management and environmental scenarios. Incorporation of cover crop modules into cropping system models would permit a more complete analysis of crop performance as well as environmental impacts.

III. Smother Crops

A smother crop will be defined as a cover crop that remains alive for part or all of the cropping season. Consequently, competition for essential resources will be a major mechanism of weed suppression by a smother crop. Smother crops may be either perennial or annual. Species that are perennials or self-seeding annuals have the advantage of eliminating the need for annual seeding and associated seedbed preparation.

A. Weed Control by Smother Crops

An established, living smother crop potentially can inhibit germination and establishment of weeds more effectively than desiccated cover crop residue. Teasdale and Daughtry (1993) showed that chlorophyll in living vegetation intercepted red light and reduced the red to far-red ratio sufficiently to convert phytochrome to an inactive form. Light transmittance and soil temperature amplitude were reduced more by living than by desiccated mulches. In addition, seedlings that emerged successfully were at a competitive disadvantage against an established smother crop.

Smother crops have been shown to reduce weed populations in cropping systems. Hartwig (1989) showed that a crownvetch (*Coronilla varia* L.) living mulch in corn reduced dandelion (*Taraxacum officinale* Weber in Wiggers) populations by 74% compared to a no-tillage control. Enache and Ilnicki (1990) found better weed control by a self-seeding subterranean clover (*Trifolium subterraneum* L.) living mulch in corn than other treatments with or without herbicides. White and Scott (1991) showed that various established perennial legumes in small grains reduced weed biomass compared to a control without mulch. De Haan et al. (1994) found that a yellow mustard (*Brassica hirta* Moench) smother crop in corn for 6 to 8 weeks reduced weed biomass by an average of 82%.

Although smother crops may have potential for controlling weeds, experience suggests that monocultures rarely exist in nature. Most reliable perennial forage species develop gaps where a diversity of species can become established. Weed species with a compatible phenology and morphology with any chosen smother crop can invade and become established. Wilkinson et al. (1987) reported that weed establishment in a tall fescue (*Festuca arundinacea* Schreb.) living mulch for corn became a major factor limiting yield after three years. Mohler (1991) reported a buildup of weeds in a white clover (*Trifolium repens* L.) living mulch that reduced sweet corn yields after four years. Although smother crops may be effective tools for controlling weeds, they require sound management, particularly to prevent the invasion and establishment of weed species over time.

B. Living Mulch Regulation

A living mulch competitive enough to suppress weeds usually will suppress the cash crop as well. A comparison of various grass or legume cover crops demonstrated that those most suppressive to weeds also were most suppressive of corn yield (Echtenkamp and Moomaw, 1989). Weed and small grain yields were reduced most when perennial legume living mulches were established and had the greatest biomass (White and Scott, 1991). Management practices that permitted yellow mustard to suppress weed growth also resulted in reduced corn yields (De Haan et al., 1994). Weed suppression and yield reduction usually are correlated with living mulch biomass. Both Elkins et al. (1983) and Nicholson and Wien (1983) showed a high correlation between living mulch biomass and yield reduction (Figure 3). Regulation of smother crop growth, therefore, is a major requirement for developing smother crop-based cropping systems. Table 3 summarizes several approaches to living mulch regulation. This section will review some of the successful approaches that have been reported.

1. Nonlethal Herbicide

Eberlein et al. (1992) reported that alfalfa (*Medicago sativa* L.) could be partially suppressed by a sublethal rate of atrazine [6-chloro-*N*-ethyl-*N*'-(1-methylethyl)-1,3,5-triazine-2,4-diamine] but would recover from buds on the lower stems. Corn yield was not reduced by partially suppressed alfalfa when irrigated but leaf area, growth, and yield were reduced without irrigation. Elkins et al. (1983) found that corn yields were not reduced by various perennial grasses and legumes when suppressed at least 50% by various herbicide treatments. In Pennsylvania, Hartwig (1987) has developed prescriptions using sublethal herbicide applications for establishing and maintaining a crownvetch living mulch in various field crop rotations.

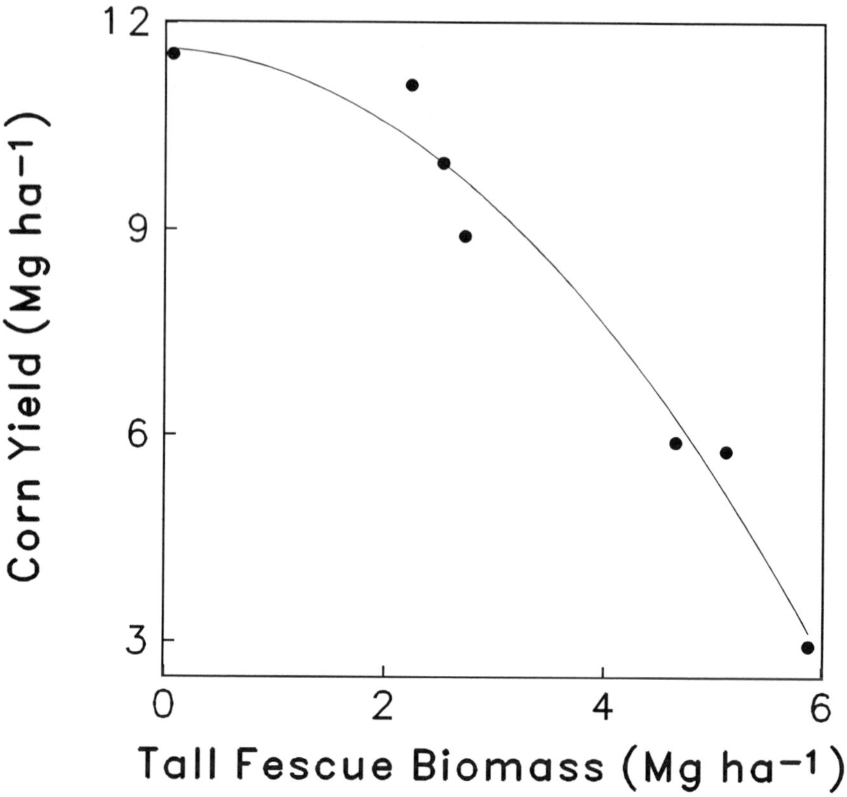

Figure 3. Relation between corn grain yield and tall fescue biomass after various herbicide treatments in 1978 at Belleville, Illinois. (Adapted from Elkins et al., 1983.)

2. Banded Herbicide

Eberlein et al. (1992) suppressed alfalfa with a band application of a lethal rate of atrazine sufficiently to prevent yield loss in irrigated corn. This approach permitted partial recovery of alfalfa between rows but recovery was less than that after the nonlethal, broadcast atrazine treatment. Elkins et al. (1983) successfully suppressed various living mulches and increased corn yields with banded herbicide treatments. Kumwenda et al. (1993) demonstrated that crimson clover killed in a 60 to 80% band over the corn row did not interfere with yield and permitted adequate reseeding of the clover.

3. Strip Tillage

Tilling a narrow band in a living mulch for planting has been practiced primarily with horticultural crops. Yields of sweet corn and cabbage (*Brassica oleracea* L. var. *capitata*) planted in a 30-cm cultivated band were not reduced by grass and white clover living mulch cultivars with limited growth (Nicholson and Wien, 1983). Pepper (*Capsicum annuum* L.) and squash (*Cucurbita maxima* L.) yields were reduced by a grass/white clover sod when grown in 0.5-m tilled strips but yields were unaffected when grown in 1.1-m tilled strips (Loy et al., 1987). However, Loy et al. (1987) found that pepper yield was reduced in 1.1-m strips in a dry year. Grubinger and Minotti (1990) showed that rototilling white clover in mid-season between 38- to 55-cm tilled strips of sweet corn did not kill the clover but suppressed it sufficiently to prevent yield reductions. Costello and Altieri (1994) reported that broccoli (*Brassica oleracea* L. ssp. *italica*) could be grown without yield loss in 10-cm cultivated strips in living mulches of various *Trifolium* species.

4. Mowing

Mowing usually is not adequate to reverse yield reduction by suppressive living mulches (Grubinger and Minotti, 1990; Loy et al., 1987). However, this approach may be useful to avoid shading during crop establishment when used in combination with other practices.

5. Row Spacing/Density

Altering row spacing, planting arrangement, or density may increase the competitiveness of the crop relative to a living mulch. Loy et al. (1987) showed that planting pepper in double rows and higher densities overcame reduced yields. Fischer and Burrill (1993) found that decreasing sweet corn row spacing from 76 to 38 cm overcame yield reduction by a white clover living mulch but doubling corn density had no effect. Use of systematic designs in which effects of both density and proportion of crop and living mulch species can be analyzed could aid understanding of intraspecific and interspecific competitive interactions (Wiles et al., 1989).

6. Relay Planting

Relay planting may be ideally suited for smother plants. The critical growth phase of a smother plant and cash crop could be staggered so they do not compete for the same resources during the same time. Additionally, early establishment of the smother crop before weed emergence would ensure better weed suppression. This approach could be applied to any situation where the phenology of the smother

Table 3. Influence of living mulches on crop yields with various regulation treatments

Regulation	Living mulch	Crop	Condition	Weed free	% Yield reduction[a]	Reference
None	Alfalfa	Corn	Irrigated	Yes	62-83	Eberlein et al., 1992
			Nonirrigated		96-99	
	Perennial legumes	Rye/wheat	Establishment year	No	10/22[b]	White and Scott, 1991
			Year after establishment	No	48/71[b]	
Broadcast, nonlethal herbicide	Alfalfa	Corn	Irrigated	Yes	None	Eberlein et al., 1992
			Nonirrigated		5-36	
	Crownvetch	Corn	Eight-year stand	?	None	Mayer and Hartwig, 1986
	White clover	Sweet corn	76 cm rows	?	12-39	Fischer and Burrill, 1993
			38 cm rows		None	
	White clover	Sweet corn	Years 1 & 2, few weeds	Yes	None	Mohler, 1991
			Years 3 & 4, many weeds	No	15-16	
	Smooth bromegrass	Corn		?	12-36	Klocke et al., 1989
	Grasses and legumes	Corn	Also strip tillage	No	25-48[c]	Echtenkamp and Moomaw, 1989
			Also band herbicide			
Band herbicide	Crimson clover	Corn		Yes	None	Kumwenda et al., 1993
	Alfalfa	Corn	--	Yes	None	Eberlein et al., 1992
			Irrigated		43-54	
	Tall fescue	Corn	Nonirrigated	No	63-84	Wilkinson et al., 1987
			20-cm band		30-70	
			41-cm band			

Table 3. continued--

Regulation	Living mulch	Crop	Condition	Weed free	% Yield reduction[a]	Reference
Strip tillage	White clover	Squash	Compost mulch in row	No	20-65	Mohler, 1995
	White clover	Sweet corn	No between-row control	Yes	37	Grubinger and Minotti, 1990
			Rototilled between row		None	
	Clovers	Broccoli	--	?	None	Costello and Altieri, 1994
	Grass/clover sod	Pepper	0.5-m strip	Yes	56	Loy et al., 1987
			1.1-m strip		8-22	
	Grasses and clovers	Sweet corn	Also herbicide in row	Yes	0-31[d]	Nicholson and Wien, 1983
		Cabbage			0-58[d]	
	Smooth bromegrass	Corn	--	?	32-54	Klocke et al., 1989

[a]Percent yield reduction compared to the closest matching treatment without a mulch. Range of values correspond to different years unless specified differently. None = no significant reduction.

[b]Average reduction across six legumes at a nitrogen rate of 56 kg ha^{-1}. The value before the "/" is rye yield reduction and that after is wheat yield reduction.

[c]Average over treatments with surviving living mulches.

[d]Range across nine grasses and clovers. The range of yield reduction values correlated with mulch biomass.

plant permitted establishment and peak growth before that of the cash crop: summer crops planted into winter annual smother crops, summer crops planted into cool-season spring-planted smother crops, or fall crops planted into summer smother crops.

Winter annuals which naturally senesce before exponential crop growth have been tested as relay planted smother crops. Corn planted into crimson clover (Kumwenda et al., 1993) or subterranean clover (Enache and Ilnicki, 1990) are successful examples of this approach. Both of these clover species naturally reestablished from seed each year thereby behaving as a living mulch despite being annuals.

More research is needed to identify species or develop improved cultivars that can serve as relay-planted smother crops. Important characteristics would include: reliable and uniform stand establishment, morphology that will maximize weed suppression but minimize crop suppression, synchronization of senescence to avoid the critical period of competition with the cash crop, and capability of reestablishment either as a perennial or self-seeding annual. De Haan et al. (1994) have explored the requirements of a spring-seeded smother crop for corn in Minnesota.

7. Intercropping

Growing a cash crop with a smother crop can be considered one approach to the broader practice of intercropping in which two or more crops are grown together. Intercropping often results in greater yields than could be achieved by each crop grown separately on the same total land area because of greater capture and/or more efficient use of resources. Many of the principles of intercropping could be applied to development of smother crop systems. Additionally, smother crops which, themselves, had an economic value could be developed to supplement income from the cash crop. Reduction in yield of the main cash crop would be acceptable if a commensurate economic benefit could be derived from the smother crop.

The capability of intercrops to improve weed suppression has been reviewed by Liebman and Dyck (1993). Generally, weed biomass was reduced more by intercrops than by, at least, one component crop alone. This generality is subject to many exceptions and conditions depending on the crops involved, crop management practices, and weather. Superior preemption of resources by the intercrop may account for weed suppression but Liebman and Dyck (1993) showed that yield advantages gained from greater resource use were not always coupled to weed suppression.

C. Resource Use by Smother Crops

Competitive interactions between weeds and crops often are subject to considerable variation depending on environmental conditions and resource availability. Interactions between smother crops, weeds, and crops would be expected to exhibit similar variation. For example, Eberlein et al. (1992) showed that corn suppression by an alfalfa living mulch could be reversed by supplemental irrigation. However, Box et al. (1980) found that corn yield was reduced by a tall fescue living mulch regardless of irrigation level. Supplementing tall fescue with excessive rates of nitrogen reduced, but did not completely reverse, corn yield reductions (Wilkinson et al., 1987).

Superior resource use in intercropping systems usually results from complementarity between the component crops grown together (Willey, 1990). This can be the result of temporal complementarity when the crops utilize resources at different times or spatial complementarity when resources are captured more completely by the intercrop. A sorghum [*Sorghum bicolor* (L.) Moench]/pigeon-pea [*Cajanus cajun* (L.) Huth] intercrop is typical of systems with temporal complementarity because higher yields result from fuller use of resources (primarily light) over time (Willey, 1990). Leach and Foale (1987) analyzed an example of spatial complementarity whereby more water was extracted by sorghum or sunflower (*Helianthus annuus* L.) with undersown cowpea [*Vigna trilobata* (L.) Verdc.] than by each component crop alone because of different rooting profiles. Willey (1990) cites greater water capture because of higher concentration and dispersion of roots as partly responsible for higher yields by a millet [*Pennisetum americanum* (L.) K. Schum.]/groundnut (*Arachis hypogaea* L.) intercrop. A physiological complementarity attributed to greater conversion efficiency of groundnut at lower light also accounted for higher yields of the millet/groundnut intercrop (Willey, 1990).

Greater knowledge of resource use by smother crops is needed before their performance in cropping systems can become more predictable and manageable. More research is needed to understand the morphological and physiological basis for the uncoupling of weed and crop suppression in successful smother cropping systems. This will make it possible to prescribe desirable traits and requirements of a smother crop. Identified traits can be used for identification or breeding of new smother crops. Modelling could permit determination of optimum edaphic and environmental conditions supporting smother crop management systems.

D. Potential Contribution of Smother Crops

Most living mulch species require herbicides to regulate growth and often require as much herbicide as would be required to control weeds in the absence of the mulch. The major contribution of these living mulches to agricultural systems is long-term soil protection and fertility enhancement, particularly on erodible land. Erosion prevention and soil stabilization is probably the highest priority for maintaining a sustainable agriculture on these sites. In this situation, herbicides should be regarded as a tool for preserving soil resources rather than an input to be eliminated.

Smother crops have potential to control weeds in herbicide-free cropping systems in the future. A smother crop that consistently can develop a uniform, competitive ground cover can effectively prevent weed establishment. Such a smother crop also must have the appropriate phenology, morphology, and physiology to complement resource use by the cash crop. Development of these smother crops represents a challenge to future weed scientists, ecophysiologists, and breeders.

IV. Research Needs

Although the utility of cover crops in cropping systems is well established and their potential for weed suppression has been demonstrated, this review has identified several areas requiring further research. Most cover crops currently researched and used were developed for other purposes, e.g., as grain or forage crops. Further improvement of cover crops probably will require breeding to develop cultivars that meet the specific requirements of cover crops. Therefore, the properties of cover crops responsible for weed suppression need to be defined precisely. Emphasis should be placed on 1) physical properties (e.g., biomass, thickness, density, surface area index) that determine radiation interception, 2) allelopathic properties, 3) phenology that complements cropping systems, and 4) adaptation to local environmental conditions. Research on smother crops should emphasize the temporal and spatial pattern of resource use. Knowledge of the salient properties of cover crops and smother plants will direct the identification and breeding of improved cultivars.

Current cover crops and smother plants require management systems that are herbicide-dependent. Refined management systems are needed to optimize weed control while minimizing herbicide inputs. Several approaches could be fruitful. Cover crops should be compatible with total postemergence herbicide programs and the feasibility of this approach needs investigation. Nonchemical methods for controlling cover crops such as rolling, chopping, and undercutting need to be developed. Several nonchemical control practices that do not provide satisfactory control alone may permit adequate weed control when combined with cover crops. Models predicting the timing and density of weed emergence could aid in determining the optimum timing and intensity of cover crop management practices.

Cover crop management must balance the requirements for weed suppression against the requirements for crop production. Properties of cover crops and smother plants that maximize weed suppression may interfere with crop production. Creative management approaches are needed that uncouple influences on weeds from those on crops. The ultimate goal of our management systems should be the replacement of a weed population that interferes with crop production by a cover crop population that enhances crop production.

References

Alm, D.M., E.W. Stoller, and L.M. Wax. 1993. An index model for predicting seed germination and emergence rates. *Weed Technol.* 7:560-569.

Barnes, J.P., A.R. Putnam, B.A. Burke, and A.J. Aasen. 1987. Isolation and characterization of allelochemicals in rye herbage. *Phytochem.* 26:1385-1390.

Blum, U., T.M. Gerig, A.D. Worsham, L.D. Holappa, and L.D. King. 1992. Allelopathic activity in wheat-conventional and wheat-no-till soils: development of soil extract bioassays. *J. Chem. Ecol.* 18:2191-2221.

Box, Jr., J.E., S.R. Wilkinson, R.N. Dawson, and J. Kozachyn. 1980. Soil water effects on no-till corn production in strip and completely killed mulches. *Agron. J.* 72:797-802.

Bristow, K.L. 1988. The role of mulch and its architecture in modifying soil temperature. *Aust. J. Soil Res.* 26:269-280.

Brown, S.M. and T. Whitwell. 1985. Weed control programs for minimum-tillage cotton. *Weed Sci.* 33:843-847.

Buhler, D.D., J.D. Doll, R.T. Proost, and M.R. Visocky. 1994. Interrow cultivation to reduce herbicide use in corn following alfalfa without tillage. *Agron. J.* 86:66-72.

Buntin, D.G., J.N. All, D.V. McCracken, and W.L. Hargrove. 1994. Cover crop and nitrogen fertility effects on southern corn rootworm (Coleoptera: Chrysomelidae) damage in corn. *J. Econ. Entomol.* 87:1683-1688.

Bussière, F. and P. Cellier. 1994. Modification of the soil temperature and water content regimes by a crop residue mulch: experiment and modelling. *Agric. and Forest Meteorol.* 68:1-28.

Carson, W.P. and C.J. Peterson. 1990. The role of litter in an old-field community: impact of litter quantity in different seasons on plant species richness and abundance. *Oecologia* 85:8-13.

Cone, J.W. and R.E. Kendrick. 1986. Photocontrol of seed germination. p. 443-465. In: R.E. Kendrick and G.H.M. Kronenberg (ed.), *Photomorphogenesis in Plants*. Martinus Nijhoff/Dr W. Junk Publishers, Dordrecht, The Netherlands.

Costello, M.J. and M.A. Altieri. 1994. Living mulches suppress aphids in broccoli. *Calif. Agric.* 48(4):24-28.

Creamer, N.G. 1994. *An Evaluation of Cover Crop-Processing Tomato (Lycopersicon esculentum L.) Production Systems, with an Emphasis on Weed Management*. Ph.D. Dissertation, The Ohio State Univ.

Crutchfield, D.A., G.A. Wicks, and O.C. Burnside. 1985. Effect of winter wheat straw mulch level on weed control. *Weed Sci.* 34:110-114.

Curran, W.S., L.D. Hoffman, and E.L. Werner. 1994. The influence of a hairy vetch cover crop on weed control and corn growth and yield. *Weed Technol.* 8:777-784.

De Haan, R.L., D.L. Wyse, N.J. Ehlke, B.D. Maxwell, and D.H. Putnam. 1994. Simulation of spring-seeded smother plants for weed control in corn. *Weed Sci.* 42:35-43.

Dyck, E. and M. Liebman. 1994. Soil fertility management as a factor in weed control: the effect of crimson clover residue, synthetic nitrogen fertilizer, and their interaction on emergence and early growth of lambsquarters and sweet corn. *Plant and Soil* 167:227-237.

Eberlein, C.V., C.C. Sheaffer, and V.F. Oliveira. 1992. Corn growth and yield in an alfalfa living mulch system. *J. Prod. Agric.* 5:332-339.

Echtenkamp, G.W. and R.S. Moomaw. 1989. No-till corn production in a living mulch system. *Weed Technol.* 3:262-266.

Elkins, D., D. Frederking, R. Marashi, and B. McVay. 1983. Living mulch for no-till corn and soybeans. *J. Soil and Water Conserv.* 38:431-433.

Enache, A.J. and R.D. Ilnicki. 1990. Weed control by subterranean clover used as a living mulch. *Weed Technol.* 4:534-538.

Facelli, J.M. and E. Facelli. 1993. Interactions after death: plant litter controls priority effects in a successional plant community. *Oecologia* 95:277-282.

Facelli, J.M. and S.T.A. Pickett. 1991a. Plant litter: its dynamics and effects on plant community structure. *Bot. Rev.* 57:1-32.

Facelli, J.M. and S.T.A. Pickett. 1991b. Plant litter: Light interception and effects on an old-field plant community. *Ecology* 72:1024-1031.

Fischer, A. and L. Burrill. 1993. Managing interference in a sweet corn-white clover living mulch system. *Amer. J. of Altern. Agric.* 8:51-56.

Forcella, F. 1993. Seedling emergence model for velvetleaf. *Agron. J.* 85:929-933.

Forcella, F., R.G. Wilson, K.A. Renner, J. Dekker, R.G. Harvey, D.A. Alm, D.D. Buhler, and J. Cardina. 1992. Weed seedbanks of the U.S. corn belt: magnitude, variation, emergence, and application. *Weed Sci.* 40:636-644.

Fortin, M.C. and A.S. Hamill. 1994. Rye residue geometry for faster corn development. *Agron. J.* 86:238-243.

Fortin, M.C. and F.J. Pierce. 1991. Timing and nature of mulch retardation of corn vegetative development. *Agron. J.* 83:258-263.

Grubinger, V.P. and P.L. Minotti. 1990. Managing white clover living mulch for sweet corn production with partial rototilling. *Amer. J. of Altern. Agric.* 5:4-12.

Hartwig, N.L. 1987. Cropping practices using crownvetch in conservation tillage. p. 109-110. In: J.F. Power (ed.), *The Role of Legumes in Conservation Tillage Systems*. Soil Conserv. Soc. Am., Ankeny, Iowa.

Hartwig, N.L. 1989. Influence of a crownvetch living mulch on dandelion invasion in corn. *Proc. Northeast. Weed Sci. Soc.* 43:25-28.

Harvey, S.J. and F. Forcella. 1993. Vernal seedling emergence model for common lambsquarters. *Weed Sci.* 41:309-316.

Hoffman, M.L., E.E. Regnier, and J. Cardina. 1993. Weed and corn responses to a hairy vetch cover crop. *Weed Technol.* 7:594-599.

Holderbaum, J.F., A.M. Decker, J.J. Meisinger, F.R. Mulford, and L.R. Vough. 1990. Fall-seeded legume cover crops for no-tillage corn in the humid east. *Agron. J.* 82:117-124.

Johnson, G.A., M.S. Defelice, and Z.R. Helsel. 1993. Cover crop management and weed control in corn. *Weed Technol.* 7:425-430.

Johnson, A.W., A.M. Goldern, D.L. Auld, and D.R. Sumner. 1992. Effects of rapeseed and vetch as green manure crops and fallow on nematodes and soil-borne pathogens. *J. Nematol.* 24:117-126.

Klocke, N.L. J.T. Nichols, P.H. Grabouski, and R. Todd. 1989. Intercropping corn in perennial cool-season grass on irrigated sandy soil. *J. Prod. Agric.* 2:42-46.

Kumwenda, J.D.T., D.E. Radcliffe, W.L. Hargrove, and D.C. Bridges. 1993. Reseeding of crimson clover and corn grain yield in a living mulch system. *Soil Sci. Soc. Am. J.* 57:517-523.

Leach, G.J. and M.A. Foale. 1987. Relations between summer crops and ground cover legumes in a subtropical environment 3. Soil water use under sorghum and sunflower. *Field Crop Res.* 16:177-191.

Liebl, R., F.W. Simmons, L.M. Wax, and E.W. Stoller. 1992. Effect of rye mulch on weed control and soil moisture in soybean. *Weed Technol.* 6:838-846.

Liebman, M. and E. Dyck. 1993. Crop rotation and intercropping strategies for weed management. *Ecol. Applic.* 3:92-122.

Loy, S.J.W., L.C. Peirce, G.O. Estes, and O.S. Wells. 1987. Productivity in a strip tillage vegetable production system. *HortScience* 22:415-417.

Mayer, J.B. and N.L. Hartwig. 1986. Corn yields in crownvetch relative to dead mulches. *Proc. Northeast. Weed Sci. Soc.* 40:34-35.

McVay, K.A., D.E. Radcliffe, and W.L. Hargrove. 1989. Winter legume effects on soil properties and nitrogen fertilizer requirements. *Soil Sci. Soc. Am. J.* 53:1856-1862.

Mohler, C.L. 1991. Effects of tillage and mulch on weed biomass and sweet corn yield. *Weed Technol.* 5:545-552.

Mohler, C.L. 1993. A model of the effects of tillage on emergence of weed seedlings. *Ecol. Applic.* 3:53-73.

Mohler, C.L. 1995. A living mulch (white clover)/dead mulch (compost) weed control system for winter squash. *Proc. Northeast. Weed Sci. Soc.* 49:5-10.

Mohler, C.L. and J.R. Teasdale. 1993. Response of weed emergence to rate of *Vicia villosa* Roth and *Secale cereale* L. residue. *Weed Res.* 33:487-499.

Monk, C.D. and F.C. Gabrielson, Jr. 1985. Effects of shade, litter and root competition on old-field vegetation in South Carolina. *Bull. Torrey Bot. Club* 112:383-392.

National Research Council. 1989. *Alternative Agriculture*. National Academy Press. Washington, DC. 448 p.

Nicholson, A.G. and H.C. Wien. 1983. Screening of turfgrasses and clovers for use as living mulches in sweet corn and cabbage. *J. Amer. Soc. Hort. Sci.* 108:1071-1076.

Putnam, A.R. 1985. Weed allelopathy. p. 131-155. In: S.O. Duke (ed.), *Weed Physiology* Vol. I. *Reproduction and Ecophysiology*. CRC Press, Inc. Boca Raton, FL.

Putnam, A.R. and J. DeFrank. 1983. Use of phytotoxic plant residues for selective weed control. *Crop Protect.* 2:173-181.

Putnam, A.R., J. DeFrank, and J.P. Barnes. 1983. Exploitation of allelopathy for weed control in annual and perennial cropping systems. *J. Chem. Ecol.* 9:1001-1010.

Rickerl, D.H., E.A. Curl, J.T. Touchton, and W.B. Gordon. 1992. Crop mulch effects on Rhizoctonia soil infestation and disease severity in conservation-tilled cotton. *Soil Biol. Biochem.* 24:553-557.

Rothrock, C.S., T.L. Kirkpatrick, R.E. Frans, and H.D. Scott. 1995. The influence of winter annual legume cover crops on soilborne plant pathogens and cotton seedling diseases. *Plant Disease* 79:167-171.

Sarrantonio, M. 1994. *Northeast Cover Crop Handbook*. Rodale Institute, Emmaus, PA.

Shilling, D.G., R.A. Liebl, and A.D. Worsham. 1985. Rye and wheat mulch: The suppression of certain broadleaved weeds and the isolation and identification of phytotoxins. p. 243-271. In: A.C. Thompson (ed.), *The Chemistry of Allelopathy*. ACS Sympos. Series 268.

Shilling, D.G., L.A. Jones, A.D. Worsham, C.E. Parker, and R.F. Wilson. 1986. Isolation and identification of some phytotoxic compounds from aqueous extracts of rye. *J. Agric. Food Chem.* 34:633-638.

Smith, M.S., W.W. Frye, and J.J. Varco. 1987. Legume winter cover crops. *Adv. Soil Sci.* 7:95-139.

Stroo, H.F., L.F. Elliott, and R.I. Papendick. 1988. Growth, survival and toxin production of root-inhibitory pseudomonads on crop residues. *Soil Biol. Biochem.* 20:201-207.

Taylorson, R.B. 1987. Environmental and chemical manipulation of weed seed dormancy. *Rev. Weed Sci.* 3:135-154.

Teasdale, J.R. 1993. Reduced-herbicide weed management systems for no-tillage corn in a hairy vetch cover crop. *Weed Technol.* 7:879-883.

Teasdale, J.R. 1994. Influence of narrow row/high population corn on weed control and light transmittance. *Weed Technol.* 9:113-118.

Teasdale, J.R. and C.S.T. Daughtry. 1993. Weed suppression by live and desiccated hairy vetch. *Weed Sci.* 41:207-212.

Teasdale, J.R. and C.L. Mohler. 1992. Weed suppression by residue from hairy vetch and rye cover crops. *Proc. First Internat. Weed Control Congr.* 2:516-518.

Teasdale, J.R. and C.L. Mohler. 1993. Light transmittance, soil temperature, and soil moisture under residue of hairy vetch and rye. *Agron. J.* 85:673-680.

Wallace, R.W. and R.R. Bellinder. 1989. Potato yields and weed populations in conventional and reduced tillage systems. *Weed Technol.* 3:590-595.

Wallace, R.W. and R.R. Bellinder. 1992. Alternative tillage and herbicide options for successful weed control in vegetables. *HortScience* 27:745-749.

White, J.G. and T.W. Scott. 1991. Effects of perennial forage-legume living mulches on no-till winter wheat and rye. *Field Crops Res.* 28:135-148.

Wicks, G.A., D.A. Crutchfield, and O.C. Burnside. 1994. Influence of wheat straw mulch and metolachlor on corn growth and yield. *Weed Sci.* 42:141-147.

Wiles, L.J., R.D. William, G.D. Crabtree, and S.R. Radosevich. 1989. Analyzing competition between a living mulch and a vegetable crop in an interplanting system. *J. Amer. Soc. Hort. Sci.* 114:1029-1034.

Wilkinson, S.R., O.J. Devine, D.P. Belesky, J.W. Dobson, Jr., and R.N. Dawson. 1987. No-tillage intercropped corn production in tall fescue sod as affected by sod control and nitrogen fertilization. *Agron. J.* 79:685-690.

Willey, R.W. 1990. Resource use in intercropping systems. *Agric. Water Management* 17:215-231.

Yenish, J.P., A.D. Worsham, and W.S. Chilton. 1995. Disappearance of DIBOA-glucoside, DIBOA, and BOA from rye cover crop residue. *Weed Sci.* 43:18-20.

Application of Micrometeorology to Weed Biology and Modeling

J.L. Hatfield

I. Introduction

Weeds are intrusions into a biological system that man tries to maintain as homogeneous as possible. There is nothing more aesthetic than a uniform field without a weed in sight. Weeds, however, invade our perfect world and distract from our perception of uniformity, and we need to understand why the weed populations within a field are variable, scattered, and comprised of a range of species. We need to understand that the act of growing a crop induces variation into the ecosystem, and weeds attempt to fit within niches of that ecosystem. Variation induced by the normal agricultural practices gives rise to variability in a number of abiotic factors, e.g., temperature, moisture, wind, and radiation. In this chapter, variation in microclimate induced by changes in micrometeorological parameters will be discussed relative to weed germination, emergence, and growth. The goal of this chapter is to provide a thought-provoking discussion of some of the variables that determine why weed populations vary across a field and to propose a physically-based model of weed seed germination.

There are three primary factors that determine how well weed seed germinate and grow in a crop canopy: availability of light, soil water availability, and soil and air temperature (Egley, 1986). Light may be necessary to induce germination and is required for photosynthesis in leaves of weed seedlings. Therefore, an understanding of the dynamics of light interception and distribution under a crop canopy at different growth stages is an important component in our understanding of weed populations. Soil water availability is a necessary prerequisite for germination and

Table 1. Extinction coefficients for solar radiation for a range of canopies

Crop	Extinction coefficient	Reference
Corn	1.1 (0.4 to 0.65 μm)	Hatfield and Carlson, 1979
Corn	0.38 (0.65 to 1.05 μm)	Hatfield and Carlson, 1979
Corn, flat leaves	0.7 (all wavelengths)	Lemeur, 1973
Soybean, upright leaves	0.5	Lemeur, 1973
Soybean, flat leaves	1.2	Hatfield and Carlson, 1979
Sunflower	1.0	Lemeur, 1973
Wheat, upright leaves	0.2	Angus et al., 1972
Wheat, long, droopy leaves	0.6	Angus et al., 1972

growth, and a soil that dries quickly will not be able to support young seedling growth and development. One can imagine a situation with dry soil between the crop rows and moist soil under the canopy as providing different environments for weed growth. Similarly, temperatures, both air and soil, are critical factors for biological system response from seed to mature plant. Temperature patterns within a canopy are a function of type of plant, time of day, and time of year.

Variations in microclimate are affected by time of year, stage of development of the crop plant, type of tillage and residue management practices, and meteorological conditions. With these dynamic variables being the critical factors in determining how well a weed might grow, it is easy to understand why the processes may be too complicated to quantify. However, within this chapter each factor will be discussed separately and then integrated to develop a microclimatic model applicable to describing weed populations within a field.

II. Light

Light penetration through a crop canopy can be described by a rather simple equation derived from Beer's Law as:

$$I = I_0 e^{-kL} \tag{1}$$

where I is the irradiance, J m^{-2} s^{-1}, at some position within a canopy below a leaf area, L, for a given amount of radiant energy impinging upon the canopy, I_0, and with a particular extinction coefficient, k (Cowan, 1968). Extinction coefficients vary with the type, leaf angle, and age of the crop (Lemeur and Blad, 1974). Extinction coefficients are larger for more dense canopies and decrease with more upright grass-type canopies (Table 1). However, within a given crop there can be a large range of extinction coefficients caused by variation in leaf angles and leaf shapes. Extinction coefficients also change with portion of the solar radiation spectrum being evaluated and sun angle. The more overhead the sun, the lower the extinction coefficient because of the chance that a ray of light will make it through the plant leaves to lower levels of the canopy. Lemeur (1973) showed that there

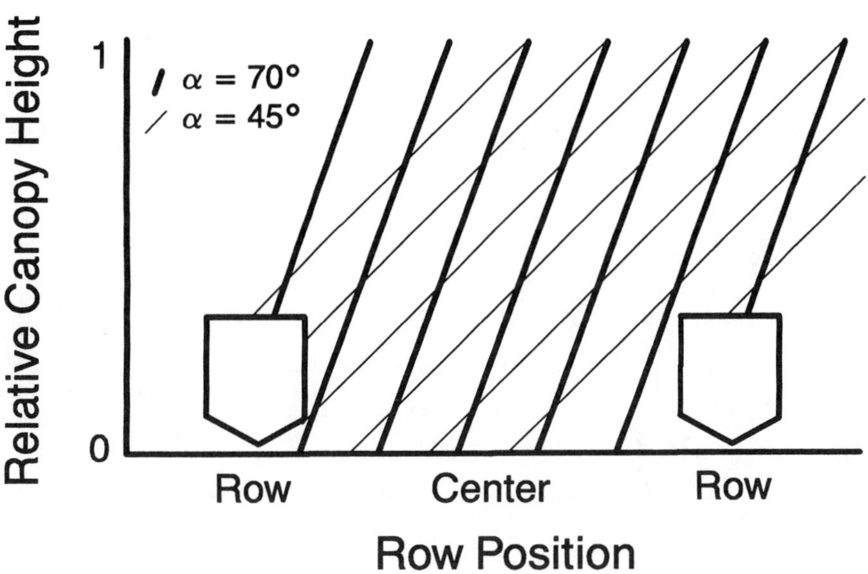

Figure 1. Schematic of direct beam solar radiation through a partial canopy of a crop.

was a rapid decrease in the extinction coefficient as the solar angle increases and for corn (*Zea mays* L.), found values above 5 when the solar elevation was less than 10° and decreased exponentially to 0.7 at solar elevations greater than 50°. There is a difference in the absorption between the photosynthetically active wavelengths of light (0.4 to 0.7 μm) compared to near-infrared wavelengths (0.7 to 1.1 μm). Leaves have a greater absorption coefficient for photosynthetically active wavelengths since these wavebands are used for the photosynthetic process while the absorption is lower in the near-infrared wavebands because these wavebands contain a large amount of energy but do not contribute to physiological processes. These aspects have been discussed by a number of researchers (Lang et al., 1985; Lemeur, 1973; and Whitfield, 1986).

Light penetration through a crop canopy is complex; however, in terms of understanding the principles, the major factors are illustrated in Figure 1. In a partial canopy that does not completely cover the space between the rows, a beam of light will reach the soil surface depending upon the sun angle and row direction of the canopy (Lemeur, 1973; Whitfield, 1986). This is coupled with the amount of solar radiation transmitted through the canopy throughout a day as shown in Figure 2. Patterns at the soil surface are complex and dependent upon the geometry of the canopy, the stage of growth, and the time of year. The latter is due to a change in angle of the direct beam radiation with the azimuth angle of the sun. Ross (1975) provided an overall summary of solar radiation penetration into crop canopies for both direct and diffuse solar radiation. Radiation at the soil surface is

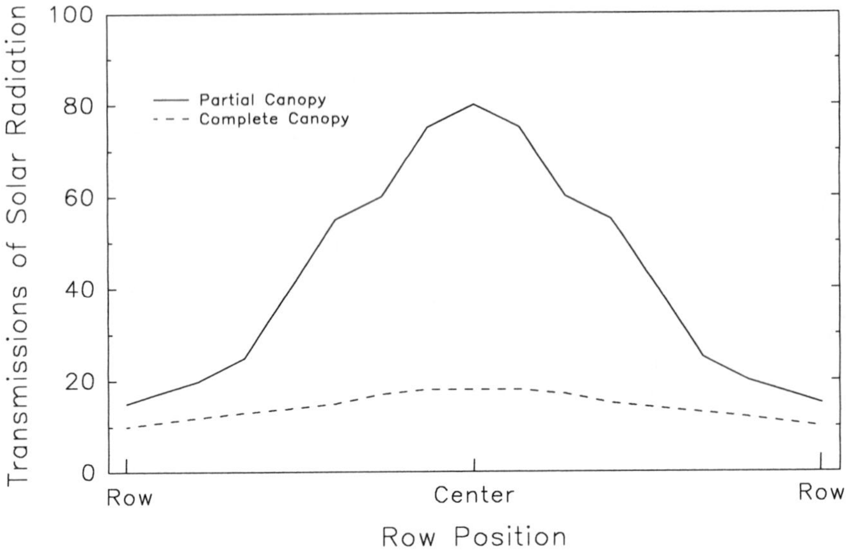

Figure 2. Transmission of solar radiation through a partial and complete canopy with rows oriented in a north-south direction.

extremely variable both temporally and spatially. The factors that contribute to this variability are:

1. Fluctuations due to the leaf flutter in the wind,
2. Fluctuations due to short periods of sun movement in the sky,
3. Fluctuations due to the diurnal movement of the sun, e.g., day and night, and
4. Fluctuations due to the annual change in the azimuth angle of the sun (Ross, 1975).

Direct beam solar radiation penetrating to a given leaf area, L, can be written as:

$$S_{bp}(L,n_s) = S_b(n_s)\tau_s(L,n_s) \quad (2)$$

where L is described as the function

$$L(z) = {}_0\int^h a_L(z)\, dz \quad (3)$$

where a_L is the vertical distribution of leaf area with height, n_s is the direction of the solar beam given as function of solar inclination and azimuth orientation, and τ_s is the representation of the fractional area of sunflecks at the soil surface. Miller and Norman (1971a,b) provided an explanation of the sunfleck theory for plant canopies and later Norman et al. (1971) developed a model for the prediction of sunflecks through plant canopies. Area of sunflecks are important since the bright spots at the soil surface may be responsible for providing sufficient energy for growth of weeds in the understory of a crop canopy and potential red and blue light for triggering the germination process. Penetration of solar radiation is more complicated when the diffuse beam is considered because there is no longer a directional angle and radiation comes from all angles above the canopy. An expression similar to Equation 2 can be used for diffuse solar radiation, $S_{dp}(L)$, and given as:

$$S_{dp}(L) = {}_0\int^{2\pi} d(n)\tau_s(L,n) \cos\theta \, d\Omega \quad (4)$$

where d(n) is the radiance of the sky in the direction n, τ_s is the penetration function in the direction n, and $d\Omega = \sin\theta \, d\theta \, d\phi$ where θ is the solar inclination angle and ϕ is the azimuth angle (Ross, 1975). Combining Equations 2 and 4 will allow us to evaluate the changes in the area of sunflecks at the soil surface. The difference between the extinction coefficient, k, and τ_s is given as:

$$\tau_s = \exp(-L\, G(n_s)/\sin\beta - L_s \cot\beta/\pi) \quad (5)$$

where $G(n_s)$ is a foliage orientation function, which is a function of solar inclination, azimuth angle, and leaf angle, β is the solar elevation, L is the leaf area, and L_s is the stem area. If we examine Equations 1 through 5 we can see that solar radiation at the bottom of the canopy changes dramatically throughout a day and over the course of a growing season. Variation among canopies in leaf orientation, size, and planting configuration will cause variation in the amount of solar radiation at the lower part of the canopy (Figure 1).

Combining Equations 3, 4, and 5 permits the estimation of the area of sunflecks at the soil surface for two different canopies of soybean to illustrate the effect of a difference in canopy density (Figure 3). The variation at each position was estimated by simulating different times of the day with the sunfleck model. Two observations are obvious: there is less variation about the area of sunfleck between the rows compared to next to the row, and there is an increase in variation when there is a decrease in canopy density. Weeds would have an increased chance of growing when there is an increase in plant skips or variations in the canopy that would lead to increased variation in solar radiation at the soil surface. Mann et al. (1980) showed that in cotton (*Gossypium hirsutum* L.) canopies, random plant spacing greatly impacted the distribution of light within the plant canopy and at the soil surface. A similar approach can be used to estimate the photosynthetically active radiation at the soil surface. These models could be used to estimate the amount of light energy required for photosynthesis and weed growth.

A complication in light availability for weed germination or growth occurs when there is crop residue on the soil surface. Sauer et al. (1996b) showed that light transmission through corn residue was dependent upon the crop residue index. They described this index as a function of the area of soil covered by a crop residue. There was little effect of different ages of crop residue on light transmission through the residue. Teasdale and Mohler (1993) showed that transmission of light through residue of hairy vetch (*Vicia villosa* Roth) and rye (*Secale cereale* L.) was dependent upon the amount and thickness of residue present. Transmittance

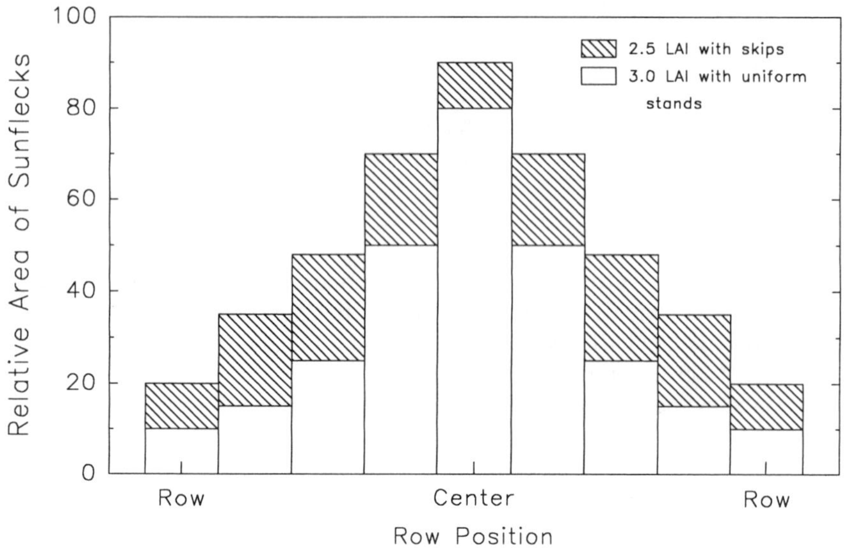

Figure 3. Area of sunflecks at the soil surface under a soybean canopy with a leaf area index (LAI) of 3 and 2.5 (represents a skip in plant rows).

through residue of both crops behaved as an exponential decay function; however, penetration of light through the hairy vetch changed more rapidly throughout the season because of the faster decay of the plant material. Teasdale and Mohler (1993) concluded that there was sufficient light transmitted through crop residue to activate phytochrome and stimulate germination. Woolley and Stoller (1978) had shown earlier that light penetration into the soil was limited to the upper 4-5 mm and this depended upon soil moisture, texture, and color. Teasdale and Mohler (1993) concluded that addition of residue to the surface would prevent germination of seed requiring red light because even under unmulched conditions, the quantity of red light available at the depths of many seeds would be marginal for promotion of germination. Addition of crop residue to the soil surface will further reduce the potential amount of red light to weed seed scattered upon the soil surface and will greatly impact those seed buried in the upper soil layer.

At the soil surface under a plant canopy there would be a shift in the red:far red ratio because of the differential adsorption of solar radiation by the plant canopy. Red light is preferentially adsorbed while far red is less adsorbed (Hatfield and Carlson, 1979). This shift and the effect on seed germination has been extensively discussed by Frankland (1981), who showed that shifts to the far red would inhibit germination. Weed seed germination would be inhibited under dense canopies because of the shifts in the red:far red ratio.

The effect of canopy shape and residue on light availability at the soil surface is a complex system. However, the components to develop a weed germination model are available since we can accurately simulate radiation through a canopy at

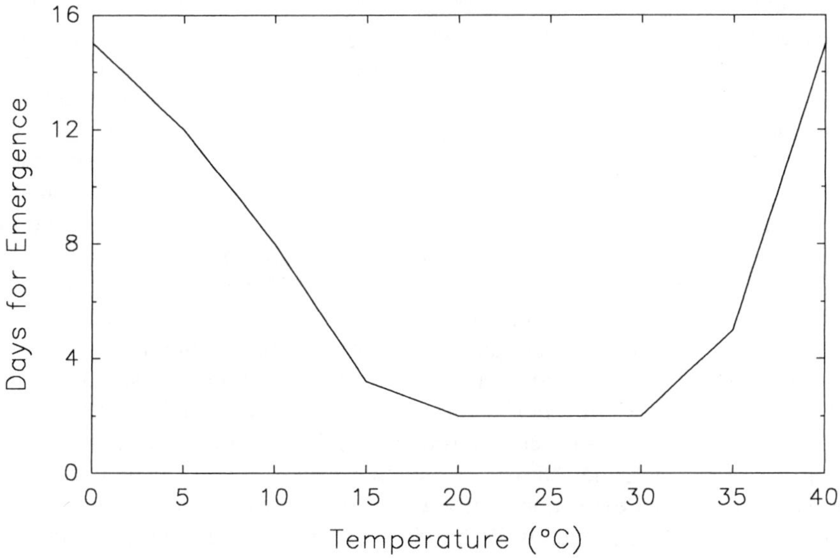

Figure 4. Temperature response function for the germination of a warm season weed seed.

different stages of growth (Lemeur and Blad, 1974) and can also estimate the impact on the radiation regime at the soil surface with the presence of crop residue. The radiation balance can be combined with the light requirements of different weed seed to provide a comprehensive model of weed seed germination.

III. Temperature

Seed germination and plant growth are functions of temperature. For the purpose of this discussion, the temperature response for germination is assumed to be similar to that shown in Figure 4. Horowitz and Taylorson (1984) and Wiese and Binning (1987) provide details on the temperature requirements of weed seed germination. Thus, we need to develop an understanding of the temperature regime of the soil surface and shallow layers of the soil if we are to understand the dynamics of temperature change relative to weed seed germination and growth. This is further complicated by the relationship between soil temperature regime and soil water status of the soil surface. Surface and near-surface soil temperature and water status are critical areas that we need to understand before we can appreciate the complexities of the temperature regime at the plant canopy-soil surface interface.

Soil temperature is a critical part of the soil microclimate and a factor in biological processes within the soil. There are several models that predict soil temperature. Luo et al. (1992) developed a simulation model that predicts soil temperature with variable crop cover and soil water content. Their model is based on evaluation of the components of the energy balance at the soil surface and produced results in close agreement to measured values. Grant et al. (1995) developed a similar framework for a model that could be used to assess soil temperature under different crop management practices. Gupta et al. (1983) and Gupta et al. (1984) showed that the upper soil boundary temperatures are dominated by the disruption of the soil surface through tillage and the addition of residue. They found the largest difference due to residue and tillage was in the nighttime period because of the greater effect of these changes on the cooling rate during radiation loss from the soil surface. McInnes et al. (1986) used a coupled heat and mass transport simulation model to estimate both soil surface water content and soil temperature. They found that soil temperatures agreed within 1.7° C for two different soils when simulated over a range of net radiation values. Thunholm (1990) evaluated a numerical model, based on air temperature and soil surface net radiation, to compare measured and predicted soil temperatures over a 22-year period. He found an underestimate of soil temperature during autumn because of the use of a constant extinction coefficient for net radiation through the canopy layer. This model performed well with the major differences due to within year variations induced by changes in the growth of the grass that influenced the surface energy balance. Novak (1991) developed a similar model for British Columbia and evaluated it for a 66 day period in the spring. He found a good agreement between measured and predicted results during the period after the soil began to thaw in the spring. Kluitenberg and Horton (1990) developed an analytical solution model to estimate heat conduction under varying amounts of surface mulch. They used this model to demonstrate that there can be a large transfer of energy horizontally within the upper layers of the soil profile. These types of models can be routinely used for the estimation of soil temperature with acceptable and repeatable results across a wide range of conditions.

One of the critical aspects that is currently missing is the temporal and spatial variation in soil properties. The amount of energy stored within the upper soil profile and released to the atmosphere is dependent upon properties that vary with farming practices. Mohanty et al. (1995) evaluated the spatial distribution of soil temperatures at 1 and 10 cm within a field. They found that soil temperatures in the direction of the crop row or the tillage operation were related to one another within a 9 m distance. However, across the rows the measurements were related to one another within a 1.5 m distance. Beyond these distances, soil temperatures varied randomly. These observations show that we need to consider different aspects of farming practices as we conduct experiments to measure the microclimate relative to weed germination and growth.

Density of the soil changes in response to how the soil is tilled; however, we don't understand the rate at which the surface soil density changes after the soil is disturbed and the role that precipitation plays in the change of the soil surface. Bradford and Huang (1992) have detailed processes affecting soil crusting and changes in surface density. They found that soil texture, antecedent moisture content, aggregate stability, surface roughness and clodiness, and climate were factors that influenced formation and stability of a surface soil crust. These variables interact as the soil is disturbed and will influence the structure of the soil surface. There are inherent differences in density among soils so that each soil will have a unique density at a given soil water content. For example, the less organic soils will tend to have a larger density because of higher fraction of minerals. Water content of the soil surface layer changes rapidly as the soil dries and influences the thermal conductivity. We have little detailed understanding of the temporal variation in water content or density and if we add the fact that soils vary across a field, there is little information on how thermal conductivity varies across a landscape. This latter issue is important since many of the large-scale models require an estimate of the available energy that is partitioned into the soil and the variation of soil heat flux. Surface thermal properties respond to changes induced by man (tillage) and nature (rainfall) and both of these factors will determine responses of the soil thermal regime. There are common properties of the soil surface that can be used to couple temperature and water components. Thermal conductivity describes the ability of the soil to conduct heat, while hydraulic conductivity describes the soil's ability to transfer water from one layer to another, and gaseous diffusivity describes how quickly different gases move through the soil layers. Each of these factors will be discussed briefly to explore the role of soil factors in energy and mass transport.

Soil temperature changes are a function of the thermal conductivity and the energy available at the soil surface. Soil heat flux, G, can be estimated using a relatively simple approach as:

$$G = -\lambda\ \partial T/\partial z \tag{6}$$

where λ is the thermal conductivity and $\partial T/\partial z$ is the temperature gradient with depth in the soil. Thermal conductivity, λ, can be considered a measure of the energy exchange for a given temperature gradient and determines the rate of temperature change (Equation 6). Thermal conductivity is a function of the type of material and ranges from 0.025 to 8.8 W m^{-1} °C^{-1} for air and quartz, respectively. In soil, this is further complicated by the fact that the water content in the surface layer changes very rapidly and there are differences in the proportion of clay and organic materials. We can imagine that as the composition of the soil changes (e.g., the amount of organic matter, air content, or water content), thermal conductivity changes. One of the most dynamic parts of soil is the surface layer, and through human intervention and cultural practices, the upper portion of the soil layer changes dramatically in both time and space. Water contents in the upper 1 cm often vary from <10% to >40%, and in a clay soil this would represent over a 600% increase in thermal conductivity from 0.25 to 1.58 W m^{-1} °C^{-1}. Changes in the density of the soil as shown as the soil texture changes from a sand to a peat affects the thermal conductivity for a given water content. This influences the rate at which the soil changes temperature throughout the day. As the mineral properties of the soil vary and the water content varies with depth, there would be a change in the thermal conductivity.

Since mineral properties and water content change in the soil, an additional term has been introduced to describe the rate of temperature change in a given soil volume. Thermal diffusivity, κ, is estimated as:

$$\kappa = \lambda / \rho c \tag{7}$$

in which κ is related to thermal conductivity and volumetric specific heat capacity of the soil, $\rho_s c$. The volumetric specific heat capacity of soil is a function of the density, ρ_s, and the specific heat, c. Both of these properties vary with soil minerals, organic matter content, and water content.

Johnson and Lowery (1985), Mielke et al. (1986), and Azooz and Arshad (1995) have shown that the soil surface properties respond to management. Johnson and Lowery (1985) found that thermal diffusivity was 20-25% higher in soils under no-till management compared to cultivated fields. They found volumetric heat capacity increased as the degree of soil disturbance decreased. Mielke et al. (1986) reported that the effect of different tillage operations on the soil physical properties was confined to the upper 75 mm of the soil profile. Their samples were collected at one time of the year so it is difficult to assess the temporal variation; however, the depth of the effect of tillage would be within the zone of weed seed deposition and potential germination. Potter et al. (1985) found soil volumetric heat capacity was similar for three different tillage systems. Their tillage systems ranged from no disturbance of the soil surface to complete disturbance. Thermal diffusivity and thermal conductivity increased when the soil surface was not disturbed under the no-tillage conditions. Increased soil water content in reduced tillage fields increased the thermal conductivity, and increased thermal conductivity may offset any negative effects due to cooler temperatures in reduced tillage practices (Azooz and Arshad, 1995). Density decreases as the soil surface is disturbed through tillage operations, and specific heat increases with increasing water content since water has the highest specific heat of any of the soil constituents. It is reasonable to assume that as we till soil, soil properties will change and create a different rate of energy exchange with the atmosphere because of the disruption of the thermal regime within the soil.

Weed germination and growth are influenced by soil temperature and although there are models to estimate soil temperature, these have not been related to weed management. There are a number of factors that influence soil temperature. Soil temperature under a plant canopy is affected by the stage of canopy development, the surface soil water content, tillage history, and time of year. Teasdale and Mohler (1993) found that soil temperature under rye and hairy vetch residue was reduced in both maximum temperatures and daily amplitude. Hatfield and Prueger (1996a) showed that corn residue influenced the rate of cooling in the fall and the diurnal amplitude throughout the year. Sauer et al. (1996a) also showed that there was an effect of corn residue on soil temperature and the presence of residue reduced the soil water evaporation rate by 40% compared to a bare soil. Addition of crop residue influences light available at the soil surface, and reducing the energy available at the soil surface in turn reduces the amount of heating and evaporation.

Soil temperature is a critical component of the microclimate influencing weed germination and emergence. However, soil temperature, like solar radiation, can be estimated with energy balance models. Horton et al. (1984) developed a model to predict soil temperature between the rows of a canopy that partially shade the soil surface. They coupled a radiation distribution model similar to that described earlier with a heat conduction model based on Equation 6. To solve for the soil temperature they used an energy balance model as shown in Equation 8:

$$R_n = G + H + LE \tag{8}$$

where R_n is the net radiation, G is the soil heat flux (Equation 6), H is the sensible heat flux, and LE is the latent heat flux with all components expressed in J m^{-2} s^{-1}. Hatfield and Prueger (1996b) showed the details of the use of the energy balance to estimate both canopy and soil parameters for a variety of conditions. Horton et al. (1984) showed that maximum soil temperatures at the 5 cm depth could range from 27° C under the canopy to 39° C between the rows for the period in mid-August. They found good agreement between the measured and estimated soil temperatures.

Positional variation of the energy balance at the soil surface under a partial canopy will determine the variation in soil temperature at the soil surface. Ham et al. (1991) showed that the energy exchanges in the lower canopy near the soil surface were dependent upon the sensible heat transport, radiation balance, and surface soil water content. Ham and Kluitenberg (1993) found large positional variation in the soil energy balance under a soybean (*Glycine max* L. Merr.) canopy. The magnitude and pattern of the soil heat flux and soil temperatures were strongly dependent upon position relative to canopy row. The primary factor influencing the temperature variation was the sun-canopy geometry as shown in Figure 1. Maximum soil surface temperatures ranged from 27° C directly under the row to 53° C between the rows when the canopy was at 50% ground cover on Day-of-Year (DOY) 204 (July 13). This magnitude of soil temperature variation would give rise to extreme differences in seed germination response if one considered a warm season species as shown in Figure 4. Using a similar approach to Ham and Kluitenberg (1993) and the energy balance model of Horton et al. (1984), estimates of the 1 cm soil temperature for early June for a partial canopy cover are shown in Figure 5. There is a large difference in the soil temperatures (>20° C) as the position changes from under the crop row to between the rows. Soil surface temperatures are more dramatic in their variation due to position. The implications for weed germination, emergence, and growth are large and could be used to describe why there is variation in weed populations under plant canopies.

Addition of crop residue and tillage will further impact the surface energy balance below the crop canopy. Crop residue will alter the radiation available at the soil surface, as shown by Teasdale and Mohler (1993) and Sauer et al. (1996b), and will impact the red light available for germination, and radiation for energy exchange. Crop residue will compound the variation of temperature under a canopy and will require the additional measurement of the amount and distribution of residue. Fortin and Hamill (1994) showed that corn development was influenced by the configuration of rye residue on the surface. They found that standing rye

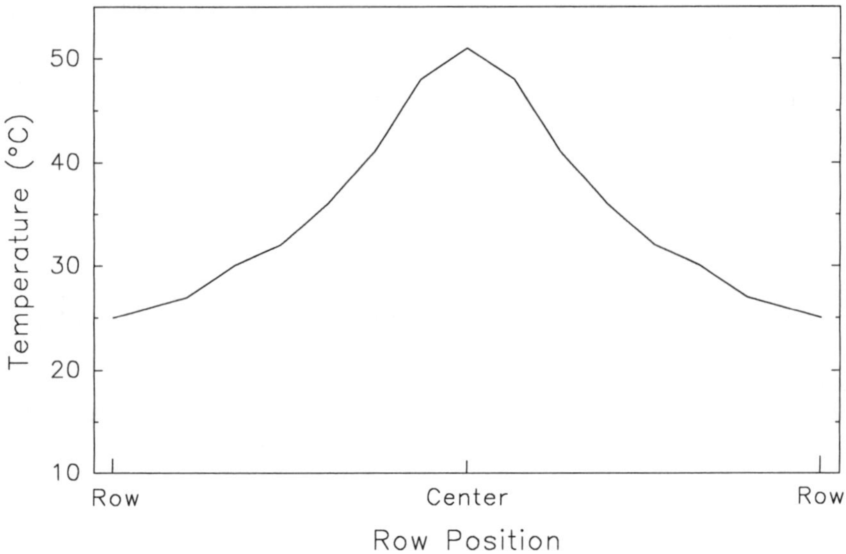

Figure 5. Simulated 1 cm soil temperatures for a partial canopy in early June with a bare, dry soil surface.

residue resulted in faster corn development than flat rye residue because of the warmer seed zone in the standing residue compared to flat residue. Configuration of residue and its effect on weed emergence and growth under plant canopies has not been evaluated. The available simulation models of the energy exchanges will have to be modified to estimate the effect of different crop canopy and residue configurations on weed germination and emergence. This will require knowledge of the temperature response curve shown in Figure 4 and light sensitivity for germination. Residue configuration and amount will change with tillage practices, and McCool et al. (1995) described how tillage interacts with residue mass and cover. This approach could provide useful parameters for incorporation into a canopy model.

IV. Moisture

Soil moisture is a critical component of microclimate required for weed germination, emergence, and growth. The dynamics of the soil water regime under a crop canopy are more complicated than light or soil temperature. Hydraulic conductivity of soil is similar to thermal conductivity and is related to the arrangement of pores within soil and water content. We can express hydraulic conductivity, K, as a function of soil water potential, which is a nonlinear function. Hydraulic conductivity values range from 10^{-5} m s^{-1} when the soil water potential is near saturation to 10^{-11} m s^{-1} when the soil is nearly air dry. Messing (1989) employed a relationship between clay content and effective porosity to relate soil water data to hydraulic conductivity. This approach is typical of the methods used to experimentally evaluate hydraulic conductivity. Porosity of a soil is a function of the number of pores within the soil and connectivity of the pores and determines how quickly water will move through the soil. Globus and Gee (1995) evaluated a column method to estimate the hydraulic conductivity of moderately dry soil since there are relatively few measurements under less than saturated soil conditions. They found that a method of using sieved soil in columns could reliably estimate hydraulic conductivity and water diffusivity. This method could be used to obtain data on soil properties that have traditionally been estimated.

Estimation and understanding of the soil hydraulic properties is important in the determination of the rate of drying of the soil surface. Changes in the soil water content near the surface are driven by the latent heat of vaporization given as:

$$LE = \rho C_p(e_s(T_s) - e_a)/(r_{av} + r_s) \quad (9)$$

where ρ is the density of air (kg m^{-3}), C_p is the specific heat of air (J kg^{-1} K^{-1}), e_s is the saturation vapor pressure (kPa) at the surface temperature (T_s), e_a is the actual vapor pressure of the air (kPa), r_{av} is the resistance for water vapor transfer in the air (s m^{-1}), and r_s is the surface resistance (s m^{-1}). Equation 9 can be used in combination with Equation 8 to estimate the rate of water lost from the soil surface and hence the rate of soil drying at different positions under a canopy. Hatfield and Prueger (1996b) summarized the various approaches available for the estimation of r_{av} and r_s for different soil surface conditions and plant canopies.

Surface resistances, r_s, for a bare soil are relatively simple and increase in complexity as the number of vegetative layers increase. Shuttleworth and Wallace (1985) were the first to describe a one-dimensional model of a partial canopy. They partitioned the available energy at the upper layer of the canopy into that intercepted by the crop and that available at the soil surface. This is a simple box type of model that assumes a distinct change between the canopy layer and soil layer. This conceptual model has been adopted by a number of different research groups with a range of results. Ham et al. (1991) applied this concept to cotton to separate LE from the soil surface, LE_s, and from the canopy, LE_c. Before irrigation, when the soil surface was dry, LE_s was approximately a quarter of the total LE, and when the soil surface was wet, immediately after irrigation, LE_s was over half of the daily LE. Allen (1990), working in northern Syria with barley, found that LE_s accounted for 67 to 77% of the total LE for the crop. These differences in the LE from the soil and crop canopy suggest that a rigorous model is required to handle the complexity of the partial canopy surface. Lascano et al. (1987) applied the concepts of the Shuttleworth and Wallace model to a cotton crop to separate LE_s from LE_c and found the soil fraction was 30% of the total LE under both irrigated and rainfed canopies on the High Plains of Texas. The cumulative LE_s values were within 1 standard deviation of the measured values throughout the season-long study (Lascano et al., 1987). Ham and Kluitenberg (1993) measured the variation beneath a soybean canopy and found that soil aerodynamic transfer coefficients were highly variable but were not related to windspeed above the canopy or position beneath the canopy. Their values were similar to other studies; however, the variation in transfer coefficients could not be evaluated with the methods of their study. They found that, when the soil surface was dry, the majority of the sensible heat was from the soil directly between the two rows and often exceeded 400 W m^{-2}.

In the desert rangeland, Nichols (1992) found the Shuttleworth and Wallace model to produce satisfactory estimates of sensible heat from the soil and latent heat from the canopy; however, latent heat estimates from the soil were less reliable. Sensible heat flux from the canopy was not predicted well from the resistance approach and was not well-behaved when a bulk aerodynamic resistance value was used in the model. In this study the aerodynamic resistances ranged from <10 s m^{-1} to 100 s m^{-1} and r_a values decreased with increasing windspeed. Soil resistance values were between 70 and 400 s m^{-1} and linearly decreased with windspeed. Differences between measured and estimated values of resistances were largest at the low windspeeds. Nichols concluded that for sparsely vegetated surfaces, a better understanding of the factors affecting energy transport would be required to improve the resistance type models beyond their current level.

Horton (1989) used a two-dimensional model to account for shading of the soil surface by a crop canopy. He used a numerical model to estimate surface energy partitioning, soil temperature, and soil water content as a function of the soil water evaporation rate. Spatial variation increased when soil between plant rows became drier than soil beneath a plant canopy. Massman and Ham (1994) proposed an alternative method to separating the canopy and soil fractions for LE estimation. They used a combination of soil surface resistance and a Bowen ratio technique. A Bowen ratio was computed for the soil based on observations of standard meteorological information. This was possible since they found that soil surface resistance and soil Bowen ratio did not change significantly during the day and that it was possible to estimate both terms from diurnal meteorological data. Their empirical approach partitioned available energy into soil and plant components for each day. Their results show a better agreement between measured and predicted transpiration rates on dry days compared to days with a wet soil surface. Villalobos and Fereres (1990) measured the soil water evaporation rate beneath three canopies and found general agreement among corn, cotton, and sunflower (*Helianthus annuus* L.) canopies. The relationship between the fraction of water evaporated from the soil surface and transpired from the crop canopy was related to the amount of leaf area. They found as LE_s decreased, LE_c increased because of the increase in microadvection from the warm bare soil to surrounding vegetation. Leuning et al. (1994) measured both the rainfall interception and the soil water evaporation under a wheat canopy for an 80 day period during the growth cycle. Combined rainfall interception and soil water evaporation accounted for 49% of the LE_t for this growth period. They found that decreasing the soil water evaporation rate by accelerating canopy closure would only increase the rainfall interception and subsequent loss via evaporation from leaves.

It is possible to estimate the evaporative flux from the soil surface under a canopy using the procedures described by Horton (1989) and the expansion of these terms contained in Shuttleworth and Wallace (1985). If we assume partial canopy

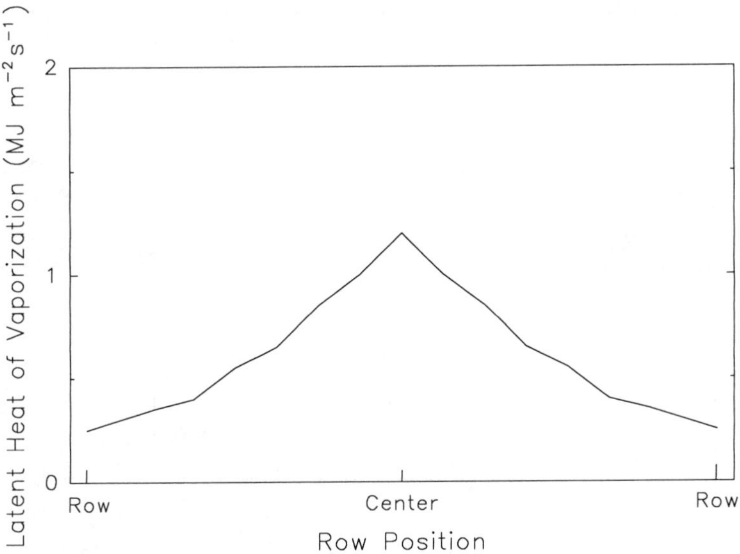

Figure 6. Simulated latent heat of vaporization fluxes for midday under a partial canopy in early June with a wet, bare soil surface.

conditions, the evaporative flux beneath a canopy will depend upon the position relative to the plant row. This is illustrated by the LE_s results shown in Figure 6 for a single day with assumptions about the changes in hydraulic conductivity of the surface with soil water content and no crop residue present on the soil surface. Crop residue adds an additional resistance in Equation 9 that further reduces water vapor transport from the soil surface to the atmosphere. This will lead to a greater soil water content and reduced rate of soil drying under a crop residue. This aspect has been illustrated by Teasdale and Mohler (1993), Horton et al. (1994), and Sauer et al. (1996a).

The approach for a partial canopy can be extended to complete canopy cover. Microclimatic conditions under a complete canopy do not exhibit the spatial variation at the soil surface compared to a partial canopy because of the more uniform radiation environment at the soil surface. To illustrate the change in the soil surface microclimate under a canopy, the change in water content of a silt loam soil without a crop residue layer was simulated. Thermal conductivity and hydraulic conductivity were spatially adjusted between the rows to simulate the effect of tillage. To represent the effect of tillage, both values were decreased by 40% for a silt loam soil for 75% of the area between the rows. To further simplify the model, solar radiation for a clear sky June day at Ames, Iowa was used as input. Atmospheric conditions with a typical diurnal variation and canopy characteristics were maintained constant among days. The number of days for the soil surface to change from a wet to dry condition (30% of saturation) was estimated and the results are shown in Figure 7. There was rapid drying between the rows and by the

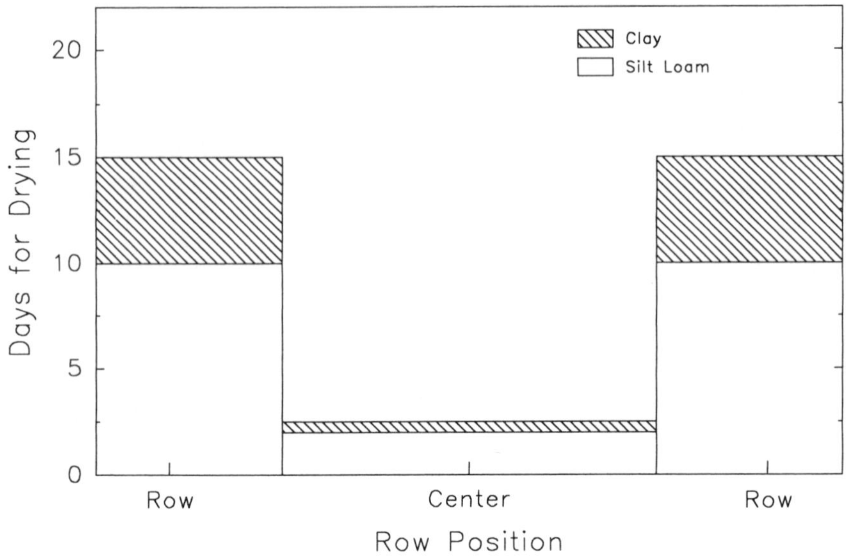

Figure 7. Number of days for a silt loam and clay soil surface to dry from a wet soil condition to 0.15 volumetric water content using a differential thermal and hydraulic conductivity between rows and typical atmospheric conditions for early June in central Iowa.

third day, the tilled area was dry while the area under the canopy did not dry for 10 days. Changing the soil parameters to represent a condition of a clay loam soil revealed that the rate of drying was more influenced for the shaded conditions under the row than between rows, and the time to dry under the rows increased to 15 days. These are relatively simple models that couple energy exchange and soil water conditions. Surface energy balance models could be applied to improving the understanding of the dynamics of weed germination and emergence under crop canopies by linking the soil microclimate with physiological requirements for weed growth. These aspects have not been combined for weeds; however, exist for several crops.

V. Application of Micrometeorology to Weed Management

In the previous sections the radiation, temperature, and soil water regimes at the soil surface have been discussed and procedures for the estimation of these regimes have been detailed. We can couple these components together as suggested by Egley (1986) to demonstrate how a weed emergence and growth model could be driven based on micrometeorological and soil processes. Combining the radiation regime with soil temperature and soil water changes, a model was constructed that

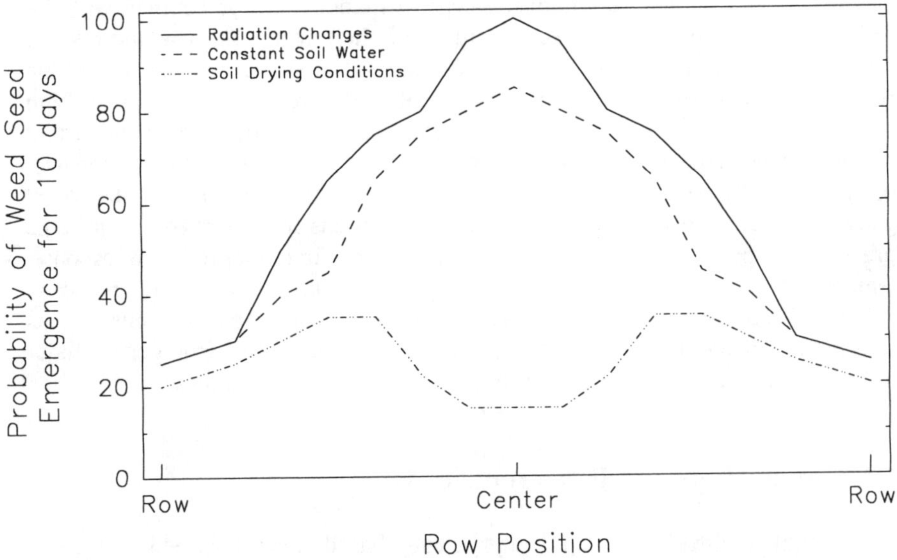

Figure 8. Probability of a weed seed emerging from the soil under partial canopy conditions with an initial wet soil surface and no surface residue. Conditions are typical of clear sky conditions in early June in central Iowa. (Simulations made with radiation changes only, constant soil water, or soil drying conditions.)

computed the probability of a weed seed emerging and growing for 10 days with an adequate soil water supply in the upper 5 cm of soil. It was assumed if the temperature of the upper 1 cm of soil exceeded 40° C for over 4 hours and the soil water content in the upper 5 cm decreased to less than 50% of the available soil water that a weed seedling would not survive. The results of the simulation are shown in Figure 8 for a partial canopy of soybean planted in 75 cm rows oriented north-south that covers 50% of the ground area. If we hold all factors constant other than radiation, e.g., temperature and soil water, there is sufficient radiation at the soil surface to support initial weed growth (Figure 8, line a). Soil water content was held constant and latent heat of vaporization proceeded at the maximum rate based on the energy balance at the surface for the second simulation. Simulated temperature in the upper 1 cm of soil showed a slight decrease in the probability of a weed growing for 10 days in the area between the rows (Figure 8, line b). If we allow the soil surface to dry as in a natural environment, the probability of a weed growing is greatly diminished between the rows and the area partially shaded under the canopy shows a more favorable area for weed growth (Figure 8, line c). Only line c on Figure 8 is realistic; however, the other changes show the interaction among microclimatic factors and the need to develop integrated models of canopy microclimate for application to weed problems.

This approach is instructive because it permits an evaluation of the effect of skips in the plant stand, addition of different residue configurations, row width, and canopy geometry. The results of this type of model could be coupled with the observations made by many weed scientists on the effects of tillage and residue management on weed population shifts (Buhler, 1996). Unfortunately, few of the required parameters needed to drive this type of model are currently available from weed population studies. Many of these factors have been suggested as important considerations in the various chapters within this volume; however, there is the lack of a physically-based biological model that can be used to describe weed germination, emergence, and growth. These simulations are illustrative and lack the precise description of various weed seed requirements for red:far red light response, temperature and water requirements for germination, photosynthetic response curves of seedlings, and effect of weed seed size on soil water response. This information could then be used to estimate potential weed emergence under different canopies and management (tillage and residue) techniques.

VI. Conclusions and Research Needs

It is possible to develop and apply a physically-based model for weed emergence and growth although there is insufficient knowledge of the relevant biological parameters for this type of model to handle different weed species. However, this type of approach shows the utility of integrating physical and biological scientists to work on weed management problems. The micrometeorological components of the model have been developed and tested at a number of sites with acceptable results; however, the coupling with weed problems remains an unknown. There are a number of research needs that can produce fruitful results through this integration. Some of these are listed below in order to produce some potential dialogue among scientists.

Evaluation of the temperature response curves for the germination of different weed species.

Determination of soil water response of different seed size of weeds.

Evaluation of the weed emergence distribution under different tillage and residue configurations.

Evaluation of the effect of tillage on the weed seed zone microclimate (upper 10 mm of soil).

Evaluation of the dynamics of the tillage response on the soil microclimate and the interaction with weed growth.

Evaluation of weed distribution within a field as related to soil water and soil temperature variation caused by soil differences.

Determination of the expected impact on changing tillage and residue management on the projected weed population shifts.

These are only a few items that can be easily tackled by a number of research groups. The micrometeorological components are fairly standard and can be adapted to a range of conditions. Coupling this model with research results from weed ecologists and weed management specialists will open new insights into an extremely challenging and perplexing problem.

References

Allen, S.J. 1990. Measurement and estimation of evaporation from soil under sparse barley crops in northern Syria. *Agric. For. Meteorol.* 49:291-309.

Angus, J.F., R. Jones, and J.H. Wilson. 1972. A comparison of barley cultivars with different leaf inclinations. *Aust. J. Agric. Res.* 23:945-957.

Azooz, R.H. and M.A. Arshad. 1995. Tillage effects on thermal conductivity of two soils in Northern British Columbia. *Soil Sci. Soc. Am. J.* 59:1413-1423.

Bradford, J.M. and C. Huang. 1992. Mechanisms of crust formation: Physical components. In: Sumner, M.E. and B.A. Stewart (eds.), *Soil Crusting: Chemical and Physical Processes. Advances in Soil Sci.*, Lewis Publishers, Boca Raton, FL. p. 55-72.

Buhler, D.D. 1996. Tillage systems and weed population dynamics and management. In: Hatfield, J.L., D.D. Buhler, and B.A. Stewart (eds.), *Weed and Soil Management, Advances in Soil Science*, CRC Press, Boca Raton, FL (this volume).

Cowan, I.R. 1968. The interception and absorption of radiation in plant stands. *J. Appl. Ecol.* 5:367-379.

Egley, G.H. 1986. Stimulation of weed seed germination in soil. *Rev. Weed Sci.* 2:67-89.

Fortin, M.-C. and A.S. Hamill. 1994. Rye residue geometry for faster corn development. *Agron. J.* 86:238-243.

Frankland, B. 1981. Germination in shade. In: Smith, H. (ed.), *Plants and the Daylight Spectrum*. Academic Press, London. p. 187-204.

Globus, A.M. and G.W. Gee. 1995. Method to estimate water diffusivity and hydraulic conductivity of moderately dry soil. *Soil Sci. Soc. Am. J.* 59:684-689.

Grant, R.F., R.C. Izaurralde, and D.S. Chanasyk. 1995. Soil temperature under different surface managements: Testing a simulation model. *Agric. For. Meteorol.* 73:89-113.

Gupta, S.C., W.E. Larson, and D.R. Linden. 1983. Tillage and surface residue effects on soil upper boundary temperatures. *Soil Sci. Soc. Am. J.* 47:1212-1218.

Gupta, S.C., W.E. Larson, and R.R. Allmaras. 1984. Predicting soil temperature and soil heat flux under different tillage-surface residue conditions. *Soil Sci. Soc. Am. J.* 48:223-232.

Ham, J.M. and G.J. Kluitenberg. 1993. Positional variation in the soil energy balance beneath a row-crop canopy. *Agric. For. Meteorol.* 63:73-92.

Ham, J.M., J.L. Heilman, and R.J. Lascano. 1991. Soil and canopy energy balances of a row crop at partial cover. *Agron. J.* 83:744-753.

Hatfield, J.L. and R.E. Carlson. 1978. Photosynthetically active radiation, CO_2 uptake, and stomatal diffusive resistance profiles within soybean canopies. *Agron. J.* 70:592-596.

Hatfield, J.L. and R.E. Carlson. 1979. Light quality distributions and spectral albedo of three maize canopies. *Agric. Meteorol.* 20:215-226.

Hatfield, J.L. and J.H. Prueger. 1996a. Microclimate effects of crop residue on biological processes. *Theor. Appl. Clim.* (in press).

Hatfield, J.L. and J.H. Prueger. 1996b. Soil surface-atmosphere interactions. In: Stefan, V. (ed.), *Research Trends in Environmental Physics*, American Institute of Physics, New York (in press).

Horowitz, M. and R.B. Taylorson. 1984. Hardseededness and germinability of velvetleaf (*Abutilon theophrasti*) as affected by temperature and moisture. *Weed Sci.* 32:111-115.

Horton, R. 1989. Canopy shading effects on soil heat and water flow. *Soil Sci. Soc. Am. J.* 53:669-679.

Horton, R., O. Aguirre-Luna, and P.J. Wierenga. 1984. Soil temperature in a row crop with incomplete cover. *Soil Sci. Soc. Am. J.* 48:1225-1232.

Horton, R., G.L. Kluitenberg, and K.L. Bristow. 1994. Surface crop residue effects on the soil surface energy balance. In: P.W. Unger (ed.), *Managing Agricultural Residues*. Lewis Publishers, Boca Raton, FL. p. 143-162.

Johnson, M.D. and B. Lowery. 1985. Effect of three conservation tillage practices on soil temperature and thermal properties. *Soil Sci. Soc. Am. J.* 49:1547-1552.

Kluitenberg, G.J. and R. Horton. 1990. Analytical solution for two-dimensional heat conduction beneath a partial surface mulch. *Soil Sci. Soc. Am. J.* 1197-1206.

Lang, A.R.G., X. Yuegin, and J.M. Norman. 1985. Crop structure and the penetration of direct sunlight. *Agric. For. Meteorol.* 35:83-101.

Lascano, R.J., C.H.M. van Bavel, J.L. Hatfield, and D.R. Upchurch. 1987. Energy and water balance of a sparse crop: Simulated and measured soil and crop evaporation. *Soil Sci. Soc. Am. J.* 51:1113-1121.

Lemeur, R. 1973. A method simulating the direct solar radiation regime in sunflower, jerusalem artichoke, corn, and soybean canopies using actual stand structure data. *Agric. Meteorol.* 12:229-247.

Lemeur, R. and B.L. Blad. 1974. A critical review of light models for estimating the shortwave radiation regime of plant canopies. *Agric. Meteorol.* 14:255-286.

Leuning, R., A.G. Condon, F.X. Dunin, S. Zegelin, and O.T. Denmead. 1994. Rainfall interception and evaporation from a soil below a wheat canopy. *Agric. For. Meteorol.* 67:221-238.

Luo, Y., R.S. Loomis, and T.C. Hsaio. 1992. Simulation of soil temperature in crops. *Agric. For. Meteorol.* 61:23-38.

Mann, J.E., G.L Curry, D.W. DeMichele, and D.N. Baker. 1980. Light penetration in a row-crop with random plant spacing. *Agron. J.* 72:131-142.

Massman, W.J. and J.M. Ham. 1994. An evaluation of a surface energy balance method for partitioning ET data into plant and soil components for a surface with partial canopy cover. *Agric. For. Meteorol.* 67:253-267.

McCool, D.K., D.E. Stoot, J.M. Laflen, and D.L. Schertz. 1995. Residue incorporation by tillage-interactions of mass and cover. *J. Soil and Water Cons*. 50:563-567.

McInnes, K.J., E.T. Kanemasu, D.E. Kissel, and J.B. Sisson. 1986. Predicting diurnal variations in water content along with temperatures at the soil surface. *Agric. For. Meteorol*. 38:337-348.

Messing, I. 1989. Estimation of the saturated hydraulic conductivity in clay soils from soil moisture retention data. *Soil. Sci. Soc. Am. J*. 53:665-668.

Mielke, L.N., J.W. Doran, and K.A. Richards. 1986. Physical environment near the surface of plowed and no-tilled soils. *Soil Tillage Res*. 7:355-366.

Miller, E.E. and J.M. Norman. 1971a. A sunfleck theory for plant canopies. I. Lengths of sunlit segments along a transect. *Agron. J*. 63:735-738.

Miller, E.E. and J.M. Norman. 1971b. A sunfleck theory for plant canopies. II. Penumbra effect: Intensity distribution along sunfleck segments. *Agron. J*. 63:739-743.

Mohanty, B.P., W.M. Klittich, R. Horton, and M. Th. van Genuchten. 1995. Spatio-temporal variability of soil temperature within three land areas exposed to different tillage systems. *Soil Sci. Soc. Am. J*. 59:752-759.

Nichols, W.D. 1992. Energy budgets and resistances to energy transport in sparsely vegetated rangeland. *Agric. For. Meteorol*. 60:221-247.

Norman, J.M., E.E. Miller, and C.B. Tanner. 1971. Light intensity and sunfleck distribution in plant canopies. *Agron. J*. 63:743-748.

Novak, M.D. 1991. Analytical solutions to predict the long-term surface energy balance components and temperature of a bare soil. *Water Resour. Res*. 27:2565-2576.

Potter, K.N., R.M. Cruse, and R. Horton. 1985. Tillage effects on soil thermal properties. *Soil Sci. Soc. Am. J*. 49:968-973.

Ross, J. 1975. Radiative transfer in plant communities. In Monteith, J.L. (ed.), *Vegetation and the Atmosphere*. Volume 1. Principles. Academic Press, New York, NY. p. 13-55.

Sauer, T.J., J.L. Hatfield, and J.H. Prueger. 1996a. Corn residue age and placement effects on evaporation and soil thermal regime. *Soil Sci. Soc. Am. J*. (submitted).

Sauer, T.J., J.L. Hatfield, and J.H. Prueger. 1996b. Over-winter changes in radiant energy exchange of a corn residue-covered surface. *Agric. For. Meteorol*. (submitted).

Shuttleworth, W.J. and J.S. Wallace. 1985. Evaporation from sparse crops-an energy combination theory. *Quart. J. R. Met. Soc*. 111:839-855.

Teasdale, J.R. and C.L. Mohler. 1993. Light transmittance, soil temperature, and soil moisture under residue of hairy vetch and rye. *Agron. J*. 85:673-680.

Thunholm, B. 1990. A comparison of measured and simulated soil temperature using air temperature and soil surface energy balance as boundary conditions. *Agric. For. Meteorol*. 53:59-72.

Villalobos, F.J. and E. Fereres. 1990. Evaporation measurements beneath corn, cotton, and sunflower canopies. *Agron. J*. 1153-1159.

Whitfield, D.M. 1986. A simple model of light penetration into row crops. *Agric. For. Meteorol*. 36:297-317.

Wiese, A.M. and L.K. Binning. 1987. Calculating the threshold temperature of development for weeds. *Weed Sci.* 35:177-179.

Woolley, J.T. and E.W. Stoller. 1978. Light penetration and light-induced seed germination in soil. *Plant Physiol.* 61:597-600.

Weed Spatial Variation and Weed Management

David A. Mortensen, J. Anita Dieleman, and Gregg A. Johnson

I. Introduction

Farmers, consultants, and land managers have always recognized spatial heterogeneity in production fields. Most agree that heterogeneity has increased as smaller fields have been incorporated into larger ones and as the complexity of rotations has decreased. Heterogeneity expresses itself visually in nonuniform plant vigor, crop yield, soil color, and pest infestations (Almekinders et al., 1995). Currently, resources and practices are uniformly applied to whole fields in spite of this recognized variation. While the technology to manage fields on sub-whole field units has become available, information processing and linking information with decision-making has lagged. For example, while we know that weed density and composition are not uniformly distributed across fields, we know relatively little about the stability of these distributions or how they are linked to field average efficacies for common weed control practices. If weed population distributions are stable from year to year, it would be possible to use last year's weed map to adjust for next year's weed management decisions. If more were known about the link among weed (composition and density, presence/absence) and soil spatial heterogeneity and weed control efficacy, it may be possible to vary rates and/or products or practices that reduce inputs. The ability to manage weed populations on less than whole field units could result in significant reductions in herbicide use, particularly if spatially variable weed management approaches were used as a part of an integrated pest management system. In such an approach, spatially variable weed management could be implemented as a preventive or therapeutic practice.

Using current practices, weeds are estimated to cause annual losses of $4.7 billion in U.S. corn (*Zea mays* L.) and $3.9 billion in soybean (*Glycine max* (L.) Merr.) crops (Bridges, 1992). A 1988 survey indicated that herbicides were used on 96% of corn and 99% of soybean crops in the north central United States compared to 57% and 49%, respectively, in 1966 (Swanson and Dahl, 1989).

Several concerns regarding herbicide use have been raised despite the fact that herbicides have been effective weed control tools to reduce weed infestations and the risk of crop loss. First, current corn and soybean production methods are responsible for the presence of atrazine, cyanazine and, to a lesser degree, metolachlor and alachlor in ground and surface water (Klaseus et al., 1988; National Research Council, 1986). Because these products are applied prior to crop and weed emergence, they are often applied without knowledge of the weed infestation. This lack of knowledge may lead to unnecessary applications. Secondly, the real value of corn and soybean production has declined. The market value of U.S. corn production per acre in 1982 dollars dropped from over $740 during most of the 1970s to $420 during 1986-1988 (Mercier, 1989). Caught in a profit squeeze, farmers need to reduce variable economic inputs.

Several recent action plans proposed by the U.S. Department of Agriculture (USDA) address the environmental concern of herbicide-based weed control. The USDA, in cooperation with the Environmental Protection Agency (EPA) and the Food and Drug Administration (FDA), has recently embarked upon a mission to significantly reduce pesticide use in the U.S. The mission commits to the implementation of integrated pest management (IPM) practices on 75% of America's crop land by the year 2000. Altered weed management systems that result in reduced herbicide input, while allowing grower flexibility within the cropping system, have the greatest promise for successful adoption of IPM.

The purpose of this paper is to outline ways in which spatially explicit weed population information could be used to enhance weed management practices. Since no one weed management approach should be seen as a silver bullet, the management practices outlined in this paper are considered as a component of an integrated system.

II. IPM Principles — The Concept of Population Management

Integrated pest management (IPM) is seen as an approach to effectively manage pest populations as well as address the myriad of environmental, economical, and sociological issues currently confronting our agricultural production systems.

Integration of several tactics to manage pest populations is central to all IPM plans (Cate and Hinkle, 1993).

IPM is built on three principles, the first two being preventive, and the third therapeutic (Mortensen et al., 1995a; Pedigo, 1995). First, strategies should focus on decreasing the pest's equilibrium density. The equilibrium density is the background vegetative or seedbank population occurring at a site. This density could be reduced by preventing seed production, enhancing rate of propagule bank turnover, or moving propagules into unsafe germination sites. Second, preventive strategies should also focus on reducing the impact of the pest on the crop, that is, decreasing the crop loss expected per pest. This can be accomplished by either making the crop more competitive or reducing weed competitiveness. Preventive approaches require research into methods of establishing a competitive hierarchy between the crop and weeds through the use of cover crops, competitive cultivars, delayed planting, reduced herbicide doses, and conservation tillage practices. The third principle, the use of therapeutic control practices, should only be used as a last resort to truncate pest population peaks when and where they are needed. Therapeutic practices include the application of postemergence herbicides or selective tillage. However, if changes in management practices are to impact herbicide use and represent a significant change from current practices, the IPM concept and relative management practices must address weed spatial heterogeneity.

Using spatial information to improve weed management decision-making is consistent with the logic of land managers who are aware of the high degree of spatial heterogeneity in their fields. The challenge will be to describe that variation and subsequently manage with rather than overcome it. Current development of site-specific weed management strategies focuses on using spatial information to optimize herbicide placement and timing. We believe a broader systems view could expand the ways in which information on field variation could be used to improve management methods. Five approaches are outlined in the *Incorporating spatial information into weed management* section of the paper. In any management approach no one method of control should be seen as a stand-alone weed control tactic. For example, benefits resulting from spatially variable herbicide application made to a field with a high weed infestation will result in little or no changes from a broadcast prophylactic herbicide application. However, if several density and/or fitness reduction practices are used, such as conservation tillage, competitive crop cultivars, or band application of reduced herbicide doses, followed by a spatial weed management practice such as intermittent herbicide application, it is possible that significant herbicide use reduction could occur while maintaining acceptable levels of weed management.

III. Weed Distribution, Density, and Composition in Agricultural Fields

To assess the impact of an agronomic practice or event on weed management, it is necessary to quantify changes in weed distribution, density, and composition over space and time. Weed density and composition are not quantifiable without knowledge of the population distribution. To date, little research has been conducted to describe field distributions of weeds. Also, weed science has lagged behind other pest management disciplines in describing, modeling, and predicting weed population dynamics with regard to changes in management practices (Navas, 1991; Doyle, 1991).

Long-term population models have proved useful for predicting effects of changes in management and thresholds on weed dynamics (Cousens, 1987). However, most population biology studies assume that weeds occur in monospecific stands, composed of genetically similar individuals that are homogeneously distributed throughout the field (Navas, 1991). On the contrary, recent research indicates the co-occurrence of multiple weed species within a stand and often in an aggregated pattern across the field (Mortensen et al., 1993; Thornton et al., 1990; Navas, 1991; van Groenendael, 1988). Recent studies have attempted to incorporate numerical and spatial distribution parameters into weed management decision models (Brain and Cousens, 1990; Doyle, 1991; Wiles et al., 1992). These models assume that the numerical distribution is described by a negative binomial function, suggesting a clumped or aggregated distribution. This was observed by Johnson et al. (1995a) for most weed species identified in a survey of 18 corn and soybean fields.

Potential exists to quantify field-scale variation in weed seed and seedling populations and to identify factors affecting this variation. To quantify farm-scale variation, information must be spatially referenced using tools such as global positioning systems (GPS). Spatially referenced information can then be used in a Geographic Information System (GIS) to identify events and processes that give rise to an aggregated spatial pattern. For example, field consultants who assist farmers with pest management recommendations commonly map weed populations using handwritten notes or GPS software (Figure 1). Such information is used to direct subsequent weekly trips to the field (Mark Hinze, personal communication).

Spatial and temporal information obtained from multiple farms can be used to identify the influence of several farm practices on weed distribution, density, and composition. Andreasen et al. (1991) recorded the frequencies of weed species in cereal grain crops in 316 Danish fields by listing the presence/absence of species in 10 randomly selected areas in each field. Weed presence was most dependent on clay content and cereal grain species (an integration of crop and associated management practices). Weed distribution is also correlated with agronomic practices and site characteristics. For example, preliminary multivariate statistical analysis separated out weed associations in central Nebraska corn fields. Large-seeded broadleaf species, velvetleaf (*Abutilon theophrasti* Medic.) and sunflower (*Helianthus annuus* L.), were correlated to areas of the field with high organic matter and low relative elevation, while higher populations of annual grasses were located in well-drained, higher areas of the field (Dieleman and Mortensen, 1997).

Few studies relating weed population distributions and site heterogeneity are available simply due to their complex nature. Andreasen et al. (1991) stated that "any attempt to relate the distribution of weed species to soil factors must be linked to several other explanatory variables that are not directly associated with the

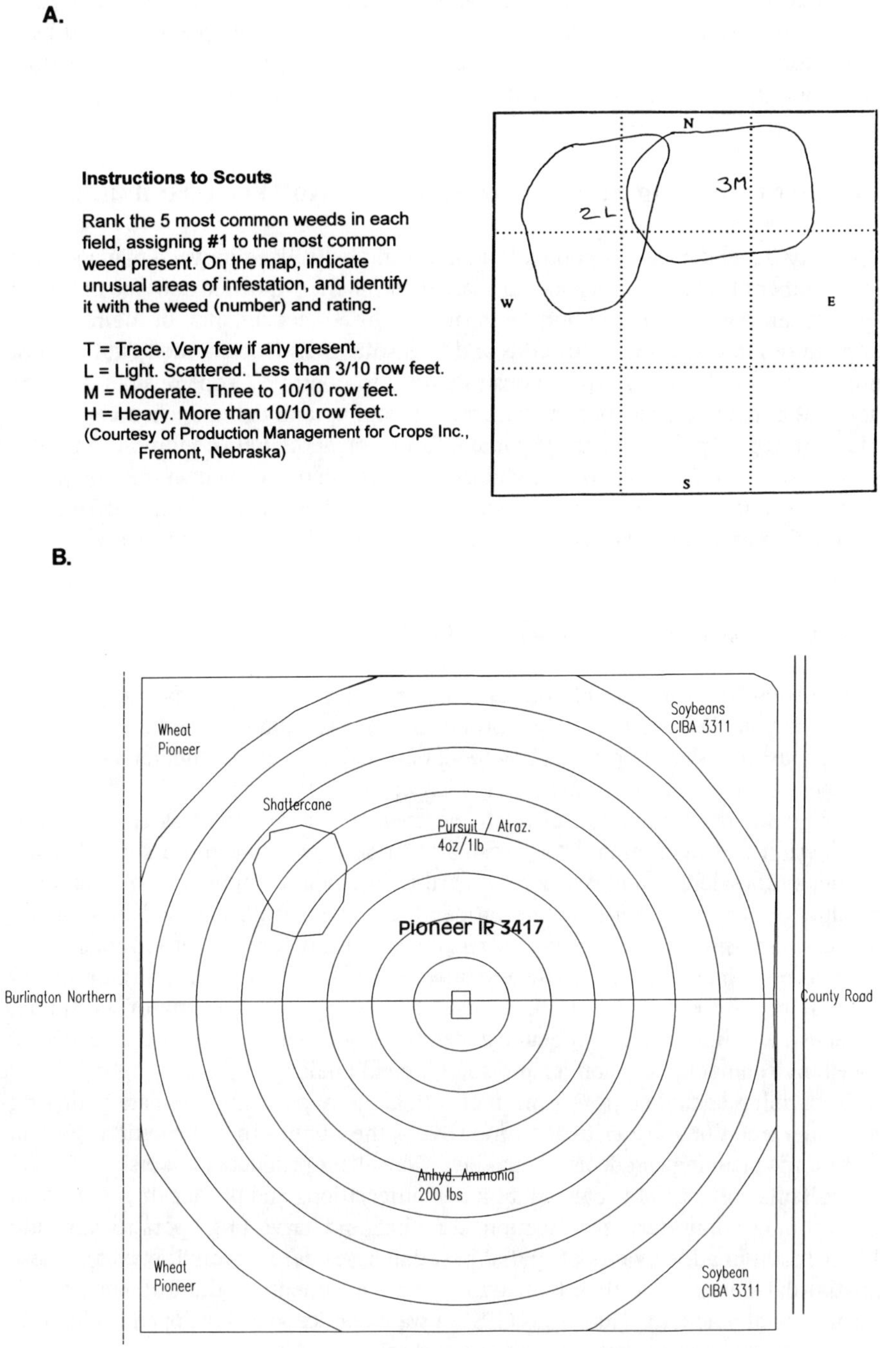

Figure 1. Hand-drawn (1a) and GIS maps (1b) are commonly used to spatially-reference pest populations.

measured soil property." Biological variables such as crop and weed species, site characteristics, efficacy of control practices, other agronomic practices, and land use history interact to influence weed species distribution, density, and composition. A method must be developed to analyze this complex system of interacting factors.

IV. Incorporating Spatial Information into Weed Management

Spatially-dependent weed population data could be utilized in weed management in a number of ways. Farmers could share their inherent knowledge of the cropping system and weed distribution with newly hired consultants or commercial applicators through the use of GPS and GIS software. Documented data on pest populations could be used to demonstrate need-to-control. Crop consultants often use spatial data to direct pest management sampling and to diagnose components of integrated cropping systems. Spatial information about field conditions and pest populations could be used to alter cultural practices in order to better manage pests. Finally, spatial variation in weed distributions could be used to direct intermittent application of herbicides. These five approaches will be discussed below.

A. Farmer Involvement — Documenting Knowledge

The current trend in agricultural land use is increased field sizes and producer specialization. For example, many producers have relinquished certain aspects of their production system to specialists (e.g., pest management recommendations are directed by consultants and pesticide application to commercial applicators), allowing for more time and effort to be spent on other activities such as crop production and marketing. As a result of these changes, it is possible that the farmer's knowledge and wisdom regarding certain components of the crop production system are lost or the farmer is removed from the decision-making process. Site-specific management offers the possibility for farmers and consultants to build upon individual expertise and talent through exchange of information and ideas using, for example, a GIS. Constant updating and enriching of the spatial database will likely lead to a better understanding of what changes are needed to meet environmental, economic, and sociological goals.

Technologies that empower producers to actively participate in data gathering and interpretations are critical to improving the quality of information used in agronomic planning and decision-making. Consultant-producer partnerships, where consultants collect field-scale data on pest infestations and producers record data on pest infestations during cultivation, spraying, and harvesting operations, would likely result in added value of spatial pest databases at relatively low cost. These spatial databases could then be used to plan adjustments in the following year's management practices. Numerous GIS software applications developed exclusively for the agricultural market are now available for home use.

B. Recordkeeping — Documenting the Need to Control

The development of GIS software for agricultural applications has been driven by the need to map crop yields, concentrations of principal nutrient elements, and applications of restricted use pesticides. With many commercially available software applications, virtually any input made to the field, including amount, location, and date of application, can be recorded. Future regulations may require that assessments of a pest infestation be performed prior to purchasing a pest control product. Such assessment strategies are under consideration in several European countries as a way to address environmental and social concerns associated with pesticide use (Matteson, 1995). Furthermore, accurate recordkeeping can help producers track the success or failure of pest management plans. In Nebraska, consultants identify ex-post evaluation of the effectiveness of crop production practices as the single greatest benefit of spatially referenced yield monitors and resulting maps. Such maps provide information for the first time on the correlation between escaped weed populations and resulting crop yield.

C. Sampling Plans — Documenting Distribution, Density, and Composition

Unlike other pest disciplines, weed sampling methodology has received little attention. Since the predominant weed management strategy has relied on preventive herbicide applications, only a few scientists or practitioners saw the need for refined weed sampling plans. While insect management has relied on scouting to determine quantitative insect density levels, weed populations are often assessed using qualitative measures of infestation level, such as high, medium, or low. With the advent of economic thresholds, effective postemergent herbicides, and increased interest in reducing inputs, the need for quantitative weed population assessment has grown (Johnson et al., 1995b; Wiles et al., 1992; Marshall, 1988).

Unlike insects that are mobile in and between fields, emerged weed seedlings are sedentary. Distribution of weeds is largely dependent on propagule dispersal and, for many species, is limited to relatively short distances. Because of greater spatial and temporal stability (Wyse-Pester et al., 1995; Johnson, 1994), mapping weed distributions could provide meaningful information for managing populations in future years. Under current scouting practices in Nebraska (Raun, 1995), it is common for field scouts to indicate the location of weed infestations, either individuals or groups of species, on hand-drawn field maps (Figure 1a). Such maps are used to alter weed management inputs in subsequent years and direct sampling to "hot spots" in the field. For example, a large shattercane patch (*Sorghum bicolor* L. Moench) resulted in a change in herbicide use in half of a 160 acre field (Figure 1b). In this field, the consultant used center pivot tracks for spatial referencing.

The highly skewed and patchy nature of weed distributions makes random, or blind, sampling strategies difficult and costly to implement. Johnson et al. (1995b) found that in order to adequately estimate weed populations, prohibitively large numbers of sample counts were required. The most effective weed sampling method may involve "directed sampling" where previous weed distribution information is used to direct sampling. After all, directed sampling is the procedure used by farmers and consultants to qualitatively describe field distributions of weeds. The challenge is to improve upon the quality of sample data for use in decision processes.

As mentioned earlier, crop scouts indicate the location of weed and insect infestations on hand-drawn field maps. This map information is refreshed with weekly field visits. Scouts have indicated that this process could be improved with easier access to previously collected data and ability to enter new data. They stress that the scouting system must be fast, hands-free, and able to withstand the rigors of in-field use (e.g., rain, dew, wind, dust, and vibration). Value of field data could be increased by distributing the cost of gathering the information across a number of other activities. For example, pest management and crop status information could be collected while a field is being cultivated or while a crop is being harvested. In addition, ground-truthing aerial images collected on individual fields would be another method of updating a field database. In any given season, 5 to 10 sets of observations at varying degrees of resolution could be gathered on an individual field. The sources of that data might include four field visits by the consultant, data gathered during cultivation and harvesting, and two aerial images prior to crop canopy closure. Such data acquisition methods are currently being evaluated (Wiles and Schweizer, 1996).

It is essential that better weed assessment methods be available as growers consider reduced input strategies and potential for increased risk. Weed management strategies such as intermittent herbicide application, reduced dose herbicide application, and increased reliance on mechanical weed control represent a willingness on the part of the producer to increase risk in their weed management plan. Therefore, it is critical that producers have access to timely and accurate information in order to minimize that risk. The use of GPS and GIS tools to log, integrate, and retrieve information regarding pest distributions will enable consultants and farmers to enrich their knowledge of fields by accessing spatially-referenced data and linking that data with management decisions. The concept of refreshing spatial databases to improve the accuracy of information is an exciting one. The agricultural consulting business is expanding and provides the expertise on the ground to facilitate data collection and synthesis in ways that could significantly influence farmer practices in the field.

D. Context for Considering Fundamental Changes in Management Practices

Adjusting cropping systems according to spatial variation across a landscape is another way in which spatially variable site data could be used. The modern agricultural model that includes spatially variable information could advance beyond just the accurate placement of herbicide inputs. It is conceivable that practices could be changed in response to weed presence. On-the-go technology offers the possibility of planting cultivars that have desirable weed suppressive traits in areas of high weed infestation. This approach is particularly attractive if the alternative practice carries a cost (lower yield, higher cost) that can't be recovered in areas of the field where pest populations are low. The same conceptual approach could be used to integrate narrow row spacing, crop seeding density, living mulches, crop species, biological control agents, or selective tillage practices into field management decisions.

Site-specific changes in management practices may not eliminate weeds but rather would offer ways of reducing the equilibrium density and per weed yield losses. An example of such an approach would be wild oat (*Avena fatua* L.) management in barley (*Hordeum vulgare* L.) and wheat (*Triticum aestivum* L.) in the high plains states and provinces of the U.S. and Canada. Wild oats is a troublesome weed in cereal grain production systems because its growth habit mimics the crop, it has evolved resistance to many of the currently used herbicides, and it is a prolific seed producer. Wild oats occur as elliptical patches ranging in size from 3 to 10 meters (Colliver and Maxwell, 1995). Suggested preventive IPM practices targeted at wild oat management include increasing barley density, which has reduced wild oat competitive losses by 60% and seed production by 70% (Maxwell, 1995). However, because of increased cost of barley seed, it is unlikely such a management approach would be widely adopted. Increasing barley density only when wild oat density exceeded some threshold population would make the practice cost effective and would provide a durable, long-term management approach.

Changing cropping practices on a site-specific basis to address weed abundance and distribution has received little attention. This lack of research reflects the relatively poor understanding that currently exists for such alternative management practices and the relative costs and risks associated with conventional versus alternative management approaches. We would argue that intermittent herbicide application has dominated the thinking and research in this area. This is because herbicide application is seen as the management system rather than as a tool in the system. In principle, the perceived increased risk in pest management failure associated with alternative, less herbicide-intensive approaches could be minimized by placing the alternative strategies where they are most likely to provide acceptable levels of pest suppression. This is an area of research needing greater attention.

E. Intermittent Herbicide Application

Two approaches can be taken to intermittent herbicide application. The first, "historic," involves collection of weed population data in the form of spatially-referenced maps. These maps are then used at a later time on an intermittent herbicide applicator as information from which herbicide application decisions are made. The second approach is "real-time," which involves sensing plants and then instantaneously applying a herbicide based on a decision algorithm.

Since the amount of spray applied intermittently is related directly to the area occupied by weeds, it is possible that less herbicide would be needed for the same level of control when compared to current broadcast methods. Of course, the extent of herbicide reduction is contingent on the weed infestation level, distribution, and performance of application equipment (Thompson et al., 1991; Mortensen et al., 1993; Stafford and Miller, 1993). Early on, Cousens (1987) recognized the utility of a spray-as-needed weed management approach when he stated that it would be the ultimate method of measuring weed populations for economic threshold decisions. Haggar et al. (1983) estimated that spot spraying could result in a 60% reduction of postemergence herbicide application. Guyer et al. (1986) estimated that use of a sensor-controlled spot sprayer could reduce wanton herbicide use in corn and soybean by twenty-four million pounds per year in the United States alone. Stafford and Miller (1993) indicated that patchiness of blackgrass (*Alopecurus myosuroides* Huds.) and wild oat resulted in 40% of winter wheat fields free of these weeds and estimated that a 35% reduction in herbicide use could be realized with patch spraying. Shearer and Jones (1991) reported a 15% reduction in herbicide use when a lateral view photoelectric sensor actuated a spray nozzle using an infrared beam interrupted by a weed. Johnson et al. (1995c) found that significant areas of corn and soybean fields were free of broadleaf and grass weeds. In surveys of 12 fields, weed-free area ranged from 30 to 70% in the interrow (preemergence band treated fields in which the interrow had received no weed control) at a time when a postemergence herbicide would be applied. When the threshold weed density was increased (above zero), the proportion of area below the threshold also increased. In 6 of 10 fields sampled, greater than 50% of the field area was below a threshold of 1 broadleaf weed per linear meter of row. When weed populations were mapped in the intrarow (where a preemergence herbicide had been applied), 71 and 94% of the sampled area, on average, was free of broadleaf and grass weeds using a zero population threshold density (Johnson et al., 1995c).

1. Historic

The historic or GIS-centered approach involves collection of spatially-referenced pest information, linking this information with knowledge (decision algorithms), then linking algorithms with a control action. The historic approach to regulating agrichemical applications has recently become commercial practice for fertilizer application. The use of spatial data to regulate nitrogen, phosphorus, and potassium applications is a commercial practice in corn, potato, soybean, and wheat crops (Mulla, 1993). For these applications, soil samples are taken in a grid configuration in the fall preceding the fertilizer application. Soil property maps are then computed through the process of linear interpolation or kriging, methods used to interpolate between unsampled locations in the field (Isaaks and Srivastava, 1989). The same application and location technologies can be used for applying herbicides (Brown et al., 1990; Thornton et al., 1990). Gerhards et al. (1995) collected and analyzed spatially-referenced weed density data in two German winter wheat fields for site-specific management using a GIS-based intermittent herbicide sprayer coupled with a GPS. Infestations of blackgrass and nine broadleaf species including catchweed bedstraw (*Galium aparine* L.) and wild chamomile (*Matricaria chamomilla* L.) were sampled and mapped prior to postemergence herbicide application in early spring. An economic threshold model (Gerhards and Kuhbauch, 1993) linked control decision algorithms with GIS weed maps, and herbicide use with this map-based approach was reduced by 40-50%.

2. Real-Time

Presently, two approaches can be taken for real-time or instantaneous plant sensing. The first involves the use of optical sensors. Here a single element optical sensor is used to measure the quantity of reflected light of a given wavelength and waveband. The optical element sensor is selected and filtered to contrast the target plant and its background. Optical contrast indices further improve the separation of the target and its background. A contrast ratio can be expressed as a difference between the reflected energy of the filtered brighter objects and darker objects within the field of view. This principle was tested in early versions of weed "patch sprayers" when reflectance ratio meters were used to measure the ratio of near infrared to red light reflected from plant and soil surfaces (Shropshire et al., 1990; Haggar et al., 1983, 1984). While these systems could effectively detect green vegetation, they were extremely sensitive to changes in light irradiance levels, and soil and plant surface characteristics. Successive sensors were improved by normalizing the reflectance signal, reducing the variability resulting from changes in light irradiance levels. Felton and McCloy (1992) reported on a commercialized vertical view optical reflectance sensor system based on Haggar's and Shropshire's earlier concepts. While this sensing approach represents the first commercial attempt at real-time sensing, several problems with the technology have persisted with the most significant being its sensitivity to variable light and ground surface environments. Patchen (1995) developed an optical sensor intermittent sprayer that largely overcomes the problems of soil and plant surface variation as well as variations in irradiance levels by using a modulated artificial light source. This commercialized system works equally well in the dark or light and is largely unaffected by variations in the ground surface. At present, it is not possible to discriminate between weed and crop using optical sensing. Research is currently underway in which such sensors could be used to assess pattern of crop and weed plant location and frequency and extent of signal to discriminate between weed and crop. With the development of crops resistant to nonselective herbicides, it is possible that such optical systems may provide acceptable plant discrimination for individual plant real-time sensing.

Another approach to real-time or instantaneous plant sensing involves the use of image analysis. In contrast to the optical element (single element) approach, image analysis involves digitizing images into an array composed of numerous picture elements (pixels), then classifying the image. Numerous approaches to image classification are currently under study and include color (Brown et al., 1990; Woebbecke et al., 1995a), shape (Guyer et al., 1986; Woebbecke et al., 1995b), and texture (Mehta, 1995).

While it is rather certain that herbicide use reduction can be achieved with the appropriate intermittent technology, caution should be taken in assessing its potential. First, the estimates of herbicide use reduction outlined in this section assume a best case-scenario with respect to the application technology. Most simulations assume that individual crop rows and small individual plants can be sensed. In the Johnson et al. (1995c) simulations, species were grouped as broadleaves and grasses and classified as something other than a broadleaf or grass crop. While the sensing technology to accomplish this task is under development (Woebbecke et al., 1995a,b), and encouraging results have been reported, field testing of such plant discrimination systems is still in the future.

The historic map-based approach assumes that information on the pest distribution is collected at some time prior to the herbicide application. Rew and Cussans (1995) identified several challenges to accurately locating and intermittently spraying a weed population. First, individual plants cannot be sprayed since either sections of or the entire spray boom is turned on and off (or rate adjusted) according to maps. Because of this coarser resolution, it is likely that a greater proportion of the field would need to be sprayed to reduce the risk of missing weeds that had gone undetected in previous surveys. In addition, spray swath overlap and location of the sprayer with respect to the weed maps is critical. If the ability to determine sprayer location is less than ideal, buffer areas would be needed. These areas would be sprayed to avoid the risk of leaving uncontrolled weeds. Finally, the feasibility of the historic approach will depend on the ability to accurately assess the weed population infestation and its spatial and temporal stability. Several recent studies have attempted to quantify the spatial and temporal stability of summer annual weeds. In general, species with long-lived seed appear more stable than short-lived species (Johnson, 1994; Wyse-Pester et al., 1995).

Because of the decreased resolution of this approach (wider swath than individual rows), the sprayer would be turned on a greater amount of the time. With the historic approach, the reduction in resolution could be overcome by coupling variable rate application with weed population infestation level (Dieleman and Mortensen, 1997). Finally, herbicide use reduction can only be achieved where the weed infestation is low-moderate. Mortensen et al. (1995b) found that this condition was met by approximately 50% of the soybean and 70% of the corn fields surveyed.

V. Future Research and Conclusions

The extent to which spatial information on weed populations will be used to manage weeds will depend on the economic feasibility of the approaches outlined in the paper, the risks associated with these approaches, and the co-evolution of knowledge and technology. Farmers will most likely adopt these spatially variable approaches if they result in net profits equal to the management system that the farmer is currently using (Reichelderfer, 1980). Risk also plays an important role in a grower's willingness to adopt a new pest management approach. In the case of farmers, "the value of controlling a pest, whose incidence varies in an uncertain way, is greater than the average loss caused to them by the pest. Risk-averse farmers are willing to pay a premium or insurance charge to reduce the risk of the uncertainty that they face" (Tisdell et al., 1984). Adopting methods that involve farmers in pest management through the use of a GIS, recordkeeping, and knowledge of pest distributions makes it likely that these methods will be readily adopted as growers associate value with them, determine that they are risk neutral, or enable farmers to meet regulatory demands.

The extent to which growers adopt spatial information to alter cropping practices on sub-field units and for intermittent herbicide application practices is less clear. Currently, risk is managed conservatively by applying higher inputs to manage for a mean field condition. The approaches outlined in this paper suggest that an alternative method of managing risk is to characterize the spatial diversity existing in a field and manage with rather than overcome it. In many ways, the development of technology has out-paced the biological understanding needed to evaluate the economic and environmental value of a spatially-based management approach. This is particularly true for intermittent herbicide and fertilizer application, and yield mapping. The value of the latter two approaches is very much dependent on the weed infestation level in a given field and the extent to which the management system is integrated beyond broadcast herbicidal weed control. Greater knowledge of the distribution of weeds that have been subjected to a range of mortality and weed suppressive events is needed. Here both mechanistic and descriptive field-scale studies are needed. Mechanistic studies aimed at understanding the causal events and processes driving weed distribution are needed in cropping systems for which this technology has promise. Descriptive on-farm studies are needed to determine farm to farm variation in weed distribution and infestation level. In addition, the stability of these populations under current and future management practices is poorly understood. The extent of spatial stability in weed populations is an important determinant in the feasibility of GIS-based weed management. Significant variation in the spatial distribution of weed populations from year-to-year would increase the risk of poor weed control. The GIS-based approach also assumes that a relationship between weed infestation level and efficacy exists. While limited data support this assumption (Winkle et al., 1981; Buhler et al., 1992), a greater understanding of the link between pre- and postemergence herbicide and nonchemical control efficacy and weed infestation level is needed.

Advances in sensing technology are also needed. Current sensors are unable to discriminate between plant species and are limited to fallow or no-till applications. More research is needed to couple nondiscriminating sensors with counting or pattern recognition algorithms in which the frequency or size of the desirable plant is sensed and plants outside of this size distribution or plant spacing are classified as weeds. In the area of machine vision, the sensing task must be simplified. Most current research focuses on creating an encyclopedia of weed shape, color, surface roughness, and edge characteristics. This approach could be simplified by recognizing the crop and sensing noncrop green plants. Neural networking is one promising approach in which a sensor would be trained-up to recognize a crop species and all green things that are not the crop. Another approach to simplifying the requirements of a real-time sensor would be to discriminate between grass and broadleaf plants. As mentioned earlier, the results of Woebbecke et al. (1995b) suggest that reducing the complexity of the sensing task; that is, grasses vs.

broadleaves, could lead to breakthroughs in machine vision sensor development. In both historic and sensor research, a greater emphasis must be placed on field evaluation. To date, most projections of herbicide use reduction have resulted from simulations based on field-derived weed distribution maps.

Finally, the development of information-intensive approaches to managing spatially variable weed populations must proceed in a systems context. In this paper, we outlined ways in which information on weed distribution could be used to: involve growers in pest management decision-making, direct pest sampling, consider fundamental changes in management practices, and perform intermittent herbicide application. All of these approaches will most likely succeed if they are conceived in an integrated pest management framework where information technologies are tools for use in weed management.

References

Almekinders, C.J.M., L.O. Fresco, and P.C. Struik. 1995. The need to study and manage variation in agro-ecosystems. *Neth. J. Agric. Sci.* 43:127-142.

Andreasen, C., J.C. Streibig, and H. Haas. 1991. Soil properties affecting the distribution of 37 weed species in Danish fields. *Weed Res.* 31:181-187.

Brain, P. and R. Cousens. 1990. The effect of weed distribution on predictions of yield loss. *J. Appl. Ecol.* 27:735-742.

Bridges, D.C. 1992. Crop Losses Due to Weeds in the United States - 1992. Weed Sci. Soc. Am., Champaign, Illinois, 403 p.

Brown, R.B., G.W. Andersen, B. Proud, and J.P. Steckler. 1990. Herbicide application control using GIS weed maps. ASAE paper No. 90-1061. Amer. Soc. Agric. Eng., St. Joseph, MI 49085.

Buhler, D.D., J.L. Gunsolus, and D.F. Ralston. 1992. Integrated weed management techniques to reduce herbicide inputs in soybean. *Agron. J.* 84:973-978.

Cate, J.R. and M.K. Hinkle. 1993. Integrated pest management: the path of a paradigm. Nat. Audubon Soc. Special Rept., Nat. Audubon Soc., Wash. D.C. 43 pp.

Colliver, C.T. and B.D. Maxwell. 1995. Detection and mapping of wild oat (*Avena fatua* L.) seedling patterns in a field. *Proc. WSSA* 35:50.

Cousens, R. 1987. Theory and reality of weed control thresholds. *Plant Prot. Quart.* 2:13-20.

Dieleman, J.A and D.A. Mortensen. 1997. Influence of weed biology and ecology on development of reduced dose strategies for integrated weed management systems. (This volume.)

Doyle, C.J. 1991. Mathematical models in weed management. *Crop Prot.* 10:432-444.

Felton, W.L. and K.R. McCloy. 1992. Spot spraying. *Agric. Eng.* 73:9-12.

Gerhards, R. and W. Kuhbauch. 1993. Dynamisches Entscheidungs-modell zur Lenkung von Unkrautkontrollmasnahmen in Wintergetreide mit Hife digitaler Bildverarbeitung. *J. Agr. & Crop Sci.* 171:329-335.

Gerhards, R., R. Hayer, M. Sokefeld, K. Schulze-Lohne, W. Kuhbauch, W. Buchner, and M. Graff. 1995. Ein Verfahren zur teilschlaggerechten Unkrautknotrolle in Winterweizen. *Mitt. Ges. Pfl.bauwissenschaften* 8:172-175.

Guyer, D.E., G.E. Miles, M.M. Schreiber, O.R. Mitchell, and V.C. Vanderbilt. 1986. Machine vision and image processing for plant identification. *Trans. ASAE* 29:1500-1506.

Haggar, R.J., C.J. Stent, and S. Issac. 1983. A prototype hand-held patch sprayer for killing weeds, activated by spectral differences in crop/weed canopies. *J. Agric. Eng. Res.* 28:349-358.

Haggar, R.J., C.J. Stent, and J. Rose. 1984. Measuring spectral differences in vegetation canopies by a reflectance ratio meter. *Weed Res.* 24:59-65.

Isaaks, E.H. and R.M. Srivastava. 1989. *An Introduction to Applied Geostatistics.* New York, USA: Oxford University Press.

Johnson, G.A. 1994. Model parameterization, parametric sequential sampling, and geostatistical analysis of weed seedling populations. Doctoral thesis, University of Nebraska, Lincoln.

Johnson, G.A., D.A. Mortensen, L.J. Young, and A.R. Martin. 1995a. The stability of weed seedling population models and parameters in eastern Nebraska corn (*Zea mays*) and soybean (*Glycine max*) fields. *Weed Sci.* 43:604-611.

Johnson, G.A., D.A. Mortensen, L.J. Young, and A.R. Martin. 1995b. Parametric sequential sampling based on multistage estimation of the negative binomial parameter k. *Weed Sci.* (in press.)

Johnson, G.A., D.A. Mortensen, and A.R. Martin. 1995c. A simulation of herbicide use based on weed spatial distribution. *Weed Res.* 35:197-205.

Klaseus, T.G., G.C. Buzicky, and E.G. Schneider. 1988. *Pesticides and Groundwater: Surveys of Selected Minnesota Wells.* Prepared for the Legislative Commission on Minnesota Resources by the Minnesota Department of Agriculture and Health, St. Paul, MN.

Marshall, E.J.P. 1988. Field-scale estimates of grass weed populations in arable land. *Weed Res.* 28:191-198.

Matteson, P.C. 1995. The "50% pesticide cuts" in Europe: a glimpse of our future? *Amer. Ent.* 41:210-220.

Maxwell, B.D. 1995. The relative importance of seed dispersal and environmental heterogeneity in determining annual weed spatial dynamics. *Proc. WSSA* 35:50.

Mehta, T. 1995. Comparison of optical, color, and textural imaging methods for distinguishing plants from soil background. Master of Science thesis, University of Nebraska, Lincoln.

Mercier, S. 1989. Corn: background for 1990 farm legislation. Commodity Economics Div., Economic Research Service - USDA Staff Report No. 89-47.

Mortensen, D.A., G.A. Johnson, and L.J. Young. 1993. Weed distribution in agricultural fields. p. 113-124. In: P. Robert and R.H. Rust (eds.), *Site Specific Crop Management*, Am. Soc. Agron.

Mortensen, D.A., D.L. Holshouser, L.G. Higley, L.D. Pedigo, J.L. Lindquist, and J.A. Dieleman. 1995a. Adopting pest management principles from a neighboring discipline. *Proc. WSSA* 35:101.

Mortensen, D.A., G.A. Johnson, D.Y. Wyse, and A.R. Martin. 1995b. Managing spatially variable weed populations. p. 397-415. In: P. Robert (ed.), *Site Specific Crop Management*, Am. Soc. Agron.

Mulla, D.J. 1993. Mapping and managing spatial patterns in soil fertility and crop yield. p. 15-26. In: P. Robert and R.H. Rust (eds.), *Site Specific Crop Management*, Am. Soc. Agron.

National Research Council. 1986. Pesticides and groundwater: issues and problems in four states. Nat. Acad. Sci., Washington, D.C.

Navas, M.L. 1991. Using plant population biology in research: a strategy to improve weed management. *Weed Res.* 31:171-179.

Patchen. 1995. The Newest WeedSeeker is a Hit! Patchen: On Target. Patchen Selective Spray Systems, Patchen, California, Inc., 22217 Old Santa Cruz Hwy., Los Gatos, CA 95030.

Pedigo, L.P. 1995. Closing the gap between IPM theory and practice. p. 209-210. In: *Proc. Second Nat. IPM Symp.*

Raun, E. 1995. Pest field scouting activities: general scouting procedures. Pest Management Comp., 3036 Prairie Rd., Lincoln, NE 68506.

Reichelderfer, K.H. 1980. Economics of integrated pest management: discussion. *Am. J. Ag. Econ.* 62:1-12-1013.

Rew, L.J. and G.W. Cussans. 1995. An economic appraisal of patch spraying quack-grass (*Elymus repens*, L.). *Proc. WSSA* 35:151.

Shearer, S.A. and P.T. Jones. 1991. Selective application of post-emergence herbicides using photoelectrics. *Trans. ASAE* 34:1661-1666.

Shropshire, G.J., K. VonBargen, and D.A. Mortensen. 1990. Optical reflectance sensor for detecting plants. In: J.A. DeShazer and G.E. Meyer (eds.), *Optics in Agriculture. Proc. SPIE* 1379:222-235.

Stafford, J.V. and P.C.H. Miller. 1993. Spatially selective application of herbicide. *Comp. and Elect. in Agric.* 9:217-229.

Swanson, J.A. and D.C. Dahl. 1989. The U.S. pesticide industry: usage trends and market development. Staff Paper P89-5. Department of Agric. Economics, Univ. of Minnesota, St. Paul.

Thompson, J.F., J.V. Stafford, and P.C.H. Miller. 1991. Potential for automatic weed detection and selective herbicide application. *Crop Prot.* 10:254-259.

Thornton, P.K., R.H. Fawcett, J.B. Dent, and T.J. Perkins. 1990. Spatial weed distribution and economic thresholds for weed control. *Crop Prot.* 9:337-342.

Tisdell, C.A., B.A. Auld, and K.M. Menz. 1984. Crop loss elasticity in relation to weed density and control. *Agric. Syst.* 13:161-166.

van Groenendael, J.M. 1988. Patchy distribution of weeds and some implications for modelling population dynamics: a short literature review. *Weed Res.* 28:437-441.

Wiles, L.J. and E.E. Schweizer. 1996. Distribution of weed seed and seedling populations in commercial corn fields in Colorado. *Proc. WSSA* 36:82.

Wiles, L.J., G.W. Oliver, A.C. York, H.J. Gold, and G.G. Wilkerson. 1992. Spatial distribution of broadleaf weeds in North Carolina soybean (*Glycine max*) fields. *Weed Sci.* 40:554-557.

Winkle, M.E., J.R.C. Leavitt, and O.C. Burnside. 1981. Effects of weed density on herbicide absorption and bioactivity. *Weed Sci.* 29:405-409.

Woebbecke, D.M., G.E. Meyer, K. VonBargen, and D.A. Mortensen. 1995a. Color indices for weed identification under various soil, residue, and lighting conditions. *Trans. ASAE* 38:259-269.

Woebbecke, D.M., G.E. Meyer, K. VonBargen, and D.A. Mortensen. 1995b. Shape features for identifying young weeds using image analysis. *Trans. ASAE* 38:271-281.

Wyse-Pester, D.Y., D.A. Mortensen, and C.A. Gotway. 1995. Statistical methods to quantify spatial stability of weed populations. *Proc. NCWSS* 50:74.

Engineering and Application of Precision Farming Technology

K.A. Sudduth

I. Introduction

Precision farming, also known as site-specific management (SSM), seeks to optimize crop production by tailoring management inputs and practices from point to point within fields. Inputs such as fertilizers and pesticides are variably applied to meet the specific needs of the crop at each location. Since over-application and under-application of agrichemicals are minimized, this strategy has the potential for increasing farmer profits and reducing environmental impacts.

In the early part of the 20th century, scholars were already studying variability in soil properties such as nutrient status and organic matter levels (Robinson and Lloyd, 1915; Waynick and Sharp, 1919), and advising farmers to map soil acidity and vary application rates of lime accordingly (Linsley and Bauer, 1929). Similarly, researchers conducting yield tests with hybrid corn were documenting spatial variations in crop yields (Fairfield Smith, 1938). Although researchers have continued to report on soil and yield variability (Reed and Rigney, 1947; Beckett and Webster, 1971; James and Dow, 1972), the mechanization of agriculture and the trend toward larger implements led agricultural production in the opposite direction. In the last decade, technological advances and the pressure of environmental concerns have revitalized the idea of defining smaller management units and applying inputs based on the individual characteristics of those units, in the concept now generally referred to as precision farming.

II. Engineering Technology for Precision Farming

The concept of precision farming is not a product of engineering technology, but rather has its foundation in the application of sound agronomic principles on a within-field, point-to-point basis. However, the practical implementation of precision farming concepts is dependent on technological developments to provide, manage, and utilize the vast quantities of data required to understand point-by-point variations in yields and in the factors that affect yields. Today, low-cost powerful computers, real-time controllers, variable rate application hardware, accurate location systems, and advances in sensor technology have combined to provide the technology to make precision farming a reality. Several technologies important to the practical implementation of precision farming are discussed below.

A. Location Determination

Knowing the location of an operation or a data collection point is an important part of precision farming implementation. The accuracy to which these locations must be known is variable, and is based upon the intended use of the information. For variable application and referencing of soil and yield data, an accuracy of one to several meters is sufficient. More accurate systems would be useful for vehicle guidance to eliminate skips and overlaps with a chemical applicator, or for precision cultivation operations (Auernhammer and Muhr, 1991). Most current precision farming efforts use Global Positioning System (GPS) technology to provide location data. GPS is a satellite-based radio navigation system developed and operated by the U.S. Department of Defense, which allows land, sea, and airborne users to determine their three-dimensional position and velocity anywhere in the world with a high degree of accuracy (Tyler, 1992).

The most common GPS positioning technique, pseudo range positioning, measures the distance to each satellite in view by comparing the coded signal transmitted from the satellite with an exact replica of the code generated in the receiver. Pseudo range measurements from at least four satellites allow for the computation of receiver position and a clock offset. When employed with a single receiver, the pseudo ranging procedure has an expected accuracy of 50 m. The accuracy can be increased substantially by placing one receiver in a fixed position and using the GPS position information from that fixed receiver to compensate for inaccuracies in the GPS position of the roving receiver. This procedure, called differential GPS (DGPS), improves accuracies to 5 m or better (Leick, 1990). A new generation of DGPS receivers with horizontal accuracy claims of as good as

1 m are now commercially available. DGPS positions can be calculated either in real-time or through post-processing of the signal. Real-time differential positioning requires a communications link, such as VHF-FM radio, to transmit the correction signal from the fixed receiver to the roving receiver.

The carrier phase positioning technique is an alternative to using the coded data imposed on the satellite signal. By directly observing the phase of the carrier wave, maximum accuracies are attainable, and this is the procedure employed in GPS surveying applications. Because of complexities in the measurement techniques, carrier phase measurements require much greater attention to detail during data collection. Therefore, although carrier phase measurements result in higher accuracies (ranging from sub-centimeter to a few centimeters), the techniques are less robust than pseudo range techniques (Leick, 1990).

Application of GPS technology to production agriculture and precision farming has been discussed by a number of researchers. For example, Shropshire et al. (1993) used post-processed DGPS to locate sample points and real-time DGPS for fertilizer application rate control. They had some difficulty with erroneous position data, which they attributed to problems with the GPS receiver software. They reported that the system they used was adequate for precision farming data acquisition and control. Muhr et al. (1994) used real-time DGPS to monitor the position of a forage chopper during harvesting and reported that position accuracy was better than 2.25 m at all times. Eliason et al. (1995) reported that they were able to obtain accuracy at the sub-meter level by applying a robust carrier phase smoothing technique to the GPS C/A code data.

B. Soil Property Measurement

Although many soil properties (such as fertility levels) can currently be quantified by more traditional methods, widespread adoption of precision farming will depend on automation to improve the efficiency of the soil property analysis process. The spatial and temporal intensity at which each property must be measured is a function of its variability. Some parameters, such as soil nitrate content and soil moisture content, can change rapidly (both spatially and temporally) and must be measured in real-time or near real-time to be useful for input control. Other parameters, such as organic matter content and topsoil depth, are more stable and can be measured off-line on a multi-year frequency. Likewise, spatial measurement intensity can be related to the spatial variation in the property being measured. Unfortunately, the necessary sampling intensity can only be determined after sampling on a small scale to determine the spatial variability, which is cost prohibitive without automated sensors.

There are two sources of error in soil testing — analysis error due to sub-sampling and analytical determination, and sampling error due to variation in soils from point to point. With traditional soil testing, analysis error is relatively low; however, sampling error can be substantial, since cost limits the sampling intensity. Real-time sensors can provide a sampling intensity several orders of magnitude greater than traditional methods. Therefore, a real-time soil sensor can tolerate much higher analysis errors while providing greater overall accuracy in mapping soil variability.

Three soil parameters that have been the subject of significant developments in sensing technology are soil nutrients, soil organic matter, and soil electrical conductivity as an indirect indicator of salinity, soil moisture, and texture.

1. Soil Nutrient Sensing

Current real-time nutrient sensor development has concentrated on nitrogen sensors, due to the economic importance of nitrogen fertilizers, the potential environmental problems associated with excessive nitrates in water, and the considerable temporal and spatial variability of nitrates. The variability of soil nitrate levels dictates that sampling occur within a short time prior to fertilizer application and at a very high sampling intensity. Since potassium and phosphorus are immobile nutrients, and soil pH changes relatively slowly, mapping of these soil properties need not be carried out every year. This allows the cost of traditional soil sampling techniques to be spread over multiple years.

Adsett and Zoerb (1991) reported on real-time nitrate sensing using ion selective electrodes. An automated field monitoring system consisting of a soil sampler, nitrate extraction unit, flow cell, and controller was laboratory and field tested. Variations in soil sample consistency affected the performance of the nitrate extraction system, and the mixing and filtration system required improvement. The nitrate extraction time and methodology were limiting factors in the system, and additional research was planned to improve the mixing and extraction phases.

Ion Selective Field Effect Transistors (ISFETs), which are based on the same chemical principles as ion selective electrodes, have several possible advantages such as small dimensions, low output impedance, high signal-to-noise ratio, fast response, and the ability to integrate several sensors on a single electronic chip. However, ISFETs have the disadvantage of greater long-term drift and hysteresis than ion selective electrodes. The use of a dynamic measurement system such as flow injection analysis (FIA) minimizes the effects of drift and hysteresis, and exploits the specific properties of ISFETs. Hummel and Birrell (1995) successfully used a multi-ISFET/FIA system to measure soil nitrates in manually-obtained soil extracts. The rapid response of the system allowed samples to be analyzed within 1.25 s, and the low sample volumes required by the multi-sensor ISFET/FIA system made it a likely candidate for use in a real-time soil nutrient sensing system.

McGrath et al. (1995) reported on the development of an integrated system of soil sample collection, automated sample preparation and chemical analysis. The sampler automatically collected samples across the field and packaged them in plastic bags that were connected to form a long band. The banded samples were then analyzed using an automated laboratory workstation set up near the field. The initial preparation of the automated workstation required 90 minutes and then samples could be analyzed at a rate of one per minute. Macro-nutrient concentrations (NO_3, NH_4, K, P, Ca, Mg), particle size, soil acidity, and several other parameters were determined (McGrath et al., 1995). The analysis results were then used in an expert system to determine fertilizer recommendations.

2. Soil Organic Matter Sensing

For many soil-applied herbicides, the application rate required for effective weed control increases as soil organic matter content increases. Organic matter variations as high as 100% were seen within an 80-m transect in a flat, apparently uniform field in Central Illinois (Sudduth and Hummel, 1993b). Even higher variations might be expected in areas with more variable topography or over longer distances.

With this amount of variability, significant opportunity exists for cost savings and input optimization of soil-applied herbicides based on soil organic matter (SOM) content. Qiu et al. (1994) used modeling to estimate that up to one-half of preemergence herbicides applied in a "typical" Kentucky corn field could be conserved using site-specific application based on variations in soil organic matter, texture, and weed competition. Their analysis was based on organic matter, soil texture, and weed pressure data obtained on a 30 m by 30 m grid in a 24 ha research field.

Soil organic matter content also affects nitrogen fertilizer needs. Many nitrogen recommendation algorithms allow a credit for the nitrogen-supplying power of the soil, which increases with increasing organic matter content. Thus, SOM sensing may also provide important data for control of variable-rate nitrogen fertilizer application.

Several authors have shown that increases in SOM content result in a decrease in soil reflectance in the visible range (Page, 1974; Griffis, 1985). Dalal and Henry (1986) used reflectances at three wavelengths in the near infrared (NIR) region to estimate SOM. Krishnan et al. (1980) correlated visible and NIR reflectance with SOM for ten Illinois soils at four moisture levels. Better correlations were obtained with information from the visible range than with NIR data.

Shonk et al. (1991) developed a real-time SOM sensor intended to be recalibrated for each new soil catena. Laboratory tests on soil samples collected from five representative midwestern U.S. fields yielded strong correlations between red light reflectance and SOM for soils obtained within a single catena and prepared to a single moisture content. Field tests showed a curvilinear relationship between sensor output and SOM, with new calibrations developed for changes in travel speed or sensing depth (Shonk et al., 1991). The sensor was licensed for commercial development, and used to control the rate of a granular herbicide applied by a pneumatic metering system (McGrath et al., 1990). The sensor probe was mounted to the front of a custom applicator and operated at a depth of 10 cm and speeds of up to 19 km/h. Soil samples were collected from each different soil catena to develop a specific sensor calibration curve. Weed control was said to be "excellent" in these tests.

Sudduth and Hummel (1993a) developed a portable spectrophotometer designed to acquire NIR soil reflectance data at a number of narrow-band wavelengths. Although this type of instrument is more complex, more expensive, and less rugged than a single-band sensor, the additional reflectance information allowed generation of accurate SOM calibrations applicable to soils obtained from multiple landscapes (Sudduth and Hummel, 1993b; Sudduth et al., 1990). Laboratory predictions of SOM were successful, across a range of soil types and moisture contents, with accuracies approaching those obtained for the same soils with a research-grade spectrophotometer. Field operation of the prototype sensor did not yield acceptable results, due at least in part to errors introduced by the movement of soil past the sensor during scanning (Sudduth and Hummel, 1993b).

NIR reflectance techniques have also been used to estimate other soil properties. Kano et al. (1985) estimated soil moisture content by NIR. Sudduth and Hummel (1993b) predicted soil moisture and cation exchange capacity for 30 Illinois soils. Dalal and Henry (1986) determined total nitrogen and moisture content of selected Australian soils.

3. Soil Conductivity Sensing

Soil conductivity is a function of soil salinity, clay content, and water content (Rhoades et al., 1989). Therefore, soil conductivity measurements have the potential for providing estimates of within-field variations in these properties. However, care must be taken to understand the effects of the other, nonestimated properties on the conductivity measurement. In areas without saline soils, spatial variation in soil moisture content is often a major factor determining variations in bulk soil conductivity. Kachanoski et al. (1988) found that soil electrical conductivity was highly correlated with soil water content stored in the top 0.5 m, within a single field and on a single measurement date. Sheets and Hendrickx (1995) measured electrical conductivity along a 1950 m transect in New Mexico over a 16 month period, and found a linear relationship between conductivity and profile soil water content. Independent measurements of soil water at several calibration points along the transect were required for each measurement date. In both of these studies, soil conductivity was measured using commercial electromagnetic induction (EM) ground conductivity meters, the EM-31[1] and EM-38 (Geonics Ltd., Mississauga, Ontario, Canada). These portable meters allow immediate, noncontact measurements of soil conductivity. They can be coupled with GPS location equipment and data loggers, and pulled across a field using an all-terrain vehicle to rapidly map conductivity variations over large areas (Jaynes et al., 1995a).

Soil conductivity measurements can provide information on soil texture, in addition to estimating soil water content. Williams and Hoey (1987) used EM measurements of soil conductivity to estimate within-field variations in soil clay content. Doolittle et al. (1994) found that EM measurements were highly correlated with the depth to a subsurface claypan horizon. An automated EM sensing system was then used to map depth to claypan over a number of fields by Kitchen et al.

[1] Mention of a trade name, proprietary product, or specific equipment does not constitute a guarantee or warranty by the USDA-Agricultural Research Service and does not imply the approval of the named product to the exclusion of other products that may be suitable.

(1995b). They found it necessary to obtain calibration measurements of claypan depth with a soil probe at a number of locations within a field to remove the effects of temporal variations in soil water content and temperature.

Since soil conductivity integrates texture and moisture availability, two characteristics that both vary over the landscape and also affect productivity, EM sensing of conductivity shows some promise in interpreting grain yield variations, at least in certain soils (Sudduth et al., 1995; Jaynes et al., 1995a; Kitchen et al., 1995b). EM data have also been used to estimate other soil properties related to clay content, including cation exchange capacity (McBride et al., 1990). Jaynes et al. (1995b) found that EM data were a good alternative to the standard method of estimating herbicide partition coefficients (K_d) from measurements of the mass fraction of soil organic carbon. Estimates of K_d based on EM data showed similar spatial patterns to measured K_d values. Varying application rates based on these estimated K_d values could potentially reduce herbicide leaching.

C. Weed and Crop Sensing

Weed identification by means of an optical sensing system mounted on an agricultural vehicle has been investigated by a number of researchers. Some projects have focused on locating green plant material on a background of bare soil, as would be needed to spray between the rows of a row crop (Merritt et al., 1994; Woebbecke et al., 1995a). Others have sought to discriminate weed species from crop species, on the basis of spectral properties (Franz et al., 1991a) or shape features (Woebbecke et al., 1995b; Franz et al., 1991b).

The high cost and complexity of machine vision systems, and the voluminous quantities of data to be processed, have led some researchers to conclude that, for the foreseeable future, site-specific weed control will be based upon the development of maps that will control sprayer output (Thompson et al., 1991). Geographic information systems (GIS) can be used to store maps of weed distributions and to predict weed densities from historical data (Brown et al., 1990; Stafford and Miller, 1993). Mapping of weeds at opportune times, such as at flowering or senescence, for control during the subsequent season, could enhance the accuracy of the map, as well as provide information on the species of the weed infestation (Thompson et al., 1991). Brown et al. (1990) stated that the use of a map to control herbicide application during the next season was valid because many weed species appeared in well-defined patches which remained relatively static from year to year. Stafford et al. (1996) developed a data logging system to facilitate the development of such weed maps by in-field scouting.

Crop nitrogen status during the growing season can be used to provide input to a variable-rate N fertilizer application system so that only the amount of N required to support crop growth is applied. Researchers in Nebraska have used chlorophyll meters to monitor crop N status and apply fertilizer N through an irrigation system (Blackmer and Schepers, 1994). Optical methods of detecting nitrogen stress in crops have included remote sensing and aerial photography, directed at measuring the “greenness” of the crop as an indicator of nitrogen stress (McMurtrey et al.,

1994). Research has also shown that 550 nm light reflectance from corn in the late milk stage was highly correlated with corn yield (Blackmer et al., 1994).

D. Crop Yield Sensing

Within-field variation of crop yield is an important input for site-specific decision making. Crop yield is an integrator of many varying crop and soil parameters, such as moisture, nutrients, and pest problems. Linking spatial information on both yield and soil properties through a GIS system allows for diagnostic determination of the predominant factor(s) controlling crop production. This information then becomes the basis for developing precision input strategies. Additionally, yield measurements can provide feedback on the effect of variable rate application of inputs, allowing refinement of the application plan for future years.

Combine-based quantification of spatial variations in grain yield requires the successful integration of several sensors and information sources. Grain flow rate must be measured at some point on the combine, usually as the material enters the grain tank. The grain transport dynamics of the combine must be modeled and used to adjust the measured grain flow in recreating the actual rate of grain entering the combine header (Searcy et al., 1989). Combine forward velocity must be measured or determined to convert from grain harvest rates per unit time to harvest rates per unit area, or yield. In drilled grains, the width of cut must be measured to provide an accurate measure of yield. Combine position within the field must be determined and that data integrated with the other data streams. Finally, all these data must be combined and processed to provide interpretable grain yield maps.

Several grain flow sensors are either on the market or in commercial development. In addition, a number of other prototype units have been developed or refined by public sector researchers (for example, Peterson et al., 1989; Vansichen and De Baerdemaeker, 1991; Pfeiffer et al., 1993; Pringle et al., 1993; Birrell et al., 1995). Researchers have also developed systems for measuring silage yield (Vansichen and De Baerdemaeker, 1993; Auernhammer and Demmel, 1995), cotton yield (Wilkerson et al., 1994), and potato yield (Campbell et al., 1994).

E. Variable Rate Technology

Variable rate technology (VRT) devices allow for differential application of crop inputs to be controlled automatically while traversing a field. A number of components and systems have been developed for VRT control of liquid and dry fertilizers, herbicides, seeds, and irrigation water (Schueller, 1992). Commercially available units range from completely integrated systems, such as the Soil-Teq fertilizer applicator (Monson, 1995), to individual controllers for varying planting or application rates, which are available from a number of manufacturers. While the Soil-Teq unit and others are targeted at the commercial applicator market, many of the individual controllers may be used by an individual producer to retrofit existing equipment for variable rate product delivery.

A number of new VRT systems for a range of products have been the subject of research and development efforts in industry and the public sector. Anderson (1995) reported on a prototype system for variable rate control of seed and fertilizer, as did Neuhaus and Searcy (1993). Camp and Sadler (1994) designed an experimental center pivot irrigation system, which they planned to use for site-specific water and nutrient application after its completion. Cahn and Hummel (1995) developed a VRT system for side-dressing liquid nitrogen fertilizer. In field testing, the applicator system exhibited a response time of 2 s or less, and an application rate accuracy of 95% or better.

1. VRT for Herbicide Application

Considerable research attention has been focused on development of VRT systems for herbicide application. Significant cost savings may be possible from turning a sprayer on and off to spot treat infested areas of fields, in an approach sometimes termed "patch spraying." Control of the sprayer could be from previously mapped weed information (Brown et al., 1990) or through on-board weed sensing. Relatively simple two-wavelength sensors can locate green plant material on a background of bare soil and control spray nozzles for fallow weed control (Felton and McCloy, 1992). Similar sensors have been used to control spray application of herbicides between the rows of a row crop (Merritt et al., 1994). Hanks (1995) reported on a sensor/nozzle unit that had been developed in cooperation with Patchen California, Inc., and tested on an 8-row toolbar for controlling weeds in row middles. Studies of this system in no-till soybeans indicated that essentially all green plant material 4 cm^2 or larger between the rows was detected and sprayed, while herbicide use was reduced by 85% as compared to conventional application.

Other VRT systems have been developed to vary herbicide application rate through a field, as opposed to providing merely on-off control. Application rate changes may be indicated by variations in weed pressure or by variations in those soil properties, such as organic matter and texture, which affect herbicide efficacy. Paice et al. (1996) provided a good review of the requirements for variable-rate herbicide sprayers. To obtain the desired application rate control, either the herbicide concentration or the total nozzle output must be varied. Variations in total nozzle output can be accomplished by changing the inlet pressure of the nozzle, which may affect droplet size and distribution, or alternatively by the use of bypass nozzles that return a portion of their inlet flow back to the supply tank (Han et al., 1986; Comino and Giles, 1988).

The variable concentration method allows spraying of a constant total volume of solution at a constant nozzle pressure, thus maintaining uniform coverage with currently available nozzle designs. This approach, referred to as chemical injection, is commercially available from several firms. Problems that have been noted with chemical injection include increased deposition variability and transient errors in application rate resulting from time lags in the system (Vidrine et al., 1975; Tompkins et al., 1988). Miller and Smith (1992) developed a direct nozzle injection system that showed promise for reducing transient errors and nozzle-to-nozzle variability. Paice et al. (1995) reported on an experimental injection sprayer in which tubing size was selected to minimize time lags, and to make the lag more constant across the spray boom.

2. Accuracy of VRT Equipment

Application accuracy is important in any system, but becomes even more important with VRT. Reliable changes in application rate can only be delivered by a system having an accuracy considerably better than the minimum level of the required rate change increments. Proper attention to issues such as delivery system calibration, nozzle-to-nozzle uniformity, and proper equipment operation is essential in providing good VRT application. To aid in uniformity testing and calibration, engineers have developed standards such as ASAE S341.2, "Procedure for measuring distribution uniformity and calibrating granular broadcast spreaders" (ASAE, 1993). Chaplin et al. (1995) proposed a revision to that standard that would make it more applicable to VRT applicators.

Transient errors in VRT application rates can be caused by the delay between the system controller commanding a rate change and the application system delivering product at that rate. Rudolph and Searcy (1994) reported on a situation where an injection sprayer traveled over 150 m at normal field speeds before the commanded final application rate was obtained. They further reported that the lag could be reduced to about 22 m by optimizing the sprayer for VRT use. Lags would vary for different types and designs of application equipment and for different operating conditions, but in many cases these lags would cause significant areas of misapplication within a field. Rudolph and Searcy (1994) described a predictive, map-based approach to applicator control, where commands were given to the sprayer based upon estimating where it would be in the field at the time the commanded product delivery rate was realized.

III. Precision Farming as an Integrated System

Successful implementation of a precision farming approach to crop production requires an integration of engineering technology and agronomic knowledge, in what can be viewed as a four-step process. An initial step in this process is spatial measurement of those factors that limit or otherwise affect crop production. These variability data are then used to develop a management plan for the variable application of inputs such as fertilizers and herbicides. Inputs are applied in precision field operations. Finally, the efficacy of the precision farming system is evaluated with respect to economics and environmental impacts. This evaluation becomes a part of the data collection process for the next cropping season. Multiple iterations through the cycle allow for refinement of the precision management plan in succeeding seasons.

A. Data Collection for Variability Measurement

Most commercially available precision farming services are centered around nutrient variability mapping by means of point-by-point grid soil sampling (Buchholz, 1991; Carr et al., 1991; Wollenhaupt et al., 1994). Fields are then divided into different manageable units based on the spatial relationship of nutrient variability (Sawyer, 1994). The optimum grid size for use in soil nutrient mapping is dependent on the level of spatial variability present and the cost of soil sampling and analysis. In a field where previous nutrient maps have been obtained, it may be possible to infer the grid size needed to adequately characterize spatial variability. However, for newly mapped fields, initial sampling should be at a small grid size, so that all significant patterns can be represented. Bullock et al. (1994) studied the effect of grid size on soil fertility maps obtained at eight Illinois sites. In comparing 0.25 ha grids to 1.0 ha grids (a standard size for commercial precision farming implementation), they sometimes found significant differences, leading to over-fertilizing or under-fertilizing of large areas of fields when using 1.0 ha grids.

On-the-go sensors could potentially reduce costs while improving the accuracy of the data collection process. However, with the exception of grain yield sensors, most sensors are not yet ready to move to the point of commercial application. Grain yield sensing has now joined grid soil sampling as a basic information source for precision farming. Aftermarket yield sensors have been available for several years in the U.S. and abroad, and grain combine manufacturers are beginning to offer such sensors as original equipment on their machines.

B. Management Planning

Successful use of the large quantities of data required for precision farming requires that an appropriate computer hardware and software system be available to efficiently manage and manipulate the data. Success also requires developing an understanding of the basic interrelationships contained within the data. These relationships between yields and other factors are quite complex, and may vary from place to place even within a single field (Drummond et al., 1995).

Systems for precision farming data management generally fall into two categories, geographic information systems (GIS), and agricultural mapping software packages (Searcy, 1995). A GIS is a generic spatial data analysis system, developed to be useful for a wide range of applications. The GIS generally has good analytical tools and flexibility, as would be required in a research setting. Since the learning curve on such systems is typically steep and the terminology used is not specific to agriculture, GIS packages may be difficult for producers and consultants to master. Agricultural mapping packages have been developed for specific purposes related to production agriculture. While they may work well at generating maps, they often do not have the analysis capabilities of a GIS. With time, GIS packages are offering interfaces that may be customized for specific uses, and mapping packages are offering more analysis capabilities. Thus, the capabilities of the two types of systems are becoming more alike (Searcy, 1995).

Previous agronomic research and knowledge provide a basis for beginning to understand the spatial relationships between yields and other factors for the development of site-specific management plans. However, the use of general fertilizer response relationships based on aggregation of data from numerous sites across a state or region for precision nutrient application is questionable. Rather, research indicates that crop nutrient response is a site-specific parameter, and may vary from region to region and even from place to place within a single field. Vanotti and Bundy (1994) proposed developing corn N fertilizer recommendations by grouping soil series with similar N response data. At an individual field level, response to fertilizer can be different based upon soil and landscape features related to soil water (Hughes and Kitchen, 1994). These site-specific response functions might be inferred by a reinterpretation of the data that were used years ago to create the general, statewide relationships. However, in many cases these original data have long since disappeared, and new research may be required to develop the best response criteria for precision fertilizer input control.

As precision farming becomes widely implemented, the need for professional agronomic advice and knowledge may well outstrip the availability of trained professionals. Implementation of computerized decision support systems (DSS) will be necessary to fill this gap. Several researchers have developed prototype decision support frameworks for site-specific management, usually based on GIS technology. Simulation modeling and expert system analysis have been incorporated to estimate potential yields and fertilizer requirements (Hoogenboom et al., 1993; He et al., 1992). Continuing progress in the area of DSS development and validation will be required to provide the tools needed for widespread commercial use of precision farming methods.

C. Precision Field Operations

The ability to precisely vary application rates while traversing a field is an essential part of precision farming. The variable rate technology (VRT) that makes this possible is probably the best developed part of the integrated precision farming system (Searcy, 1995). Application control with VRT equipment can be implemented either on-line or off-line. In the on-line approach, the controlled equipment (herbicide applicator, for example) incorporates on-board sensors, and the sensor data are used immediately for automatic control. In the off-line, or map-based approach, sensor data are collected and stored in one operation, and the controlled equipment uses the information in a separate field operation. The map-based approach allows more flexibility in data manipulation and preprocessing but requires that the location of equipment in the field be precisely defined, as with GPS. Most systems currently available are map-based, but more on-line systems will likely become available as real-time sensing technologies become more mature.

Much of the VRT equipment developed to date has been targeted at the custom application market, where the costs of the technology can be spread over larger acreages. More recently, farm level controllers for VRT fertilizer application and seeding have become available. Significant potential markets also exist for VRT aerial application and irrigation delivery systems (Searcy, 1995).

D. Evaluation

1. Yield and Profitability

Research results documenting the effects of precision farming methods on yield and profitability are mixed and based on only a few studies. In the Palouse region of the Pacific Northwest, Fiez et al. (1994) reported improvements in wheat yield and profitability due to variable application of nitrogen. They also stated that variable nitrogen applications should take into account the fact that N use efficiencies varied across the landscape. Hughes and Kitchen (1994) also reported that significantly different yield responses to N fertilizer occurred among different soil and landscape areas within a field. Carr et al. (1991), working on dryland wheat and barley in Montana, reported mixed results when comparing variable-rate fertilizer treatments using application varied by soil map units to treatments based on conventional, whole-field recommendations.

Wibawa et al. (1993) increased small grain yield in North Dakota using grid soil sampling and yield goals based on soil map unit to guide VRT application of fertilizers. However, net returns were reduced, even though yields were increased, due to the additional costs of implementing their program. They noted that precision application would be most profitable when there were large differences observed within a field. Buchholz (1991) reported that economic improvement with variable rate P and K fertilization was dependent on the level of within-field fertility variation, the yield response to these variations, and the costs of implementing a precision farming program.

Precision application of crop inputs may be more profitable in high-value crops. Hilde (1994) reported on precision N management in sugar beets, where departure from the optimum N level can result in dramatic yield and quality decreases. Preliminary results indicated potential net returns of over $350/ha for variable rate fertilization based on soil nitrate measurements. Hammond (1993) reported increased net profits of over $900/ha in commercial potato production.

Audsley (1993) analyzed the theoretical benefits of patch spraying for U.K. cereal weeds, such as wild oats and black grass. Patch spraying was found to be more profitable than blanket treatment, with increased value reported for smaller weed patches, lower seed densities, and more effective herbicides. Much of the economic advantage of patch spraying was lost if the size of the smallest treatable patch was increased above 12 m^2. This small patch size indicates the need for herbicide application equipment that can detect weed levels and vary rates with a high degree of spatial precision.

2. Environmental Impacts

Improving the environment is a benefit often attributed to the site-specific application of fertilizers and pesticides. While water quality benefits are expected, only a few studies document measurements quantifying the impact of site-specific management relative to conventional management (Sawyer, 1994). For example, Kitchen et al. (1995a) found that in some cases residual soil nitrate levels could be reduced by a variable rate nitrogen application strategy, as compared to a single rate "best areas of the field" approach.

IV. Summary

Precision farming, or site-specific management, is a management system in which the application of crop production inputs is optimized for crop needs on a location-by-location basis within a field. Although this type of intensive management is not a new idea, newly developed engineering technologies are now making it feasible on a widespread basis. Location systems, electronic sensors for crop and soil properties, and electronic control systems for varying application rates of production inputs are examples of some of the important enabling technologies for precision farming.

Successful implementation of a precision farming approach to crop production requires a combination of engineering technology and agronomic knowledge. An integrated precision farming system can be viewed as a cycle composed of data collection, management planning, precision field operations, and evaluation. Multiple iterations through the cycle allow for refinement of the precision management plan in succeeding seasons.

To date, most precision farming developments have concentrated on variable rate application of fertilizer nutrients and/or variable rate seeding. Great potential exists for the application of this technology to weed control, due to the relatively higher costs for herbicide inputs and the greater potential for environmental impact, when compared to fertilizer inputs. Precision weed management may be implemented at various levels of complexity. A simple system might be based on visual identification, mapping, and spot herbicide treatment of weed patches. A more futuristic system might include automatic sensing of weed density and species, soil organic matter content, and soil conductivity as an indicator of herbicide degradation potential. Subsequent variable rate (perhaps coupled with variable product) herbicide application would then use decision algorithms based on all these sensed inputs.

Technologies under development and those now becoming commercially available will provide many of the necessary tools for precision weed management. Soil organic matter can be sensed as an input to herbicide rate control, and soil electrical conductivity sensing has shown promise as an indicator of herbicide degradation rates. Several weed sensing systems are in various stages of development. Units which discriminate green plant material from soil or residue and can be used for precision weed control between crop rows are becoming commercially available. Some basic research has been conducted on machine vision systems that could discriminate weed species from crop plant species, but much work remains to make this technology usable for in-field sensing. Systems for variable rate application of herbicides are available now and improved systems are being developed.

Although many of the first-generation tools necessary for precision crop management are available or in development, there are significant opportunities for development of new and improved sensing technologies to improve the data collection process for measurements that are now difficult or costly to obtain. Also important will be research to develop decision rules and algorithms optimized for this spatial information. Current precision farming recommendations are usually based on applying long-standing "blanket" decision rules to site-specific data. However, there is research to indicate that the response relationship of yield to nutrient levels or weed densities is often variable within a given field. Understanding and applying these site-specific response relationships for input control will be an important refinement to precision farming systems. Long-term research to quantify the economic and environmental benefits of precision farming is also critical. The few economic evaluations in the literature have reported mixed results, and have generally only looked at returns over a few years. Although positive environmental impacts are often attributed to precision farming systems, very little hard evidence has been reported to support this intuitive benefit.

Technology transfer activities are very important and must be linked closely with precision farming research. Producers and agribusiness professionals are demanding answers now to their questions about precision farming. Those of us involved in research are "playing catch-up," since many of the tools we needed to research this topic, such as GPS and combine yield monitors, became available to producers and researchers at the same time. We have been, and will continue to be, conducting research concurrently with the production implementation of many of these technologies, techniques, and systems. Because of this, timely extension of research results to the user community is crucial.

References

Adsett, J.F. and G.C. Zoerb. 1991. Automated field monitoring of soil nitrate levels. p. 326-335. In: Automated Agriculture for the 21st Century. ASAE Public. No. 1191. ASAE, St. Joseph, MI.

Anderson, N.W. 1995. The Terranova Vrs Variable Rate System. ASAE Paper No. 951754. ASAE, St. Joseph, MI.

ASAE. 1993. Procedure for Measuring Distribution Uniformity and Calibrating Granular Broadcast Spreaders. Standard S341.2. ASAE, St. Joseph, MI.

Audsley, E. 1993. Operational research analysis of patch spraying. *Crop Prot.* 12:111-119.

Auernhammer, H. and M. Demmel. 1995. Yield Measurement on Self Propelled Forage Harvesters. ASAE Paper No. 951757. ASAE, St. Joseph, MI.

Auernhammer, H. and T. Muhr. 1991. Gps in a Basic Rule for Environment Protection in Agriculture. p. 394-402. In: Automated Agriculture for the 21st Century. ASAE Public. No. 1191. ASAE, St. Joseph, MI.

Beckett, P.H.T. and R. Webster. 1971. Soil variability: A review. *Soils and Fertilizers* 34:1-15.

Birrell, S.J., K.A. Sudduth, and S.C. Borgelt. 1995. Crop yield mapping: comparison of yield monitors and mapping techniques. p. 15-31. In: P.C. Robert, R.H. Rust, and W.E. Larson (eds.), Site-specific management for agricultural systems. Proc. 2nd Intl. Conf., Minneapolis, MN. 27-30 March 1994. ASA, CSSA, and SSSA, Madison, WI.

Blackmer, T.M. and J.S. Schepers. 1994. Techniques for monitoring crop nitrogen status in corn. *Commun. Soil Sci. Plant Anal.* 25:1791-1800.

Blackmer, T.M., J.S. Schepers, and G.E. Varvel. 1994. Light reflectance compared with other nitrogen stress measurements in corn leaves. *Agron. J.* 86:934-938.

Brown, R.B., G.W. Anderson, B. Proud, and J.P. Steckler. 1990. Herbicide Application Control Using Gis Weed Maps. ASAE Paper No. 901061. ASAE, St. Joseph, MI.

Buchholz, D.D. 1991. Missouri grid soil sampling project. p. 6-12. In: Proc. 21st North Central Extension Industry Soil Fertility Conference. St. Louis, Missouri. 28-29 Oct. 1991. Potash and Phosphate Institute, Manhattan, KS.

Bullock, D.G., R.G. Hoeft, P. Dorman, T. Macy, and R. Olson. 1994. Nutrient management with intensive soil sampling and differential fertilizer spreading. *Better Crops* 78:10-12.

Cahn, M.D. and J.W. Hummel. 1995. Variable rate system for side-dressing liquid N fertilizer. p. 683-689. In: P.C. Robert, R.H. Rust, and W.E. Larson (eds.), Site-specific management for agricultural systems. Proc. 2nd Intl. Conf., Minneapolis, MN. 27-30 March 1994. ASA, CSSA, and SSSA, Madison, WI.

Camp, C.R. and E.J. Sadler. 1994. Center Pivot Irrigation System for Site-Specific Water and Nutrient Management. ASAE Paper No. 941586. ASAE, St. Joseph, MI.

Campbell, R.H., S.L. Rawlins, and S. Han. 1994. Monitoring Methods for Potato Yield Mapping. ASAE Paper No. 941584. ASAE, St. Joseph, MI.

Carr, P.M., G.R. Carlson, J.S. Jacobsen, G.A. Nielsen, and E.O. Skogley. 1991. Farming soils, not fields: a strategy for increasing fertilizer profitability. *J. Prod. Agric.* 4:57-61.

Chaplin, J., E. Roytburg, and J. Kaplan. 1995. Measuring the spatial performance of chemical applicators. p. 651-669. In: P.C. Robert, R.H. Rust, and W.E. Larson (eds.), Site-specific management for agricultural systems. Proc. 2nd Intl. Conf., Minneapolis, MN. 27-30 March 1994. ASA, CSSA, and SSSA, Madison, WI.

Comino, J.A. and D.K. Giles. 1988. Evaluation of a Variable Flow Controller for Spray Nozzles. ASAE Paper No. 881590. ASAE, St. Joseph, MI.

Dalal, R.C. and R.J. Henry. 1986. Simultaneous determination of moisture, organic carbon, and total nitrogen by near infrared reflectance spectrophotometry. *Soil Sci. Soc. Am. J.* 50:120-123.

Doolittle, J.A., K.A. Sudduth, N.R. Kitchen, and S.J. Indorante. 1994. Estimating depths to claypans using electromagnetic induction methods. *J. Soil Water Cons.* 49:572-575.

Drummond, S.T., K.A. Sudduth, and S.J. Birrell. 1995. Analysis and Correlation Methods for Spatial Data. ASAE Paper No. 951335. ASAE, St. Joseph, MI.

Eliason, M., D. Heaney, T. Goddard, M. Green, C. McKenzie, D. Penney, H. Gehue, G. Lachapelle, and M.E. Cannon. 1995. Yield measurement and field mapping with an integrated GPS system. p. 49-58. In: P.C. Robert, R.H. Rust, and W.E. Larson (eds.), Site-specific management for agricultural systems. Proc. 2nd Intl. Conf., Minneapolis, MN. 27-30 March 1994. ASA, CSSA, and SSSA, Madison, WI.

Fairfield Smith, H. 1938. An empirical law describing the heterogeneity in the yield of agricultural crops. *J. Agric. Sci.* 28:1-23.

Felton, W.L. and K.R. McCloy. 1992. Spot spraying. Microprocessor-controlled, weed-detecting technology helps save money and the environment. *Agric. Engr.* 73:9-12.

Fiez, T.E., B.C. Miller, and W.L. Pan. 1994. Assessment of spatially variable nitrogen fertilizer management in winter wheat. *J. Prod. Agric.* 7:86-93.

Franz, E., M.R. Gebhardt, and K.B. Unklesbay. 1991a. The use of local spectral properties of leaves as an aid for identifying weed seedlings in digital images. *Trans. ASAE* 34:682-687.

Franz, E., M.R. Gebhardt, and K.B. Unklesbay. 1991b. Shape description of completely visible and partially occluded leaves for identifying plants in digital images. *Trans. ASAE* 34:673-681.

Griffis, C.L. 1985. Electronic sensing of soil organic matter. *Trans. ASAE* 28:703-705.

Hammond, M. 1993. Cost analysis of variable fertility management of phosphorous and potassium for potato production in central Washington. p. 213-228. In: P.C. Robert, R.H. Rust, and W.E. Larson (eds.), Proc. first workshop on soil specific crop management. Minneapolis, MN. 14-16 April 1992. ASA, CSSA, and SSSA, Madison, WI.

Han, Y.J., L.E. Bode, and J.W. Hummel. 1986. Controlling chemical application rate with bypass spray nozzles. *Trans. ASAE* 29:1221-1227.

Hanks, J.E. 1995. Sensor-controlled sprayers for herbicide application. p. 176-179. In: Proc. Nat. Conf. on Pesticide Application Technology. University of Guelph, Guelph, Ontario, Canada.

He, B., C.L. Peterson, and R.L. Mahler. 1992. An Expert System Linked with a Gis Database for Spatially Variable Fertilizer Application. ASAE Paper No. 923556. ASAE, St. Joseph, MI.

Hilde, D. 1994. Fine-tuning sugar beet fertility management in the Red River valley. *Better Crops* 78:26-27.

Hoogenboom, G., H. Lal, and D.D. Gresham. 1993. Spatial Yield Prediction. ASAE Paper No. 933550. ASAE, St. Joseph, MI.

Hughes, D.F. and N.R. Kitchen. 1994. Corn response to nitrogen fertilizer in distinct soil management units. p. 13-17. In: Missouri Soil Fertility and Fertilizers Research Update 1993. Agron. Misc. Publ. #94-03. Univ. of Missouri, Columbia, MO.

Hummel, J.W. and S.J. Birrell. 1995. Real-time soil nitrate sensing. p.125-136. In: G. Rehm (ed.), 25th North Central Extension-Industry Soil Fertility Conference Proceedings. St. Louis, Missouri. 15-16 Nov. 1995. Potash & Phosphate Institute, Manhattan, KS.

James, D.W. and A.I. Dow. 1972. Source and degree of soil variability in the field - the problem of sampling for soil tests and estimating soil fertility status. Washington Agric. Exp. Sta. Bull. 749. Washington State Univ., Pullman, WA.

Jaynes, D.B., T.S. Colvin, and J. Ambuel. 1995a. Yield mapping by electromagnetic induction. p. 383-394. In: P.C. Robert, R.H. Rust, and W.E. Larson (eds.), Site-specific management for agricultural systems. Proc. 2nd Intl. Conf., Minneapolis, MN. 27-30 March 1994. ASA, CSSA, and SSSA, Madison, WI.

Jaynes, D.B., J.M. Novak, T.B. Moorman, and C.A. Cambardella. 1995b. Estimating herbicide partition coefficients from electromagnetic induction measurements. *J. Environ. Qual.* 24:36-41.

Kachanoski, R.G., E.G. Gregorich, and I.J. Van Wesenbeeck. 1988. Estimating spatial variations of soil water content using noncontacting electromagnetic inductive methods. *Can. J. Soil Sci.* 68:715-722.

Kano, Y., W.F. McClure, and R.W. Skaggs. 1985. A near infrared reflectance soil moisture meter. *Trans. ASAE* 28:1852-1855.

Kitchen, N.R., D.F. Hughes, K.A. Sudduth, and S.J. Birrell. 1995a. Comparison of variable rate to single rate nitrogen fertilizer application: corn production and residual soil NO_3-N. p. 427-441. In: P.C. Robert, R.H. Rust, and W.E. Larson (eds.), Site-specific management for agricultural systems. Proc. 2nd Intl. Conf., Minneapolis, MN. 27-30 March 1994. ASA, CSSA, and SSSA, Madison, WI.

Kitchen, N.R., K.A. Sudduth, S.J. Birrell, and S.T. Drummond. 1995b. Spatial prediction of crop productivity using electromagnetic induction. p. 299. In: Agronomy Abstracts. ASA, Madison, WI.

Krishnan, P., J.D. Alexander, B.J. Butler, and J.W. Hummel. 1980. Reflectance technique for predicting soil organic matter. *Soil Sci. Soc. Am. J.* 44:1282-1285.

Leick, A. 1990. *GPS Satellite Surveying*. John Wiley and Sons, New York, NY.

Linsley, C.M. and F.C. Bauer. 1929. Test your soil for acidity. Circ. 346. College of Agric. and Agric. Exp. Sta., Univ. of Illinois, Champaign, IL.

McBride, R.A., A.M. Gordon, and S.C. Shrive. 1990. Estimating forest soil quality from terrain measurements of apparent electrical conductivity. *Soil Sci. Soc. Am. J.* 54:290-293.

McGrath, D.E., J.P. Ellingson, and A.O. Leedahl. 1990. Variable Application Rates Based on Soil Organic Matter. ASAE Paper No. 901598. ASAE, St. Joseph, MI.

McGrath, D.E., A.V. Skotnikov, and V.A. Bobrov. 1995. A site-specific expert system with supporting equipment for crop management. p. 619-635. In: P.C. Robert, R.H. Rust, and W.E. Larson (eds.), Site-specific management for agricultural systems. Proc. 2nd Intl. Conf., Minneapolis, MN. 27-30 March 1994. ASA, CSSA, and SSSA, Madison, WI.

McMurtrey III, J.E., E.W. Chappelle, M.S. Kim, J.J. Meisinger, and L.A. Corp. 1994. Distinguishing nitrogen fertilization levels in field corn with actively induced fluorescence and passive reflectance measurements. *Rem. Sens. Environ.* 47:36-44.

Merritt, S.J., G.E. Meyer, K. Von Bargen, and D.A. Mortensen. 1994. Reflectance Sensor and Control System for Spot Spraying. ASAE Paper No. 941057. ASAE, St. Joseph, MI.

Miller, M.S. and D.B. Smith. 1992. A direct nozzle injection controlled rate spray boom. *Trans. ASAE* 35:781-785.

Monson, R.J. 1995. The Falcon Control System for Soilection. ASAE Paper No. 951753. ASAE, St. Joseph, MI.

Muhr, T., H. Auernhammer, M. Demmel, and K. Wild. 1994. Inventory of Fields and Soils with Dgps and Gis for Precision Farming. ASAE Paper No. 941583. ASAE, St. Joseph, MI.

Neuhaus, P.E. and S.W. Searcy. 1993. Variable Planting Density and Fertilizer Rate Application System. ASAE Paper No. 931554. ASAE, St. Joseph, MI.

Page, N.R. 1974. Estimation of organic matter in Atlantic coastal plain soils with a color-difference meter. *Agron. J.* 66:652-653.

Paice, M.E.R., P.C.H. Miller, and J.D. Bodle. 1995. An experimental sprayer for the spatially selective application of herbicides. *J. Agric. Engrg. Res.* 60:107-116.

Paice, M.E.R., P.C.H. Miller, and W. Day. 1996. Control requirements for spatially selective herbicide sprayers. *Comp. Elect. Agric.* 14:163-177.

Peterson, C.L., J.C. Whitcraft, K.N. Hawley, and E.A. Dowding. 1989. Yield Mapping Winter Wheat for Improved Crop Management. ASAE Paper No. 897034. ASAE, St. Joseph, MI.

Pfeiffer, D., J.W. Hummel, and N.R. Miller. 1993. Real-Time Corn Yield Sensor. ASAE Paper No. 931013. ASAE, St. Joseph, MI.

Pringle, J.L., M.D. Schrock, R.T. Hinnen, K.D. Howard, and D.L. Oard. 1993. Yield Variation in Grain Crops. ASAE Paper No. 931505. ASAE, St. Joseph, MI.

Qiu, W., S.A. Shearer, and G.A. Watkins. 1994. Modeling of Variable-Rate Herbicide Application Using GIS. ASAE Paper No. 943522. ASAE, St. Joseph, MI.

Reed, J.F. and J.A. Rigney. 1947. Soil sampling from fields of uniform and nonuniform appearance and soil types. *J. Am. Soc. Agron.* 39:26-40.

Rhoades, J.D., N.A. Manteghi, P.J. Shouse, and W.J. Alves. 1989. Soil electrical conductivity and soil salinity: New formulations and calibrations. *Soil Sci. Soc. Am. J.* 53:433-439.

Robinson, G.W. and W.E. Lloyd. 1915. On the probable error of sampling in soil surveys. *J. Agric. Sci.* 7:144-153.

Rudolph, W.W. and S.W. Searcy. 1994. Strategies for Prescription Application Using the Chemical Injection Control System with Computer Commanded Rate Changes. ASAE Paper No. 941585. ASAE, St. Joseph, MI.

Sawyer, J.E. 1994. Concepts of variable rate technology with considerations for fertilizer application. *J. Prod. Agric.* 7:195-201.

Schueller, J.K. 1992. A review and integrating analysis of spatially-variable control of crop production. *Fert. Res.* 33:1-34.

Searcy, S.W. 1995. Engineering systems for site-specific management: opportunities and limitations. p. 603-612. In: P.C. Robert, R.H. Rust, and W.E. Larson (eds.), Site-specific management for agricultural systems. Proc. 2nd Intl. Conf., Minneapolis, MN. 27-30 March 1994. ASA, CSSA, and SSSA, Madison, WI.

Searcy, S.W., J.K. Schueller, Y.H. Bae, S.C. Borgelt, and B.A. Stout. 1989. Mapping of spatially variable yield during grain combining. *Trans. ASAE* 32: 826-829.

Sheets K.R. and J.M.H. Hendrickx. 1995. Noninvasive soil water content measurement using electromagnetic induction. *Water Resources Res.* 31:2401-2409.

Shonk, J.L., L.D. Gaultney, D.G. Schulze, and G.E. Van Scoyoc. 1991. Spectroscopic sensing of soil organic matter content. *Trans. ASAE* 34:1978-1984.

Shropshire, G., C. Peterson, and K. Fisher. 1993. Field Experience with Differential GPS. ASAE Paper No. 931073. ASAE, St. Joseph, MI.

Stafford, J.V. and P.C.H. Miller. 1993. Spatially selective application of herbicide to cereal crops. *Comp. Elect. Agric.* 9:217-229.

Stafford, J.V., J.M. Le Bars, and B. Ambler. 1996. A hand-held data logger with integral GPS for producing weed maps by field walking. *Comp. Elect. Agric.* 14:235-247.

Sudduth, K.A. and J.W. Hummel. 1993a. Portable near infrared spectrophotometer for rapid soil analysis. *Trans. ASAE* 36:187-195.

Sudduth, K.A. and J.W. Hummel. 1993b. Soil organic matter, CEC, and moisture sensing with a portable NIR spectrophotometer. *Trans. ASAE* 36:1571-1582.

Sudduth, K.A., J.W. Hummel, and R.C. Funk. 1990. Geographic Range Testing of an NIR Soil Sensor. ASAE Paper No. 901063. ASAE, St. Joseph, MI.

Sudduth, K.A., N.R. Kitchen, D.F. Hughes, and S.T. Drummond. 1995. Electromagnetic induction sensing as an indicator of productivity on claypan soils. p. 671-681. In: P.C. Robert, R.H. Rust, and W.E. Larson (eds.), Site-specific management for agricultural systems. Proc. 2nd Intl. Conf., Minneapolis, MN. 27-30 March 1994. ASA, CSSA, and SSSA, Madison, WI.

Thompson, J.F., J.V. Stafford, and P.C.H. Miller. 1991. Potential for automatic weed detection and selective herbicide application. *Crop Prot.* 10:254-259.

Tompkins, F.D., K.D. Howard, C.R. Mote, and R.S. Freeland. 1988. Boom Flow Characteristics with Direct Chemical Injection. ASAE Paper No. 881593. ASAE, St. Joseph, MI.

Tyler, D.A. 1992. Positioning technology (GPS). p. 159-166. In: P.C. Robert, R.H. Rust, and W.E. Larson (eds.), Proc. first workshop on soil specific crop management. Minneapolis, MN. 14-16 April 1992. ASA, CSSA, and SSSA, Madison, WI.

Vanotti, M.B. and L.G. Bundy. 1994. Corn nitrogen recommendations based on yield response data. *J. Prod. Agric.* 7:249-256.

Vansichen, R. and J. De Baerdemaeker. 1991. Continuous Wheat Yield Measurements on a Combine. p. 346-355. In: Automated Agriculture for the 21st Century. ASAE Public. No. 1191. ASAE, St. Joseph, MI.

Vansichen, R. and J. De Baerdemaeker. 1993. A measurement technique for yield mapping of corn silage. *J. Agric. Engr. Res.* 55:1-10.

Vidrine, C.G., C.E. Goering, C.L. Day, M.R. Gebhardt, and D.B. Smith. 1975. A constant pesticide application rate sprayer model. *Trans. ASAE* 18:439-443.

Waynick, D.D. and L.T. Sharp. 1919. Variability in soils and its significance to past and future soil investigations: II -- variations in nitrogen and carbon in field soils and their relation to the accuracy of field trials. *Agric. Sciences* 4:121-139.

Wibawa, W.D., D.L. Dludlu, L.J. Swenson, D.G. Hopkins, and W.C. Dahnke. 1993. Variable fertilizer application based on yield goal, soil fertility, and soil map unit. *J. Prod. Agric.* 6:255-261.

Wilkerson, J.B., J.S. Kirby, W.E. Hart, and A.R. Womac. 1994. Real-Time Cotton Flow Sensor. ASAE Paper No. 941054. ASAE, St. Joseph, MI.

Williams, B.G. and D. Hoey. 1987. The use of electromagnetic induction to detect the spatial variability of the salt and clay contents of soils. *Aus. J. Soil Res.* 25:21-27.

Woebbecke, D.M., G.E. Meyer, K. Von Bargen, and D.A. Mortensen. 1995a. Color indices for weed identification under various soil, residue, and lighting conditions. *Trans. ASAE* 38:259-269.

Woebbecke, D.M., G.E. Meyer, K. Von Bargen, and D.A. Mortensen. 1995b. Shape features for identifying young weeds using image analysis. *Trans. ASAE* 38:271-281.

Wollenhaupt, N.C., R.P. Wolkowski, and M.K. Clayton. 1994. Mapping soil test phosphorus and potassium for variable-rate fertilizer application. *J. Prod. Agric.* 7:441-448.

Influence of Weed Biology and Ecology on Development of Reduced Dose Strategies for Integrated Weed Management Systems

J. Anita Dieleman and David A. Mortensen

I. Introduction

The single strategy weed management approach relying on full dose herbicide applications has led to concerns over high weed control input costs, environmental contamination, and development of herbicide resistant weed populations. Development of reduced herbicide dose strategies requires an alternative approach to applying a single herbicide expected to provide a high level of control of a weed population. Reduced herbicide dose strategies embody an integrated weed management (IWM) system that incorporates multiple weed control tactics into crop production systems and must be economically, environmentally, and socially acceptable (Mortensen et al., 1995a,b; Swanton and Weise, 1991; Walker and Buchanan, 1982). The cumulative effect of multiple control tactics should enhance the competitiveness of the crop over the weed population.

Weeds need to be managed for a variety of reasons. Weeds are known to cause crop yield losses, hamper harvest, reduce quality of the harvested product, and perhaps harbor insects and diseases that may harm the crop (Orson, 1990). Yield losses are of the greatest concern and have been predicted using early season assessments of the weed population such as weed seedling density, relative time of emergence, weed pressure, and relative leaf area (Coble and Mortensen, 1992; Cousens, 1985; Cousens et al., 1987; Dieleman et al., 1995; Harvey and Wagner, 1994; Kropff, 1988; Kropff and Spitters, 1991; Kropff et al., 1993; Weaver, 1991). These estimates have been used to determine single year economic thresholds (Coble and Mortensen, 1992; Dieleman et al., 1996; Pannell, 1990a,b; Weaver, 1991) and multi-year economic optimum thresholds that take into account weed seed production (Bauer and Mortensen, 1992; Lindquist et al., 1995). Determination of economic thresholds has been based on the application of recommended label doses of herbicides with a high level of efficacy. However, if the objective is to place the crop at a competitive advantage over the weed, such thresholds could be adjusted to allow for weed suppression rather than control in an agricultural field. This could be accomplished with the use of reduced dose strategies in combination with other weed control tactics.

Many growers already use herbicide doses below that recommended on the product label (Baandrup and Ballegaard, 1989; Defelice et al., 1989; Hartzler et al., 1993; Jones et al., 1995). This decision is based on the grower's accumulated knowledge of the weed spectrum and its location in a particular field, environmental conditions at the time of application, growth or developmental stage of the weeds, and the desire to reduce input costs and protect the environment. However, use of such reduced dose strategies is not supported by the herbicide manufacturers and any risk associated with the strategy, such as reduced efficacy or uncontrolled species, is assumed by the grower. Expected levels of control could be improved, or the variability of control reduced, with better understanding of weed species response to reduced herbicide dose strategies.

A few definitions will clarify methods of selecting a herbicide dose other than that recommended on the product label. *Reduced dose* refers to a blanket reduction such as 1/2X or 1/4X of the recommended label dose (1X). The herbicide is selected based on the weed spectrum present in the field and the dose is adjusted according to the growth or developmental stage of susceptible weeds. For example, if weeds have just emerged and are at the 1- to 2-leaf stage, 1/4X could be applied as compared to an application of 1/2X for weeds in the 2- to 3-leaf stage and 1X after the 3-leaf stage (Anon., 1995). *Biologically-effective dose* refers to the selection of a dose based on an individual weed species, its susceptibility to a product, and an expected level of control (Dieleman et al., 1996; Whiting et al., 1991). *Factor-adjusted dose* refers to the adjustment of the dose based on several key factors present at the time of application including weed species, their growth or developmental stage, and environmental conditions (Baandrup and Ballegaard, 1989; Kudsk, 1989).

The objective of this review is to integrate scientific knowledge of weed biology and ecology with the development and application of reduced dose strategies. This paper will describe the response of weed populations to reduced herbicide doses, the development of reduced dose strategies, and the incorporation of these strategies into integrated weed management systems.

II. Response of Weed Populations to Reduced Herbicide Doses

Weed populations express a variety of responses to the application of a herbicide. Characteristic responses include death, reduction in weed pressure (a qualitative descriptor of weed biomass, density, and leaf area), and reductions in weed seed production and viability. Typically, an estimate of herbicide efficacy is used to describe the response of a weed population. Herbicide efficacy is an estimate of the performance of a product based on visual estimates of weed biomass in treated plots relative to the biomass in an untreated check plot (Martin et al., 1994). From this estimate, label doses are derived which provide 90 to 95% control of the tested weed species under a broad range of environmental conditions.

There is the potential to reduce herbicide doses below that recommended on the label and still provide adequate reductions in weed pressure. It is important to assess the amount of reduction, determine how this will change the crop-weed interaction, and attempt to predict crop yield losses rather than simply expect nearly complete weed control. The expected response varies with weed spectrum, selected herbicide dose level, weed infestation level, weed growth or developmental stage at application time, environmental conditions, and application timing [preplant incorporated (PPI), preemergence (PRE), postemergence (POST), etc.]. It is necessary to determine the level of response required to place the crop at a competitive advantage over the weed.

A. Reduction in Weed Pressure

Reduction in weed pressure is the first response to any herbicide application. Reductions in weed biomass, density, and leaf area can be used to qualitatively and quantitatively assess changes in weed pressure. For example, reduced doses of bentazon caused reductions in both measured biomass and density of common cocklebur (*Xanthium strumarium* L.) 60 days after planting (DAP) (Table 1; Buhler et al., 1993). Common cocklebur biomass and density in 1989 were reduced from 815 g m^{-2} and 73 plants m^{-2} in the weedy control to 710 g m^{-2} and 43 plants m^{-2} with two inter-row cultivations, and further reduced to 222 g m^{-2} and 27 plants m^{-2} with a banded application of 0.6 kg bentazon ha^{-1} and two inter-row cultivations, respectively. Level of common cocklebur infestation differed significantly among the three years of the study, with lower biomass and densities on average in 1990 and 1991 as compared to 1989 (Table 1). Higher population levels decreased the success of reduced dose applications and inter-row cultivation (Buhler et al., 1993).

Weed infestation level, or overall mean field density, is an important variable influencing the success of reduced dose applications. Low weed infestation levels are expected to be adequately controlled with reduced herbicide doses compared to higher weed infestation levels. Winkle et al. (1981) evaluated the effects of weed density on herbicide absorption and bioactivity in the greenhouse and field. In the greenhouse, with increasing weed densities, higher doses of atrazine for white mustard (*Brassica hirta* Moench) and alachlor for foxtail millet (*Setaria italica* (L.) Beauv.) were required to cause equivalent fresh weight per pot reductions. Results

Table 1. Effect of bentazon dose, application method, and between-row cultivation on common cocklebur density and biomass 60 days after planting at Rosemount, MN in 1989, 1990 and 1991[a]

Bentazon treatment				Common cocklebur density		Common cocklebur biomass	
Rate	Time	Method	Cultivation	1989	1990-1991	1989	1990-1991
kg ha^{-1}				----------no. m^{-2}----------		----------g m^{-2}----------	
0			0	73*	51*	815*	820*
			1	63*	33*	773*	710*
			2	43*	21*	710*	590*
0.3	POST1	Band	0	77*	34*	838*	665*
			1	50*	7	734*	326*
			2	30*	7	383*	180*
0.6	POST1	Band	0	66*	26*	726*	524*
			1	45*	11	404*	202*
			2	27*	4	222*	46
0.3	POST1	Broadcast	0	76*	12*	789*	285*
			1	51*	8	606*	126*
			2	33*	3	384*	31
0.6	POST1	Broadcast	0	70*	7	568*	57*
			1	44*	4	470*	43
			2	25*	1	258*	6
0.3 + 0.3	POST1 + POST2	Broadcast + Broadcast	0	25*	3	262*	31
0.6 + 0.6	POST1 + POST2	Broadcast + Broadcast	0	1	2	59	21
LSD (0.05) for comparing bentazon treatments within a cultivation level				13	9	191	80
LSD (0.05) for comparing cultivation levels within a bentazon treatment				13	8	154	59

[a]Means followed by an * differ from bentazon at 0.6 kg ha^{-1} POST1 plus 0.6 kg ha^{-1} POST2 using Dunnet's procedure at P = 0.05. (From Buhler et al., 1993.)

Table 2. Forage sorghum fresh weight at four seeding densities and four alachlor rates grown at Lincoln, Nebraska during 1977 and 1979

Year	Alachlor rate	Forage sorghum seeding rate in seeds/71 m^2				
		0	1000	10,000	50,000	100,000
	kg ha^{-1}	--------- sorghum fresh weight (kg 71 m^{-2}) -------				
1977	0	3 ab[a]	3 ab	10 d	--	7 cd
	1.7	0 a	2 ab	8 cd	--	10 d
	3.4	0 a	1 ab	6 bc	--	10 d
	6.7	0 a	2 ab	4 abc	--	10 d
		-------- sorghum fresh weight (kg 57 m^{-2}) --------				
1979	0	--	28 fg	30 g	28 fg	18 cd
	1.7	--	12 b	20 d	19 cd	22 de
	3.4	--	13 b	16 c	26 f	21 de
	6.7	--	7 a	7 a	23 ef	22 de

[a] Numbers followed by the same letters within a year are not significantly different at the 5% level using Duncan's multiple range test.
(From Winkle et al., 1981.)

of the field experiment indicated that more alachlor was required for equal control of a high than a low density of forage sorghum [*Sorghum bicolor* (L.) Moench], a substitute for shattercane, interseeded in corn (*Zea mays* L.) (Table 2; Winkle et al., 1981). The proportion of ryegrass (*Lolium rigidum*) killed at a particular dose of diclofop-methyl was affected by pretreatment weed density (Figure 1, Pannell, 1990b). At low herbicide doses, weed survival was greatest at high weed densities. As herbicide dose increased, weeds surviving the application were weakened and made more susceptible to crop competition.

One could speculate as to the factors responsible for differences in herbicide efficacy observed at various infestation levels. Application of low PPI or PRE doses to high infestation levels could be unsuccessful in part due to a large number of root systems attempting to absorb a limited amount of herbicide and not obtaining a lethal concentration, or dilution in the soil solution out of the weed seed germination zone. Reduced POST dose applications may have lower efficacy if weed canopies overlap and reduce interception of spray, small weeds survive, and/or new seedlings emerge as part of a future infestation. High weed infestations may compete for soil moisture, become stressed, and therefore, less able to absorb a lethal concentration from the soil or have developed cuticular barriers to POST herbicide applications. More research is required to determine the mechanisms responsible for the interaction between infestation level and herbicide efficacy.

Weed infestation level varies across locations and years as a result of environmental and management variability. This will influence the success of reduced dose strategies at a given location (Eadie et al., 1992; Salonen, 1992a,b). For example, good control resulted with an application of MCPA/mecoprop in 1986 to spring cereals in Finland because of a low weed infestation level and the presence of susceptible species (Salonen, 1992a). In 1988, a high infestation level and a more

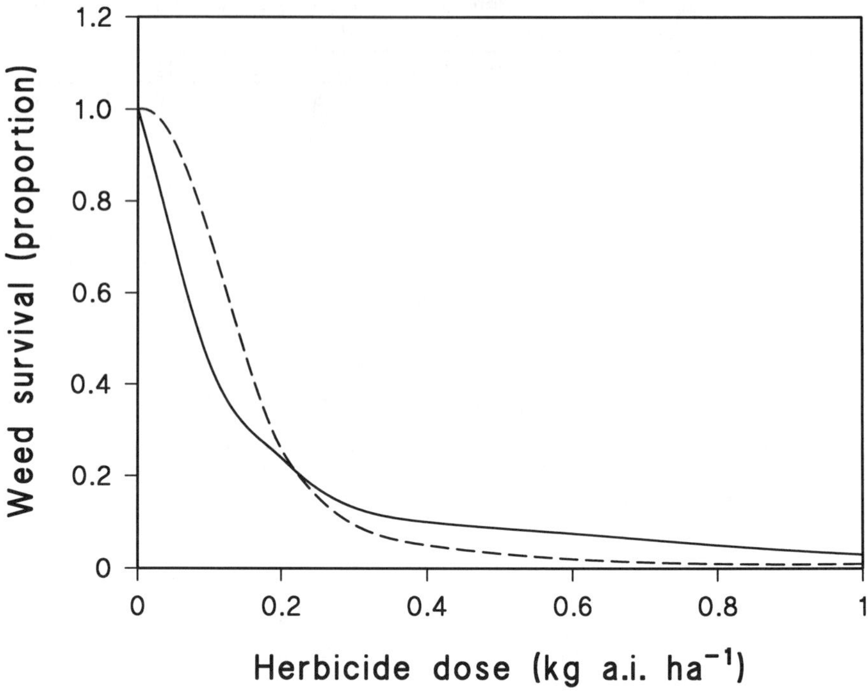

Figure 1. Effect of initial ryegrass density (— 100 m^{-2}; ---- 500 m^{-2}) on relationship between herbicide dose of diclofop-methyl and weed survival. (From Pannell, 1990b.)

tolerant weed spectrum reduced the efficacy of the herbicide. Weed biomass was reduced by all herbicide doses applied. At the high infestation level, reduced doses provided 70 to 90% efficacy and higher relative yields than the highest dose, which suppressed the weed infestation by 90% (Salonen, 1992b). High weed infestation levels reduced the benefits of switching from a chemically- to a mechanically-intensive system (Buhler et al., 1992). In 1989, with a lower weed infestation level, several mechanical weed control systems resulted in soybean [*Glycine max* (L.] Merr.) yields similar to the weed-free control. However, in 1990, under greater weed densities and delayed planting with moist conditions, mechanical weed control systems resulted in lower soybean yields compared to systems that included herbicides.

In order to successfully reduce weed pressure and overcome environmental variability, observations of weed emergence relative to time of application, assessment of weed spectrum with selection of appropriate herbicides, and recognition of current environmental conditions are important considerations (Defelice et al., 1989; Devlin et al., 1991; Prostko and Meade, 1993). Early sequential applications or single application with a timely cultivation are often suggested. Sequential 1/4X applications applied early to small velvetleaf (*Abutilon*

Table 3. Influence of reduced doses applied two weeks after planting on broadleaf weed control and velvetleaf densities with and without cultivation at Powhattan, KS in 1990, assessed at eight weeks after planting

		Broadleaf weed control	------Velvetleaf density------	
		Cultivation		
Herbicide	Rate	No	No	Yes
	kg ha^{-1}	%	---- plants m^{-2} ----	
None		0	16.4	11.6
Acifluorfen	0.14	50	17.0	6.8
	0.28	75	9.6	3.6
	0.56	89	4.0	1.6
Bentazon	0.21	90	2.8	0.9
	0.42	97	1.3	0
	0.84	99	1.1	0
Acifluorfen +	0.14 + 0.21	84	7.0	2.1
Bentazon	0.28 + 0.42	93	4.1	0.9
	0.56 + 0.84	96	1.5	0.5
Chlorimuron	0.0022	35	12.1	9.6
	0.0045	48	9.8	5.7
	0.0090	58	9.8	8.4
LSD (0.05)		16	5.8	5.8

(From Devlin et al., 1991.)

theophrasti Medic.), common cocklebur, pitted morningglory (*Ipomoea lacunosa* L.), and giant foxtail (*Setaria faberi* Herrm.) seedlings provided equivalent control to a single 1X application to larger weeds (Defelice et al., 1989). Prostko and Meade (1993) also determined that two sequential 1/4X or a single 1X application controlled redroot pigweed (*Amaranthus retroflexus* L.) and common lambsquarters (*Chenopodium album* L.) populations more effectively than a single 1/2X dose of either fomesafen, bentazon plus acifluorfen, or chlorimuron, in combination with sethoxydim. Devlin et al. (1991) observed that velvetleaf densities declined with increasing doses of acifluorfen, bentazon, chlorimuron, or a tank-mix of acifluorfen plus bentazon applied two weeks after planting and with an inter-row cultivation at four weeks after planting (Table 3). Velvetleaf density was reduced from 16.4 plants m^{-2} in the weedy control to 1.3 and 1.1 plants m^{-2} with 1/2X and 1X bentazon applications without cultivation, respectively, and none were left after cultivation. However, inadequate reduction of velvetleaf density occurred when the same herbicides were applied in combination with sethoxydim for grass control (Table 4, Defelice et al., 1989). It was suggested that antagonism between the broadleaf and grass control products resulted in poor velvetleaf control (Anon., 1995; Defelice et al., 1989). It was important that weed emergence coincided with the time of POST application and a timely cultivation appeared to improve control (Devlin et al., 1991).

Table 4. Effect of reduced dose applications on velvetleaf density and seed production at Columbia, MO in 1986 and 1987

		Plants m^{-2}		Seed m^{-2}	
Treatment	Rate	1986	1987	1986	1987
	kg ha^{-1}	no.			
None		9.6 a[a]	8.9 d	2160 a	3340 def
Hand-weeded		0 b	0 f	0 b	0 f
Bentazon +	0.8 + 0.2	0.1 b	0 f	0 b	0 f
Sethoxydim	0.6 + 0.1	0.2 b	19.7 b	0 b	7560 bc
	0.6 + 0.1	0.1 b	15.1 bc	13 b	4970 cd
	0.3 + 0.1 & 0.3 + 0.1	0 b	2.4 ef	6 b	410 f
Acifluorfen +	0.4 + 0.2	0.1 b	3.2 ef	0 b	650 ef
Sethoxydim	0.3 + 0.1	0.5 b	20.0 b	0 b	9800 ab
	0.14 + 0.1	0.3 b	26.2 a	11 b	12200 a
	0.14 + 0.1 & 0.14 + 0.1	0 b	5.4 de	4 b	2220 def
Chlorimuron +	0.009 + 0.2	0 b	1.9 ef	0 b	0 f
Sethoxydim	0.005 + 0.1	0.1 b	14.3 c	3 b	4420 cde
	0.002 + 0.1	0.1 b	18.9 bc	12 b	7990 bc
	0.002 + 0.1 & 0.002 + 0.1	0 b	0 f	0 b	0 f

[a]Means within a column followed by the same letter are not significantly different according to Fisher's Protected LSD test (a=0.05).
(From Defelice et al., 1989.)

Reduction in weed leaf area is concomitant with the reduction in weed biomass. How much leaf area reduction is required to put the weed at a competitive disadvantage with the crop? Weaver (1991) observed that common cocklebur, velvetleaf, and jimsonweed (*Datura stramonium* L.) seedlings that escaped a PPI application of metribuzin (0.4 kg ha^{-1}) were shorter and had fewer leaves at three weeks after planting compared to plants in untreated control plots. At equal densities, escaped weeds caused less soybean yield loss than untreated control plants (Figure 2; Weaver, 1991). Zhang et al. (1994) evaluated the response of common cocklebur seedling characteristics to various doses of bentazon in the field. Leaves of sprayed plants began to show injury 1 week after spraying, percentage leaf damage increased, and a measure of plant size, crown diameter at maturity, decreased with increasing bentazon doses.

Physiological responses of weeds to reduced doses could be targeted for reductions in competitiveness. As a result of exposure to a herbicide, escaped weeds could possess growth characteristics and competitive abilities differing from untreated weeds (Adcock and Banks, 1991; Adcock et al., 1990; Hume and Shirriff, 1989; Weaver, 1991). This is important information to be considered when modeling the effects of crop-weed competition (Adcock and Banks, 1991; Adcock

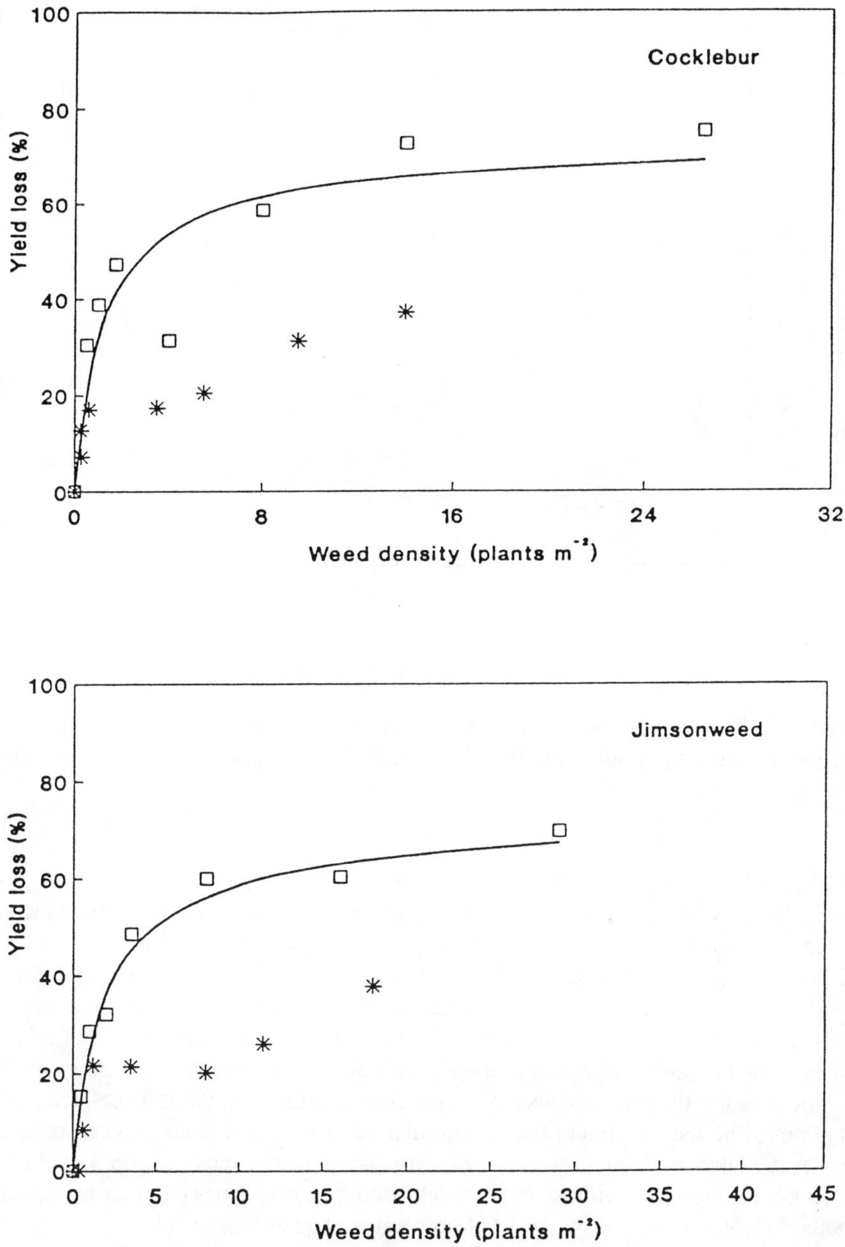

Figure 2. Soybean seed yield losses as a function of weed density of common cocklebur in 1988 and jimsonweed in 1987 on control plots (□) with fitted lines (—) and on plots treated with soil-applied metribuzin (*). (From Weaver, 1991.)

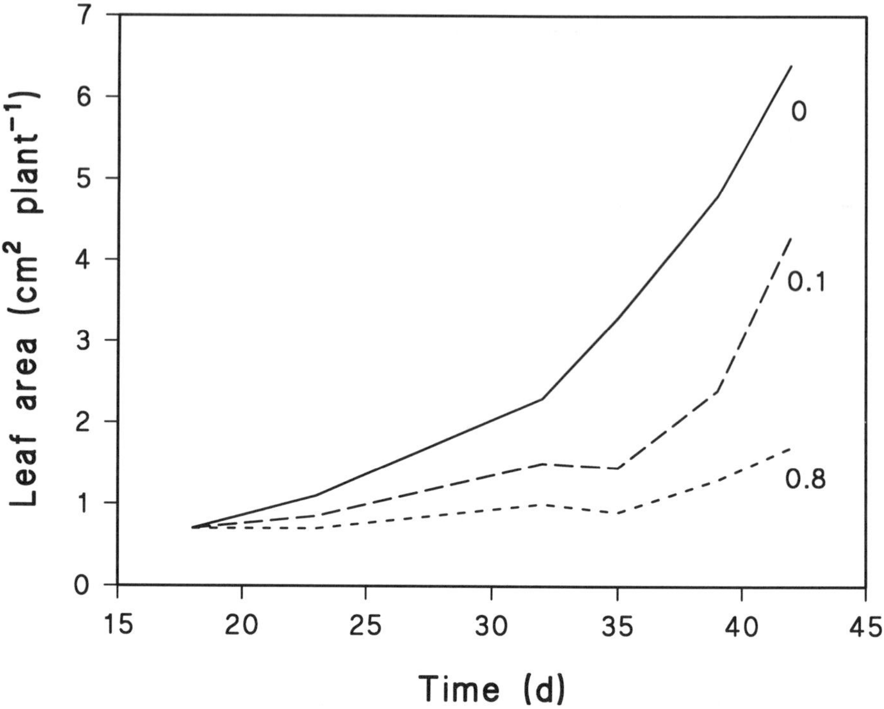

Figure 3. The development of leaf area of common lambsquarters seedlings with different herbicide treatments: 0, 0.1 and 0.8 kg ha^{-1}. (single dose metamitron) (From Kropff et al., 1993.)

et al., 1990; Lindquist et al., 1995; Kropff et al., 1993; Weaver, 1991). For example, early season relative leaf area growth rates of common lambsquarters were decreased with increasing doses of metamitron applied POST (Figure 3; Kropff et al., 1993). Ecophysiological model simulation and sensitivity analysis found that small changes in the relative growth rate of leaf area had the greatest impact on expected crop yield losses. Such ecophysiological models could serve as a tool for quantifying the impact of a range of herbicide doses, used to assess changes in the competitive ability of the weeds, and to predict potential crop yield losses. Models also could be used to integrate the cumulative effects of sublethal weed control strategies such as biological control (insects, diseases, cover crops), cultural practices (conservation tillage, cultivar selection, planting density), and mechanical methods (rotary hoe, cultivation) (Mortensen et al., 1995a,b).

B. Influence on Weed Seed Production and Viability

A concern of many growers is that lower weed control levels with reduced doses and resultant seed production will result in unmanageable weed populations in the

Table 5. Mean (± S.E.) percentage emergence and percentage viable seeds under 0 or 540 g ha^{-1} bentazon treatment to the mother plant

	Emergence (%)		Remaining viable seeds (%)	
Population	0	540	0	540
Arner	72.80±3.79	61.90±2.71	5.60±2.39	3.30±0.51
Grand Bend	39.10±2.16	38.00±3.39	17.10±2.68	15.00±3.58
Harrow	65.70±1.68	61.70±1.19	7.60±1.30	3.20±1.25
Highgate	40.10±2.60	36.60±1.28	25.80±3.52	22.60±3.10
Leamington	63.30±4.27	55.60±2.54	8.50±2.00	6.80±0.82
Morpeth	59.60±2.83	49.40±1.41	11.30±0.85	8.60±2.17
Talbotville	34.00±1.41	36.40±0.64	42.00±0.95	36.60±2.20
Point Pelee	50.80±2.11	50.40±1.63	12.30±1.52	6.70±0.90
Rondeau Park	52.50±5.12	33.90±2.77	15.10±3.74	15.50±2.17
UWO Campus	39.90±0.81	45.60±3.73	29.20±2.21	26.00±2.92

(From Zhang and Cavers, 1994b.)

future (Defelice et al., 1989; Hartzler et al., 1993; Orson, 1990; O'Sullivan and Bouw, 1993; Steckel et al., 1990). The extent of a grower's concern regarding weed seed return to the soil would depend on whether it could be easily controlled in a rotational crop or whether it is difficult or expensive to control (Orson, 1990).

Is weed seed production and viability affected by the application of reduced herbicide doses to the parent plant? Application of 1/2X bentazon to common cocklebur seedlings reduced the number of fruits produced, reduced the number of fruits with at least one developed seed, and increased the proportion of empty fruits (Zhang and Cavers, 1994a). Mean seed mass was also reduced while fruit coat thickness increased. These qualitative changes could affect future seed germination, seedling emergence, and establishment. Bentazon carryover effects on common cocklebur progeny reduced percentage of seedlings that emerged and reduced percentage of viable seeds left in the soil seedbank (Table 5; Zhang and Cavers, 1994b). Common lambsquarters that survived applications of reduced 2,4-D doses took longer to flower and mature and, thereby, reduced the likelihood of seed production before harvest or a killing frost (Hume and Shirriff, 1989). Actual seed production decreased from 1051 to 106 seeds per plant with increased 2,4-D dose from 0.1 to 0.6 kg ha^{-1}.

Application timing could be targeted to increase reductions in seed production (Andersson, 1995; Fawcett and Slife, 1978; Medd and Pandey, 1993). A 70% reduction in seed production was observed for black bindweed [*Fallopia convolvulus* (L.) A. Love] treated at the one true-leaf stage (1/2X tribenuron-methyl or MCPA) (Andersson, 1995). Applications at the beginning of flowering reduced seed production by 45% for bedstraw (*Galium spurium* L.) with 1/2X tribenuron-methyl, 83% for field pennycress (*Thlaspi arvense* L.) with 1/16X tribenuron-methyl or 100% with 1/16X MCPA. These plants were grown without crop competition which almost certainly would further suppress seed production of treated weeds. A flowering, natural stand of common lambsquarters, giant foxtail,

Table 6. Toxic effects of herbicides and growth regulators on treated *Arabidopsis thaliana* (L.) Heynh. plants and their progeny

		Parent plants		Progeny	
Compound	Conc.	Approx. shoot length	Approx. pods	Nongerm. seeds	Chlorotic & bleached seedlings
	μM	cm	no plant^{-1}	-------- %	---------
Amitrol	5.94	15	4	0	91
	2.97	20	8	3	67
	1.49	20	8	2	34
Diuron	0.11	10	4	35	46
	0.06	15	6	21	16
	0.03	20	8	0	4
Atrazine	0.23	10	2	79	20
	0.12	15	4	0	65
	0.06	20	7	0	27
Bromacil	0.075	10	3	7	35
	0.038	20	4	0	9
	0.019	20	7	0	24
Control		20	8	1	11

(From Henzell et al., 1985.)

redroot pigweed, and jimsonweed was treated with various doses of dalapon or 2,4-D to assess impact on seed production (Fawcett and Slife, 1978). Common lambsquarters and redroot pigweed were eliminated by 2,4-D and jimsonweed had reduced seed production. Giant foxtail increased seed production by 307 and 381% (low and high doses of 2,4-D, respectively) since broadleaf weed competition was removed. Dalapon inhibited flowering of giant foxtail while common lambsquarters and redroot pigweed seed production were not affected. Seed progeny of mouseear-cress [*Arabidopsis thaliana* (L.) Heynh.] was assessed after parent plant root uptake of various herbicides and growth regulators present in growth medium (Henzell et al., 1985). Parent plants were shorter and had fewer pods with increasing doses while the number of nongerminating seeds and chlorotic and bleached seedlings increased with dose (Table 6).

Seed germination and seedling emergence are influenced by parental exposure to herbicides (Fawcett and Slife, 1978; Henzell et al., 1985; Hume and Shirriff, 1989; Zhang and Cavers, 1994b). Germination of common lambsquarters seed increased with increasing 2,4-D doses from 0.1 to 0.6 kg ha^{-1} (Hume and Shirriff, 1989). The reduced number of seeds produced per plant increased the proportion of vegetative dry weight per seed and may have provided a growth advantage to the larger seed. Seedling vigor of seeds grown from parents treated with 2,4-D and dalapon depended on species (Fawcett and Slife, 1978). Vigor of common lambsquarters was less than untreated plants, not different for giant foxtail, slightly less for redroot pigweed as seed was produced before herbicide treatment, and

Table 7. Vigor ratings of weed seedlings grown from seeds from 2,4-D and dalapon-treated plants

Parent plant		Seedling vigor rating			
Treatment	Rate	Common lambsquarters	Giant foxtail	Redroot pigweed	Jimson-weed
	kg ha^{-1}	% of control			
Control	0.0	100	100	100	100
2,4-D	0.6	50	100	100	33
2,4-D	1.1	40	100	97	--
Dalapon	2.2	53	--	97	--
Dalapon	4.5	43	--	93	97
LSD (0.05)		13	0	8	9

(From Fawcett and Slife, 1978.)

variable for jimsonweed (Table 7). The magnitude to which parental exposure to reduced herbicide doses influences seed germination and seedling emergence must be studied in order to understand long-term dynamics of weed populations.

Late-season weed populations, resultant seed production, and buildup of seedbanks in areas treated with reduced doses have caused concern (Defelice et al., 1989; Steckel et al., 1990). It was found that velvetleaf plants m^{-2} and seeds produced m^{-2} were not significantly less than untreated plots after application of reduced doses of bentazon, acifluorfen and chlorimuron in combination with sethoxydim in soybean (Table 4; Defelice et al., 1989). However, a late season cultivation together with reduced POST herbicide doses in wide-row soybean reduced velvetleaf seeds $plant^{-1}$ and plants m^{-2} as well as other weeds in a two year study (Table 8; Steckel et al., 1990). Although immediate soybean yield may not be increased, a timely cultivation after POST application may have value by significantly reducing late-season weed populations and seed production. A resultant weed infestation was assessed by Salonen (1992a) one year after a 3-year trial which evaluated spring cereal competitive ability, crop rotation, and reduced doses. Measurements of the weed seedbank revealed no increase with continuous use of reduced herbicide doses for three years compared to untreated plots. However, the soil reservoir of seeds was higher in the wheat-dominated versus the barley-dominated rotation, indicating that a barley crop could reduce weed seed production to a greater extent than wheat (Salonen, 1992a).

Two opposing views have been expressed regarding application of reduced doses and selection within a weed population for herbicide resistance. Reduced herbicide doses could either enhance or delay the onset of resistance. One view is that sublethal herbicide doses could allow the naturally tolerant proportion of the weed population to recover from herbicide application and produce weed seed. This could shift future weed populations to those that possess greater tolerance to the herbicide. This variability is evident in comparing population responses, such as growth reduction and mortality, to applications of reduced herbicide doses (Morrison and Devine, 1994; Hume and Shirriff, 1989; Zhang and Cavers, 1994a,b;

Table 8. Influence of herbicide, application rate, and application timing on velvetleaf density and seed production at Columbia, MO in 1987 and 1988

Treatment	Rate	Plants m^{-2}		Seed plant^{-1}		Seed m^{-2}	
		1987	1988	1987	1988	1987	1988
	g ha^{-1}	----------	----------	----------	no.	----------	----------
None		1.1NS[a]	1.8NS	160NS	420NS	320NS	740NS
Hand-weeded		0.2	1.2	100	80	200	80
Bentazon	840	0.2	0.2	10	60	10	130
	280	1.6	1.5	100	510	240	500
	280 + 280	0.9	2.0	30	340	60	1090
Chlorimuron	8	0.4	2.2	10	440	30	640
	2	1.0	1.5	110	210	140	530
	2 + 2	1.4	2.0	110	150	260	110
Imazaquin	140	0.6	0.8	90	330	450	530
	40	0.5	3.2	120	460	200	510
	40 + 40	1.1	1.0	130	320	210	240
Imazethapyr	110	0.6	0.2	90	50	300	70
	30	0.4	1.0	50	10	100	20
	30 + 30	0.5	0.5	220	110	480	430
Cultivation		0.2a	0.5a	34a	222a	36a	251a
No-cultivation		1.3b	2.1b	156b	285b	230b	479b

[a]Means within a column followed by the same letter are not significantly different according to Fisher's Protected LSD Test (a=0.05). NS = non-significant F-test. (From Steckel et al., 1990.)

Zhang et al., 1994). From the opposing viewpoint, use of reduced dose strategies could also maintain a susceptible population to slow the enrichment of the weed population with resistant plants (Gressel and Segel, 1982, 1990; Pannell, 1990b). Current simulation models could be used to assess the benefits of reduced doses in resistance management schemes (Gressel and Segel, 1990; Jasieniuk and Maxwell, 1994; Maxwell et al., 1990). A mixed, general selection pressure, incorporating reduced dose strategies and other control tactics, would benefit resistance management.

III. Reduced Dose Strategies

Once we have an understanding of the responses a weed population exhibits to the application of reduced herbicide doses, it will be necessary to integrate these responses in a comprehensive manner. Our desire is to determine the needed dose for a given level of reduction in weed pressure and weed seed production while maintaining crop yield, reducing risk, and managing the variability of weed control.

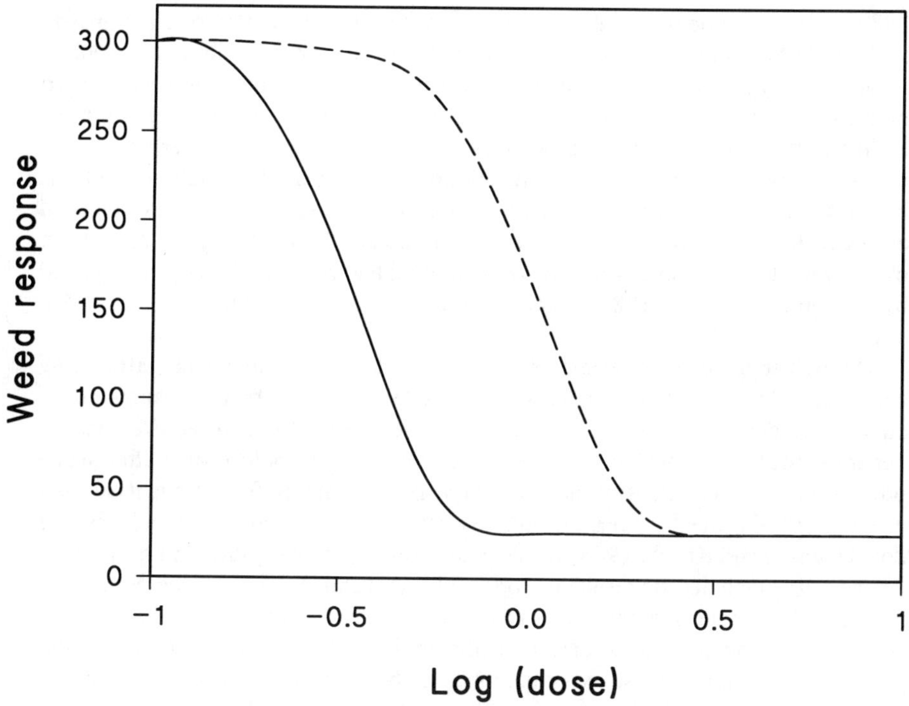

Figure 4. Examples of dose-response curves for two herbicides. (From Streibig, 1989.)

What is a reduced dose strategy? The strategy involves determination of the herbicide dose while accounting for the known responses of the weed population to the herbicide and the many interactions occurring in the field between the weed, crop, environment, and control strategies. Strategic steps include determination of the weed spectrum in the field and its growth or developmental stage, selection of appropriate herbicides, consideration of other weed control strategies, calculation of the reduction in weed pressure and seed production desired to economically maintain crop yield and quality, and finally selection of the dose necessary to accomplish this task.

A. Herbicide Dose-Response Curves

Herbicide dose-response curves are derived from evaluating the response of weeds (e.g., dry weight reduction) to the application of a wide range of herbicide doses (Figure 4; Jensen and Streibig, 1994; Streibig, 1980, 1988, 1989). When weed response is plotted against the logarithm of herbicide dose, a sigmoidal relationship often exists. The upper limit of the curve is an estimate of weed pressure without herbicides and the lower limit at high doses is approximately zero (Streibig, 1988,

1989). Variations on this relationship exist, such as the stimulation of plant growth with sublethal doses (Brain and Cousens, 1989) or nonzero biomass of weeds treated with high POST herbicide doses. Herbicide dose-response curves are based on the mechanism of action of the herbicide and provide a quantitative assessment of herbicide performance across a range of doses. Activity of a herbicide with different weed species, growth or developmental stages, or climatic conditions could be described as a parallel displacement of the dose-response curve (Jensen and Kudsk, 1988). Methods of determining the fit of such functions have been extensively described in a book edited by Streibig and Kudsk (1993) and reviewed by Jensen and Kudsk (1988), Seefeldt et al. (1995), and Streibig (1980, 1988, 1989).

Through appropriate experimental design and data collection, factors influencing herbicide efficacy could be quantified and used to derive herbicide dose-response curves and determine appropriate reduced doses. Specific data requirements to generate these herbicide dose-response curves include before and after weed population biomass, density, and leaf area measurements from field trials that evaluate specific weed species exposed to a range from very low to label herbicide doses (Jensen and Kudsk, 1988; Jensen and Streibig, 1994). Other important data for modeling includes the resultant crop yield losses and long-term weed population dynamics. Herbicide manufacturers and university researchers conduct experiments across locations and years to determine the appropriate label dose for a particular product. The value of these experiments could be enhanced with a slight increase in effort such that appropriate data are collected to generate herbicide dose-response curves (Seefeldt et al., 1995).

B. Factors to Adjust Dose

Recommended doses are often higher than required for efficient weed control. The label dose is sufficient to control the most difficult species present, at larger growth stages than required, and adjusted to reduce risk of nonperformance in a grower's field. However, key factors observed at the time of application can be the basis for refinement in herbicide dose: weed spectrum, growth or developmental stage, and environmental conditions (Jensen and Kudsk, 1988; Kudsk, 1989; Streibig, 1988, 1989). These key factors cause the herbicide dose-response curve to shift horizontally (Figure 5) as there is no change in the mechanism of action of the herbicide, only a change in the amount of response of the weed population to a particular dose of the herbicide (Baandrup and Ballegaard, 1989; Jensen and Kudsk, 1988; Kudsk, 1989). Level of weed infestation, PRE or PPI herbicide applications, rotary hoeing, inter-row cultivation, and cultural conditions such as narrow crop rows or competitive cultivars will also influence weed population response. Assessment of these key factors allow for the effective adjustment to reduced herbicide doses.

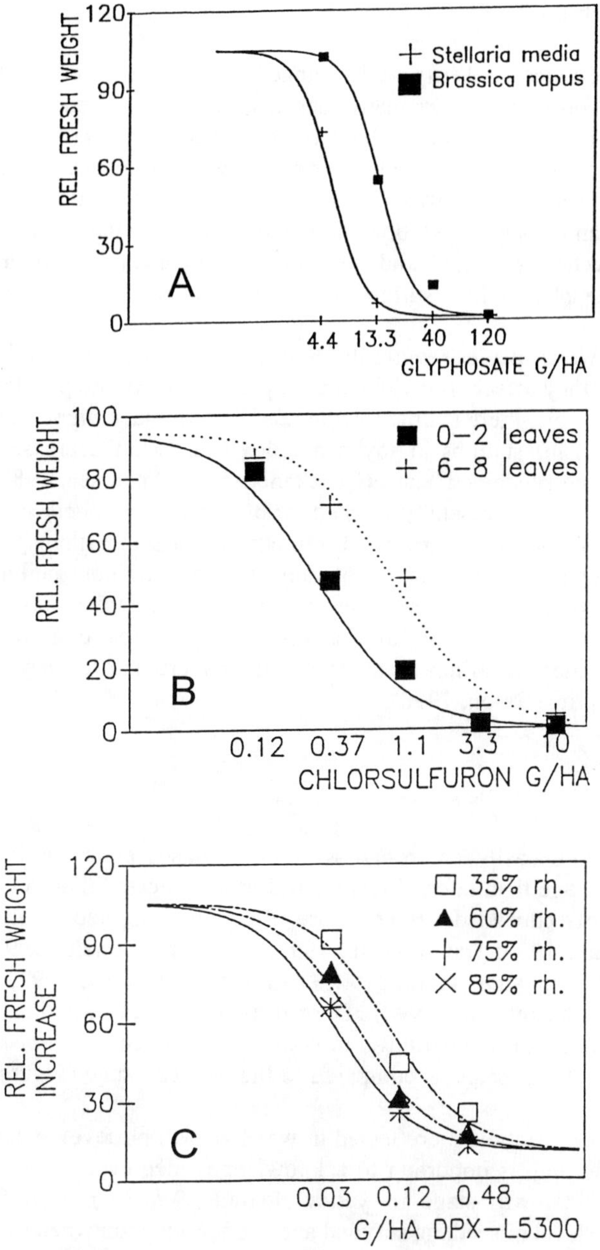

Figure 5. Examples of how factors influence the herbicidal activity of a) glyphosate on two weed species, b) chlorsulfuron on two growth stages of common chickweed and c) DPX-L5300 on white mustard at a constant temperature of 22.2° C and different relative humidities. (From Kudsk, 1989.)

1. Weed Spectrum

A herbicide product label lists a wide range of weed species controlled with the recommended dose. However, many species may be controlled with lower doses and dose-response curves need to be derived for each species (Figure 5a; Kudsk, 1989). For example, in an experiment determining the mechanisms of tolerance for thifensulfuron in soybean, it was observed that redroot pigweed required less than 0.5 g ha^{-1} for an 80% growth reduction, whereas common lambsquarters required 1.2 g ha^{-1}, velvetleaf 2.9 g ha^{-1}, and soybean tolerated at least 125 g ha^{-1} (Brown et al., 1990). The labeled thifensulfuron dose for broadleaf weed control in soybean is 6 g ha^{-1}.

Individual weed species require different levels of weed control depending on how seriously they affect crop yield and crop quality or create problems in future years. For example, there is little tolerance of eastern black nightshade (*Solanum ptycanthum* L.) infestations in soybean fields because of decreased harvesting efficiency and berry-stained beans (Quackenbush and Andersen, 1984). Reduced doses have not been successfully adopted for blackgrass (*Alopecurus myosuroides* Huds.) and cleavers (*Galium aparine* L.) in winter wheat in Britain (Orson, 1990). Blackgrass produces a large quantity of viable seed and available herbicides are not conducive to the use of reduced doses based on herbicide dose-response curves. Application of reduced doses to cleavers, a very competitive weed, are only successful if the prevailing weather conditions are particularly suitable for herbicidal activity (Orson, 1990).

2. Growth or Development Stage

Weeds are more easily controlled with lower doses at an earlier growth or development stage than later. Most studies have indicated that POST herbicide applications of reduced doses are more effective if applied to smaller weeds (Defelice et al., 1989; Devlin et al., 1991; Khan and Donald, 1992; King and Oliver, 1992; Klingaman et al., 1992; O'Sullivan and Bouw, 1993; Rathjen and Martin, 1994; Steckel et al., 1990). Dose-response curves shifted to a lower dose for an application of chlorsulfuron to common chickweed [*Stellaria media* (L.) Vill.] at the 0-2 leaf stage as compared to the 6-8 leaf stage (Figure 5b; Kudsk, 1989).

Timing of application, as reflected in weed growth or developmental stage or days after planting, is important to acknowledge when evaluating reduced dose applications. Growth stage of yellow foxtail [*Setaria glauca* (L.) Beauv.] influenced the level of seedling survival and seed production when treated with 10 g ha^{-1} of chlorsulfuron (Khan and Donald, 1992). None survived treatment at the 2-leaf stage, while 3-28% survived application at the 4-leaf stage. Reductions in seedling survival of 40-43% and seed production of 2-24% occurred with treatment at the 6-leaf stage. Poor velvetleaf and sunflower (*Helianthus annuus* L.) control in corn resulted from applications at 25 DAP of 1/4X dicamba, dicamba plus atrazine or bromoxynil plus atrazine as compared to applications at 15 or 20 DAP (Rathjen and Martin, 1994).

A suggested method to improve dose selection of imazethapyr was the development of contour graphs predicting levels of control for individual weed species based upon weed leaf number (Figure 6; Klingaman et al., 1992) or weed age (Figure 7; King and Oliver, 1992). In general, to provide 90% control, it was recommended that reduced doses be applied within 6-12 days after weed emergence (King and Oliver, 1992). Growth or development stage indicates the level of susceptibility and the best application timing of a reduced herbicide dose for an individual species.

3. Environmental Conditions

What influence do environmental conditions at the time of herbicide application have on the success of reduced doses? Conditions such as temperature, irradiance, rainfall, soil moisture, and relative humidity greatly influence herbicide activity (Gerber et al., 1983; Jensen and Kudsk, 1988). These conditions should not be assessed in isolation when evaluating herbicide performance in the field. The effect of one climatic factor, such as humidity, will be greatest when other factors, such as temperature or soil moisture, are optimal (Jensen and Kudsk, 1988). Temperature and irradiance influence the growth rate of the plant and its vigor. Rainfall influences soil-applied herbicides by increasing their availability in the soil while the activity of foliar-applied herbicides is influenced by changes in relative humidity (Gerber et al., 1983). These environmental conditions alter the rate of plant transpiration and subsequently, the rate and duration of herbicide uptake by the roots, and rate of foliar uptake and translocation (Gerber et al., 1983).

With lower doses, it may take longer for a weed to absorb a lethal concentration of the herbicide compared to higher doses. Therefore, necessary conditions near the time of POST application include adequate rain-free periods, high irradiance, and hot, humid conditions (Gerber et al., 1983). Increased humidity, soil moisture, and temperature have been reported to increase the activity of many POST herbicides, including acifluorfen on spurred anoda [*Anoda cristata* (L.) Schlecht.], velvetleaf, morningglory, common cocklebur, and pigweed (Meeusen et al., 1983) and thiameturon-methyl on white mustard (Kudsk et al., 1990). Increasing relative humidity from 35 to 85% shifted the dose response curve of DPX-L5300 to lower doses for equivalent control of white mustard (Figure 5c; Kudsk, 1989).

4. Mechanical and Cultural Strategies

The success of reduced dose strategies on a weed population will also be influenced by the use of other weed control strategies. These include the use of mechanical controls, such as rotary hoeing and inter-row cultivation, and cultural practices, such as conservation tillage and residue management, increased crop seeding rates, competitive cultivars, cover crops, and fertility and water management. The use of

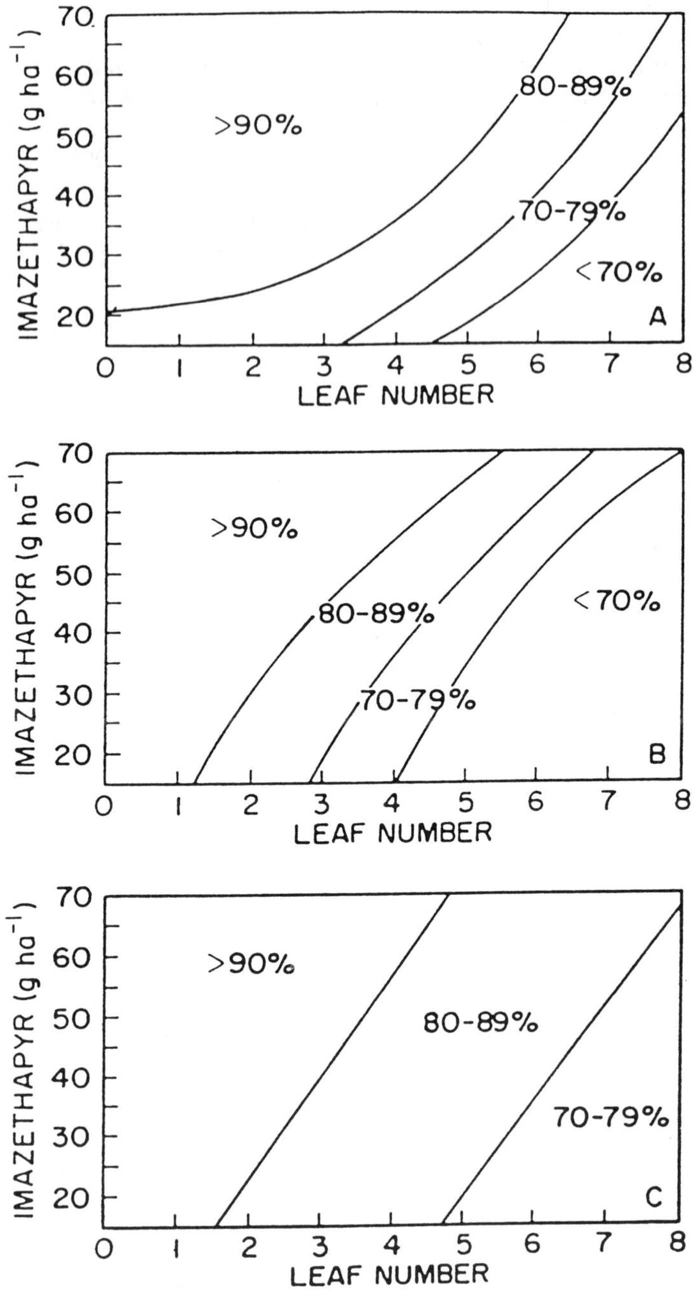

Figure 6. Predicted percent control of weeds most susceptible to imazethapyr in 1987 and 1988 at Fayetteville, AR. A = common cocklebur, B = smallflower morningglory, and C = smooth pigweed. The 0 = cotyledon growth stage. (From Klingaman et al., 1992.)

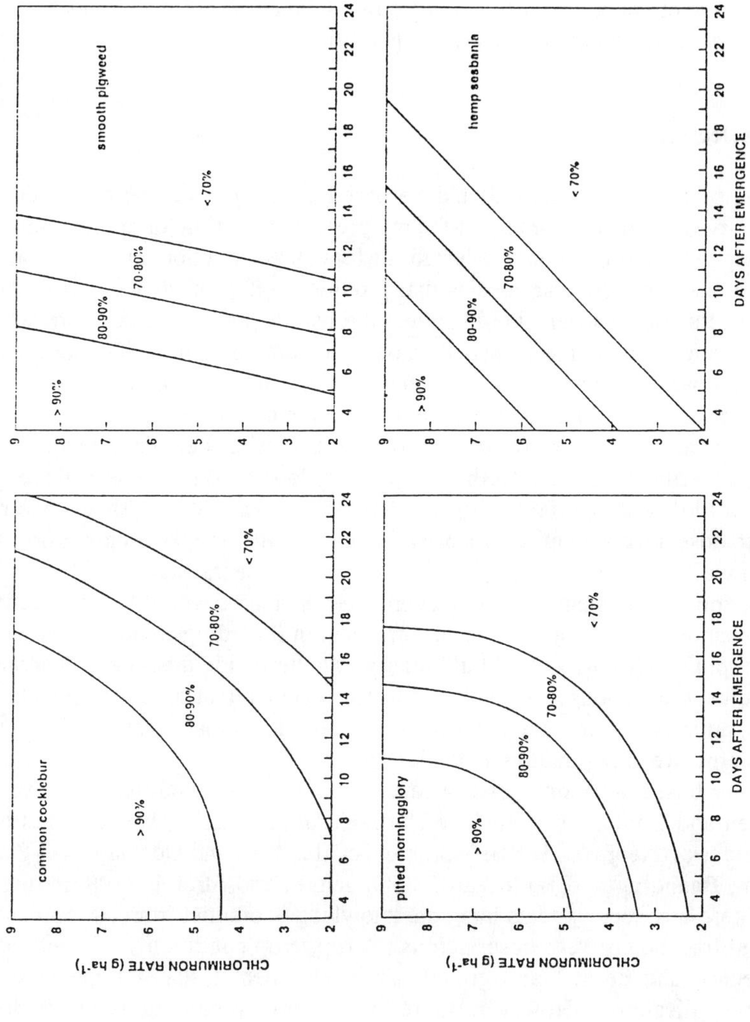

Figure 7. Predicted percent weed control from chlorimuron doses applied at 3 to 24 days after emergence for common cocklebur, pitted morningglory, smooth pigweed, and hemp sesbania (From King and Oliver, 1992).

several weed control strategies reduces the reliance on herbicides. If inter-row cultivation is already part of the crop production practice, the dose required to provide a certain level of weed pressure reduction can be significantly reduced. One inter-row cultivation provided 45 and 55% control of annual broadleaf weed species in chisel-plow and no-tillage experiments, respectively (Buhler et al., 1995). Cultivation may be more successful resulting from an interaction between reduced competitiveness or enhanced susceptibility of herbicide-treated weeds and tillage. The cumulative effects of mechanical and cultural weed control strategies and the application of reduced doses needs to be quantified.

C. Dose Selection

The actual selection of a dose should be based on the derived herbicide dose-response curves and an assessment of factors present at the time of application. A level of response must be initially selected, such as 50% reduction, or biologically-effective doses for particular species that provide 70, 80, or 90% control. The amount of reduction chosen should reflect the use of previous and future weed control strategies, such as mechanical methods or cultural practices. To make such a reduced dose strategy reasonable and useful, database construction and development of information management models are critical.

Databases need to be provided to growers to assist them in selecting the appropriate herbicide dose. Herbicide product labels could be expanded by including a biologically-effective dose by weed species matrix. Such a matrix would list dose levels required to provide 50, 70, 80 or 90% suppression of individual weed species at different growth or developmental stages.

Information management models and computer databases would be an effective method to summarize the multitudes of information regarding weed responses to herbicide application. Models should incorporate herbicide dose-weed response curves, adjust the response based on key factors occurring at time of application, and select appropriate doses to achieve both short- (immediate crop yield losses) and long-term (weed population dynamics) goals.

Computer-based advisory systems are being developed around the world (Mortensen and Coble, 1991) such as North Carolina (HERB; Wilkerson et al., 1991), Nebraska (NebraskaHERB; Mortensen et al., 1994) and Denmark (PC-Plant Protection; Baandrup and Ballegaard, 1989; Jensen and Streibig, 1994; Kudsk, 1989). These advisory systems integrate knowledge obtained from experimental data and utilize the grower's observations of crop, crop condition, soil condition, weed species, and density and growth or developmental stage at the time of herbicide application. However, more information regarding reduced dose strategies should be incorporated. Field testing of reduced dose strategies and implementation of information management models are necessary. Baandrup and Ballegaard (1989) reported on field experience with the Danish advisory system in spring barley and winter wheat. Over all three years, herbicides selected by the advisory system provided higher percentage weed control (ave. 82%) than the standard mixture of MCPA/dichlorprop (ave. 72%) although reduced dose treatments did lower yields 2.4% compared to the standard treatment.

IV. Integrated Weed Management Systems

Integrated weed management is defined as the use of multiple control tactics, including cultural, genetic, mechanical, biological, and chemical means of weed control, to enhance the competitiveness of the crop over weeds (Linker and Barfield, 1985; Mortensen et al., 1995a; Swanton and Weise, 1991; Walker and Buchanan, 1982). The development of an IWM system takes into account all aspects of the grower's cropping system, where each component of the system provides a measure of weed suppression, not necessarily complete control, and allows for the economically and environmentally sustainable application of weed control strategies.

Reduced dose strategies are one component of an IWM system. Their development requires incorporation of scientific knowledge of weed biology and ecology, grower's knowledge of crop production practices, and intensive and timely management of information. Successful integration of reduced doses with other weed control components has been accomplished at the research station level (Bicki et al., 1991; Buhler et al., 1992; Forcella et al., 1992; Mulder and Doll, 1993). However, little information exists about its success on-farm despite the fact that numerous growers employ reduced dose strategies (Hartzler et al., 1993). Success on-farm may be greater since weed infestation levels are usually low, with average densities for broadleaf and grass weeds of 11.1 and 3.8 plants per linear meter of row, respectively (Johnson et al., 1995) compared to extraordinarily high levels found on research station farms, such as common cocklebur densities of 51 to 73 plants m^{-2} (Buhler et al., 1993) and average broadleaf densities of 96 to 142 plants m^{-2} (O'Sullivan and Bouw, 1993).

Incorporating reduced doses with rotary hoeing or inter-row cultivation has resulted in 50 to 75% reductions in herbicide use without sacrificing crop yield (Buhler et al., 1992; Eadie et al., 1992; Mulder and Doll, 1993). One or two between-row cultivations reduced the amount of PPI or PRE herbicide used (reduced doses and banded over the crop row) without reducing weed control or soybean yield (Buhler et al., 1992). Maximum corn yields were observed in a PRE treatment of 1/2X applied in a 38 cm-band over 76 cm-crop rows with two inter-row cultivations (Mulder and Doll, 1993). Improved consistency of control is possible with combinations of 1/4X herbicide dose and a timely inter-row cultivation (Rathjen and Martin, 1994).

Several band widths (0, 15, 30, 45, 60, 75 cm wide or broadcast) and numbers of inter-row cultivations (0, 1, 2, or 3 times) were compared for control of sicklepod (*Cassia obtusifolia* L.) and pitted morningglory in soybean (Poston et al., 1992). The results indicated that applications of alachlor + imazaquin PRE plus imazaquin POST both in a 15 cm-band together with two cultivations maintained soybean yields. In coarse-textured soils in Illinois, herbicide use was reduced by 73% using a 20 cm-band over the crop row and two inter-row cultivations, providing equivalent weed control to the broadcast application of high large crabgrass [*Digitaria sanguinalis* (L.) Scop.] and low common lambsquarters, smooth pigweed (*Amaranthus hybridus* L.) and common ragweed (*Ambrosia artemisiifolia* L.) populations in both corn and soybean (Bicki et al., 1991).

Weed management in corn was evaluated on-farm where combinations of rotary hoeing, cultivation, and band- or broadcast-applied herbicides were compared at 64 locations in Iowa from 1987 to 1991 (Hartzler et al., 1993). Cultivation alone resulted in denser weed populations than herbicide treatments at 41% of the locations; however, corn yields were reduced at only 22% of these locations. Survey results indicated that 75 and 21% of the corn acres in Iowa were cultivated and rotary hoed, respectively. However, only 20% of the corn acres received a band-applied herbicide, which indicates a potential area for herbicide use reduction. Such regional surveys provide valuable information regarding the extent of adoption of IWM and indicate needed areas of future research and extension.

The number of research studies investigating the interactions of enhanced crop competitiveness and use of reduced doses is increasing. High crop seeding density and selection of competitive cultivars may aid in the suppression of weed growth (Christensen, 1994; Courtney, 1991; Lindquist and Kropff, 1995; Richards and Davies, 1991; Salonen, 1992a,b; Whiting et al., 1991). This strategy does not provide complete control but, together with the prevailing environmental conditions and the presence of a susceptible weed spectrum, acceptable control has been demonstrated (Christensen, 1994; Salonen, 1992a). Christensen (1994) determined a crop varietal calibration factor to adjust the herbicide dose-response curve by comparing weed biomass production in untreated plots among various crop varieties. Early crop ground cover correlated well with reduced weed growth in cereal cultivar trials in Scotland (Richards and Davies, 1991; Whiting et al., 1991). The necessary herbicide dose to provide adequate weed control in combination with cultivar competitiveness was 1/8X of mecoprop-P plus metsulfuron-methyl alone or with the addition of thifensulfuron-methyl (Richards and Davies, 1991). The use of narrower crop rows, which increases crop seeding density, has allowed 1/4 and 1/2X doses to provide consistently high levels of weed control (82 to 99%) in corn, soybean, and sunflower compared to variable control (42 to 99%) in wide rows (Forcella et al., 1992). Defelice et al. (1989) also found that reduced and label doses of POST herbicides were more efficacious in narrow row soybean crops.

The use of IWM systems requires intensive management by the grower. This management is in response to both the demonstrated risks associated with these strategies and the need for appropriate information management technologies to implement these strategies. With the need to combine various strategies in a timely fashion, several risks arise including variable field and environmental conditions, increased liability for lack of performance, and other demands on a grower's time such as livestock or off-farm employment. Field conditions may delay rotary hoeing, cultivation, or herbicide application and reduce their effectiveness. Environmental conditions such as wet soils, compaction, seedling diseases, frost, or sandblasting and reduced crop stands alter the susceptibility of weed populations to control strategies. The reduction of herbicide doses to half or even less has been shown to be possible in well-defined weed/herbicide situations.

V. Future Research and Conclusions

Development of reduced herbicide dose strategies requires scientific knowledge of weed biology and ecology to improve expected levels of control and reduce the variability of that control. Weed species response to reduced dose applications include reduction in weed pressure and weed seed production. The desire is to place the crop at a competitive advantage over weeds and not necessarily to obtain complete weed control. The influence of weed spectrum, selected herbicide dose level, weed infestation level, growth or developmental stage, environmental conditions, application timing, and crop production system on weed species response must be further evaluated. The value of current experimental designs and data collection by herbicide manufacturers and university researchers could be enhanced. Information to be collected includes before and after weed population biomass, density, and leaf area measurements, as well as environmental conditions, from field trials that evaluate specific species treated with a range of low to label herbicide doses. More information is needed on the influence of reduced doses on weed seed production and whether qualitative changes in seed viability and germination, and seedling vigor and emergence occur. Also, the tremendous concern over weed seedbank buildup with regular use of reduced doses, as well as opportunities for resistance development, require more study and indicate the need to use multiple control strategies.

The potential of using reduced dose strategies is enhanced when considered as a component of an integrated weed management system. The magnitude of information available and required to select appropriate reduced doses indicates the need for information management models and databases. For example, herbicide product labels could be a database incorporating a biologically-effective dose by weed species matrix for dose selection. Management models should incorporate scientific knowledge of weed biology and ecology and the grower's knowledge of crop production practices, and allow for intensive and timely management of information. There is a need for on-farm assessment of reduced dose strategies and their interaction with other control tactics, and to address environmental, social, and economic concerns of both the grower and the public through integrated weed management systems.

Acknowledgments

The authors thank Alex Martin, David Holshouser, John Lindquist, and the editors for valuable comments on the manuscript.

References

Adcock, T.E. and P.A. Banks. 1991. Effects of preemergence herbicides on the competitiveness of selected weeds. *Weed Sci.* 39:54-56.

Adcock, T.E., P.A. Banks, and D.C. Bridges. 1990. Effects of preemergence herbicides on soybean (*Glycine max*):weed competition. *Weed Sci.* 38:108-112.

Andersson, L. 1995. Effects of dose and application timing on the seed production of three weed species treated with MCPA or tribenuron-methyl. *Weed Res.* 35:67-74.

Anonymous. 1995. *A 1995 Guide for Herbicide Use in Nebraska.* Nebraska Cooperative Extension EC 95-130-D. University of Nebraska, Lincoln, NE.

Baandrup, M. and T. Ballegaard. 1989. Three years field experience with an advisory computer system applying factor-adjusted doses. p. 555-560. In: *Proc. Br. Crop Prot. Conf.-Weeds.* Lavenham, Suffolk, UK.

Bauer, T.A. and D.A. Mortensen. 1992. A comparison of economic and economic optimum thresholds for two annual weeds in soybeans. *Weed Technol.* 6:228-235.

Bicki, T.J., L.M. Wax, and S.K. Sipp. 1991. Evaluation of reduced herbicide application strategies for weed control in coarse-textured soils. *J. Prod. Agric.* 4:516-519.

Brain, P. and R. Cousens. 1989. An equation to describe dose responses where there is stimulation of growth at low doses. *Weed Res.* 29:93-96.

Brown, H.M., V.A. Wittenbach, D.R. Forney, and S.D. Strachan. 1990. Basis for soybean tolerance to thifensulfuron methyl. *Pest. Biochem. Phys.* 37:303-313.

Buhler, D.D., J.L. Gunsolus, and D.F. Ralston. 1992. Integrated weed management techniques to reduce herbicide inputs in soybean. *Agron. J.* 84:973-978.

Buhler, D.D., J.L. Gunsolus, and D.F. Ralston. 1993. Common cocklebur (*Xanthium strumarium*) control in soybean (*Glycine max*) with reduced bentazon rates and cultivation. *Weed Sci.* 41:447-453.

Buhler, D.D., J.D. Doll, R.T. Proost, and M.R. Visocky. 1995. Integrating mechanical weeding with reduced herbicide use in conservation tillage corn production systems. *Agron. J.* 87:507-512.

Christensen, S. 1994. Crop weed competition and herbicide performance in cereal species and varieties. *Weed Res.* 34:29-36.

Coble, H.D. and D.A. Mortensen. 1992. The threshold concept and its application to weed science. *Weed Technol.* 6:191-195.

Courtney, A.D. 1991. The role of competition in developing an appropriate rate strategy for weed control in spring barley. p. 1217-1224. In: *Proc. Br. Crop Prot. Conf.-Weeds.* Lavenham, Suffolk, UK.

Cousens, R. 1985. A simple model relating yield loss to weed density. *Ann. Appl. Biol.* 107:239-252.

Cousens, R., P. Brain, J.T. O'Donovan, and P.A. O'Sullivan. 1987. The use of biologically realistic equations to describe the effects of weed density and relative time of emergence on crop yield. *Weed Sci.* 35:720-725.

Defelice, M.S., W.B. Brown, R.J. Aldrich, B.D. Sims, D.T. Judy, and D.R. Guethle. 1989. Weed control in soybeans (*Glycine max*) with reduced rates of postemergence herbicides. *Weed Sci.* 37:365-374.

Devlin, D.L., J.H. Long, and L.D. Maddux. 1991. Using reduced rates of postemergence herbicides in soybeans (*Glycine max*). *Weed Technol.* 5:834-840.

Dieleman, J.A., A.S. Hamill, S.F. Weise, and C.J. Swanton. 1995. Empirical models of pigweed (*Amaranthus* spp.) interference in soybean (*Glycine max*). *Weed Sci.* 43:612-618.

Dieleman, J.A., A.S. Hamill, G.C. Fox, and C.J. Swanton. 1996. Decision rules for postemergence control of pigweed (*Amaranthus* spp.) in soybean (*Glycine max*). *Weed Sci.* 44:126-132.

Eadie, A.G., C.J. Swanton, J.E. Shaw, and G.W. Anderson. 1992. Banded herbicide applications and cultivation in a modified no-till corn (*Zea mays*) system. *Weed Technol.* 6:535-542.

Fawcett, R.S. and F.W. Slife. 1978. Effects of 2,4-D and dalapon on weed seed production and dormancy. *Weed Sci.* 26:543-547.

Forcella, F., M.E. Westgate, and D.D. Warnes. 1992. Effect of row width on herbicide and cultivation requirements in row crops. *Am. J. Alt. Agric.* 7:161-167.

Gerber, H.R., A. Nyfleler, and D.H. Green. 1983. The influence of rainfall, temperature, humidity, and light on soil- and foliage-applied herbicides. *Asp. Appl. Biol.* 4:1-14.

Gressel, J. and L.A. Segel. 1982. Interrelating factors controlling the rate of appearance of resistance: The outlook for the future. p. 325-347. In: J.M. LeBaron and J. Gressel (eds.), *Herbicide Resistance in Plants*, John Wiley and Sons, New York, NY.

Gressel, J. and L.A. Segel. 1990. Modelling the effectiveness of herbicide rotations and mixtures as strategies to delay or preclude resistance. *Weed Technol.* 4:186-198.

Hartzler, R.G., B.D. Van Kooten, D.E. Stoltenberg, E.M. Hall, and R.S. Fawcett. 1993. On-farm evaluation of mechanical and chemical weed management practices in corn (*Zea mays*). *Weed Technol.* 7:1001-1004.

Harvey, R.G. and C.R. Wagner. 1994. Using estimates of weed pressure to establish crop yield loss equations. *Weed Technol.* 8:114-118.

Henzell, R., J. Phillips, and P. Diggle. 1985. Influence of sublethal concentrations of herbicides and growth regulators on mouseearcress (*Arabidopsis thaliana*) progeny. *Weed Sci.* 33:430-434.

Hume, L. and S. Shirriff. 1989. The effects of 2,4-D on growth and germination of lamb's-quarters (*Chenopodium album* L.) plants having different degrees of tolerance. *Can. J. Plant Sci.* 69:897-902.

Jasieniuk, M. and B.D. Maxwell. 1994. Population genetics and the evolution of herbicide resistance in weeds. *Phytoprot.* 75(Suppl.):25-35.

Jensen, P.K. and P. Kudsk. 1988. Prediction of herbicide activity. *Weed Res.* 28:473-478.

Jensen, J.E. and J. Streibig. 1994. Herbicide dose-response curves and sustainable agriculture. p 15-33. In: *Quantitative Methods for Sustainable Agriculture*, EU HARMA Concerted Action Workshop, May 6-7, 1994, Edinburgh, UK.

Johnson, G.A., D.A. Mortensen, and A.R. Martin. 1995. A simulation of herbicide use based on weed spatial distribution. *Weed Res.* 35:197-205.

Jones, D.S., H. Lin, M.V. Kane, and L.J. Grabau. 1995. Reduced preplant-incorporated imazaquin rates for broadleaf weed control in soybean. *Agron. J.* 87:498-502.

Khan, M. and W.W. Donald. 1992. Sulfonylurea herbicides reduce survival and seed production of green and yellow foxtails (*Setaria* spp.). *Weed Technol.* 6:284-290.

King, C.A. and L.R. Oliver. 1992. Application rate and timing of acifluorfen, bentazon, chlorimuron and imazaquin. *Weed Technol.* 6:526-534.

Klingaman, T.E., C.A. King, and L.R. Oliver. 1992. Effect of application rate, weed species, and weed stage of growth on imazethapyr activity. *Weed Sci.* 40:227-232.

Kropff, M.J. 1988. Modelling the effects of weeds on crop production. *Weed Res.* 28:465-471.

Kropff, M.J. and C.J.T. Spitters. 1991. A simple model of crop loss by weed competition from early observations on relative leaf area of the weeds. *Weed Res.* 31:97-105.

Kropff, M.J., L.A.P. Lotz, and S.E. Weaver. 1993. Practical applications. p. 149-167. In: Kropff, M.J. and H.H. van Laar (eds.), *Modelling Crop-Weed Interactions,* CAB International, Wallingford, UK.

Kudsk, P. 1989. Experiences with reduced herbicide doses in Denmark and the development of the concept of factor-adjusted doses. p. 545-554. In: *Proc. Br. Crop Prot. Conf.-Weeds.* Lavenham, Suffolk, UK.

Kudsk, P., T. Olesen, and K.E. Thonke. 1990. The influence of temperature, humidity and simulated rain on the performance of thiameturon-methyl. *Weed Res.* 30:261-269.

Lindquist, J.L. and M.J. Kropff. 1996. Applications of an ecophysiological model for irrigated rice (*Oryza sativa*) - *Echinochloa* competition. *Weed Sci.* 44:52-56.

Lindquist, J.L., B.D. Maxwell, D.D. Buhler, and J.L. Gunsolus. 1995. Modeling the population dynamics and economics of velvetleaf (*Abutilon theophrasti*) control in a corn (*Zea mays*)-soybean (*Glycine max*) rotation. *Weed Sci.* 43:269-275.

Linker, H.M. and C.S. Barfield. 1985. Integrating tactics into soybean IPM programs. p. 652-662. In: R.E. Frisbie and P.L. Adkisson (eds.), *CIPM: Integrated Pest Management on Major Agricultural Systems.* Texas A&M University, TX.

Martin, A.R., D.A. Mortensen, and N. Burchell. 1994. Visual and photographic assessment of herbicide efficacy trials for use in bioeconomic modeling. *Proc. WSSA* 34:117.

Maxwell, B.D., M.L. Roush, and S.R. Radosevich. 1990. Predicting the evolution and dynamics of herbicide resistance in weed populations. *Weed Technol.* 4:2-13.

Medd, R.W. and S. Pandey. 1993. Compelling grounds for controlling seed production in *Avena* species (wild oats). p. 769-776. In: *Proc. 8th EWRS Symposium*, Braunschweig, Germany.

Meeusen, R.M., S.M. Lokey, and R.Y. Yih. 1983. Effects of environmental factors on the herbicidal activity of acifluorfen. *Asp. Appl. Biol.* 4:253-263.

Morrison, I.N. and M.D. Devine. 1994. Herbicide resistance in the Canadian prairie provinces: Five years after the fact. *Phytoprot.* 75(Suppl.):5-16.

Mortensen, D.A. and H.D. Coble. 1991. Two approaches to weed control decision-aid software. *Weed Technol.* 5:445-452.

Mortensen, D.A., D.L. Holshouser, L.G. Higley, and L.D. Pedigo. 1995a. Adopting pest management principles from a neighboring discipline. *Proc. WSSA* 35:301.

Mortensen, D.A., A.R. Martin, and J.A. Dieleman. 1995b. Reducing herbicide use through developments in methods and technologies for herbicide application. *Proc. WSSA* 35:363.

Mortensen, D.A., A.R. Martin, F.W. Roeth, T.E. Harvill, R.W. Klein, G.A. Wicks, R.G. Wilson, D.L. Holshouser, and J.W. McNamara. 1994. *NebraskaHERB, Version 4.0.*, University of Nebraska, Agricultural Research Division, Publication 94-114.

Mulder, T.A. and J.D. Doll. 1993. Integrating reduced herbicide use with mechanical weeding in corn (*Zea mays*). *Weed Technol.* 7:382-389.

Orson, J.H. 1990. Population thresholds as an aid to weed control. p. 41-48. In: *Proc. Br. Crop Prot. Conf.-Weeds.* Lavenham, Suffolk, UK.

O'Sullivan, J. and W.J. Bouw. 1993. Reduced rates of postemergence herbicides for weed control in sweet corn (*Zea mays*). *Weed Technol.* 7:995-1000.

Pannell, D.J. 1990a. An economic response model of herbicide application for weed control. *Aust. J. Agric. Econ.* 34:223-241.

Pannell, D.J. 1990b. Model of wheat yield response to application of diclofop-methyl to control ryegrass (*Lolium rigidum*). *Crop Prot.* 9:422-428.

Poston, D.H., E.C. Murdock, and J.E. Toler. 1992. Cost-efficient weed control in soybean (*Glycine max*) with cultivation and banded herbicide applications. *Weed Technol.* 6:990-995.

Prostko, E.P. and J.A. Meade. 1993. Reduced rates of postemergence herbicides in conventional soybeans (*Glycine max*). *Weed Technol.* 7:365-369.

Quackenbush, L.S. and R.N. Andersen. 1984. Effect of soybean (*Glycine max*) interference on eastern black nightshade (*Solanum ptycanthum*). *Weed Sci.* 32:638-645.

Rathjen, M.E. and A.R. Martin. 1994. Weed control in corn as influenced by rate and timing of postemergence herbicides. *Proceedings, NCWSS* 49:49.

Richards, M.C. and D.H.K. Davies. 1991. Potential for reducing herbicide inputs/rates with more competitive cereal cultivars. p. 1233-1240. In: *Proc. Br. Crop Prot. Conf.-Weeds.* Lavenham, Suffolk, UK.

Salonen, J. 1992a. Efficacy of reduced herbicide doses in spring cereals of different competitive ability. *Weed Res.* 32:483-491.

Salonen, J. 1992b. Yield responses of spring cereals to reduced herbicide doses. *Weed Res.* 32:493-499.

Seefeldt, S.S., J.E. Jensen, and E.P. Fuerst. 1995. Log-logistic analysis of herbicide dose-response relationships. *Weed Technol.* 9:218-227.

Steckel, L.E., M.S. Defelice, and B.D. Sims. 1990. Integrating reduced rates of postemergence herbicides and cultivation for broadleaf weed control in soybeans (*Glycine max*). *Weed Sci.* 38:541-545.

Streibig, J.C. 1980. Models for curve-fitting herbicide dose response data. *Acta Agriculturae Scandinavica* 30:59-64.

Streibig, J.C. 1988. Herbicide bioassay. *Weed Res.* 28:479-484.

Streibig, J.C. 1989. The herbicide dose-response curve and the economics of weed control. p. 927-935. In: *Proc. Br. Crop Prot. Conf.-Weeds.* Lavenham, Suffolk, UK.

Streibig, J.C. and P. Kudsk. 1993. *Herbicide Bioassays.* CRC Press, Inc., Boca Raton, FL.

Swanton, C.J. and S.F. Weise. 1991. Integrated weed management: the rationale and approach. *Weed Technol.* 5:657-663.

Walker, R.H. and G.A. Buchanan. 1982. Crop manipulation in integrated weed management systems. *Weed Sci.* 30 (Suppl. 1):17-24.

Weaver, S.E. 1991. Size-dependent economic thresholds for three broadleaf weed species in soybeans. *Weed Technol.* 5:674-679.

Whiting, A.J., D.H.K. Davies, H. Brown, and G.P. Whytock. 1991. The field use of reduced doses of broad-leaved weed herbicides in cereals. p. 1209-1216. In: *Proc. Br. Crop Prot. Conf.-Weeds.* Lavenham, Suffolk, UK.

Wilkerson, G.G., S.A. Modena, and H.D. Coble. 1991. HERB: Decision model for postemergence weed control in soybean. *Agron. J.* 83:413-417.

Winkle, M.E., J.R.C. Leavitt, and O.C. Burnside. 1981. Effects of weed density on herbicide absorption and bioactivity. *Weed Sci.* 29:405-409.

Zhang, J. and P.B. Cavers. 1994a. Effect of herbicide application on fruit characters of *Xanthium strumarium* L. populations. *Weed Res.* 34:319-326.

Zhang, J. and P.B. Cavers. 1994b. Seedling emergence after maternal bentazon application to 10 common cocklebur (*Xanthium strumarium*) populations. *Can. J. Plant Sci.* 74:863-866.

Zhang, J., P.B. Cavers, and M. Jasieniuk. 1994. Response of *Xanthium strumarium* populations to sublethal applications of bentazone. *Weed Res.* 34:55-61.

Decision Support Models for Weed Management: In-Field Management Tools

Alex R. Martin, David A. Mortensen, and John L. Lindquist

I. Introduction

Decision support models have a role as tools for weed management. The evolution of decision support models for weed management, including the principles upon which they are based and the limitations of their accuracy and adoption, are discussed in this paper. Economic and environmental constraints on weed management strategies coupled with numerous available herbicides present the grower with a large volume of information to evaluate before reaching a weed management decision. Computer models are well suited for evaluating these quantities of information and therefore may facilitate weed management decision making. Currently available weed management models range in sophistication from herbicide selection models based on efficacy to threshold based bioeconomic models, featuring cost/benefit analysis.

II. Models

Initial development of weed management decision support models focused on herbicide efficacy, often with consideration of treatment cost, crop rotation, and soil properties. Recent more comprehensive models feature a cost/benefit analysis of various strategies based on treatment cost and predicted crop loss. The crop loss relationship may be based on weed seedbank or seedling populations.

A. Herbicide Efficacy Based Models

The forerunner of efficacy based models were the herbicide use guides published by the Cooperative Extension Service at many universities. These publications provided herbicide information with respect to: 1) efficacy by weed species and in some cases weed growth stage, 2) soil interactions, 3) carryover potential, 4) treatment cost, and 5) compatibility with tillage systems. As the number of herbicide products and the complexity of their use specifications increased, a computer based selection model became a logical application. A number of these models were developed beginning in the early 1980s (Mortensen and Coble, 1991), most enjoying modest adoption. Examples of these models include, "COTNWEED"-University of Georgia, "Field-Crop Herbicide Selection"-University of Illinois, "Weeding with Chemicals"-Purdue University, and "Herbicides"-University of Nebraska. These models did not provide any assessment of the impact of uncontrolled weeds or the economics associated with various weed management strategies. They assisted in selecting the most efficacious treatment that met selected criterion without consideration of the economic impact of the selection. The user informally provided the cost/benefit assessment of various scenarios.

B. Threshold Based Models

Bioeconomic models link the concepts of herbicide selection based on efficacy, cost of management, and weed-crop interference (King et al., 1986; Lybecker et al., 1991; Swinton and King, 1994; Wilkerson et al., 1991). The nature of this linkage is defined by threshold theory. The weed density at which the value of crop loss equals treatment cost is commonly referred to as the single year economic threshold (Coble and Mortensen, 1992; Lindquist et al., 1995). The weed population estimate is based on the soil seedbank for certain models and weed seedling population for others. Seedbank based models utilize weed seed numbers in the soil obtained by soil sampling to estimate a resultant seedling population upon which crop loss functions are based. Seedling based models utilize field sampling of emerged weeds to estimate weed population upon which crop loss functions are based. Weed-crop interference research is the basis for the crop yield loss functions used in these models. Threshold based models developed at North Carolina State University first became available in 1989 (Mortensen and Coble, 1991). When an economic loss associated with a predicted crop yield loss is equal to the cost of treatment (adjusted for treatment efficacy), that treatment will be identified as a potential management option. These models represent an evolutionary advancement over herbicide efficacy based models in terms of the level of information utilized in the decision process.

Threshold based models support the IPM concept in that treatments are utilized only in circumstances where economics justify treatment. However, many models rely primarily on chemical control strategies and do not provide complete integrated weed management (IWM) options. Use of the threshold based models would provide growers an economic basis for treatment decisions. This would support a reduction in unjustified treatments resulting in economic benefit to the grower and perhaps environmental benefits to society as a whole. Thus, threshold based models conceptually would move weed management in a direction deemed desirable by growers and society.

III. Obstacles to Model Adoption

Limitations to the widespread development and adoption of threshold based models are both biological and socioeconomic in nature. These are factors in weed management decisions in general, and especially with threshold based weed management. Biological shortcomings of current models from the perspective of the grower include: 1) the single year nature of the assessment, 2) scouting/sampling time requirement, 3) spatial distribution of weeds as related to sampling, and 4) accuracy of yield predictions. The primary socioeconomic obstacles to model adoption deal with costs versus benefit and land tenure relationships.

A. Biological Limitations

Growers are concerned about the future cost of allowing even a low population of weeds to produce seed even if current year economics would not justify treatment. Bauer and Mortensen (1992) demonstrated that the single year economic threshold for velvetleaf (*Abutilon theophrasti* Medik) and common sunflower (*Helianthus annuus* L.) in soybean (*Glycine max*) was 7.5 and 3.5 fold greater than the multi-year economic optimum threshold. This difference is due to economic optimum threshold analysis addressing future weed management costs as a result of weed seed production that are not addressed in a single year economic threshold. Thus, growers' concerns about allowing uncontrolled weeds to produce seed are valid to some degree. Weed seed production with and without controls and its impact on subsequent weed populations needs further study. This information needs to be incorporated into our models to add a multi-year dimension to the analysis.

Crop consultants and growers object to the detailed weed density data required of most threshold based models and currently are unwilling to invest the time required to obtain the information. Seedbank sampling and seed enumeration are considered too labor intensive for widespread adoption. A more rapid, less intensive sampling procedure over that currently required by most models would be highly desirable and improve acceptance of models.

Distribution of weeds in many fields is not uniform (Johnson et al., 1992). This nonuniform weed distribution is frustrating when sampling fields. Compounding the problem, yield loss is overestimated if a mean weed density calculated from samples that include very high and very low densities is used to calculate yield (Wyse et al., 1993). Uneven weed distribution undermines a growers' confidence in both the sampling process and also in model predictions. As a result, growers question whether the models' predictions are any more precise than their own judgement. Expressing weed presence as competitive load or in terms of percent yield reduction rather than an absolute density would largely overcome the overestimation of yield loss that occurs when absolute densities are used to calculate yield losses in highly variable fields. If spatially aggregated weed populations were stable over time and information managed over time, it would be possible to reduce sampling time required in any one year. Additional research is needed in this area with the results useful in developing more appropriate, less labor intensive inputs for threshold based models.

Crop response to a given weed population may not be constant across environments. Corn and soybean response to velvetleaf has been shown to vary among environments (Bauer et al., 1991; Lindquist et al., 1995; Lindquist et al., 1996); however, soybean response to common sunflower and common waterhemp was relatively constant (Bauer et al., 1991). Use of environmental variables to explain the variation in crop-weed interference relationships across environment has received little attention to date.

Threshold densities vary with variation in the crop-weed size relationship (Weaver, 1990). Yield loss algorithms could be modified to reflect environmental conditions by including coefficients appropriate for various environments, thereby making the calculation sensitive to the environment. Use of a yield loss-weed density model to define the threshold weed density implies that all weeds are identical in their competitiveness with the crop. In many cases this assumption is not valid due to variability in weed size within the population. Refinements in models based on research bringing this effect into consideration in calculating yield response would result in more accurate predictions. While ecologists have developed theoretical models to deal with this "competitive asymmetry" (Pacala and Weiner, 1991), little work has been done to apply them in bioeconomic models.

Weeds that survive a soil active herbicide may be less competitive than untreated weeds (Weaver, 1986; Weaver, 1990). The yield loss functions used in most models are based on competition experiments conducted under herbicide-free conditions. Postemergence models may often be used for weeds surviving in fields previously treated with a soil-active herbicide. In this case, the yield impact of these weeds may be overestimated. The impact of a previous herbicide treatment on weed competitiveness and seed production may be substantial (Steckel et al., 1990; Weaver, 1990). Improvements in model predictions could be achieved by addressing the crop-weed response to specific environmental factors, by refining the crop-weed interactions as influenced by the growth stage of each species, and by assessing the effect of a previous treatment on the competitiveness of escaping weeds.

B. Socioeconomic Constraints

Although the biological and socioeconomic constraints are interrelated, the socioeconomic issues are a greater impediment to the adoption of threshold based weed management than biological issues. Ultimately the issue is one of economics. The greatest single obstacle to widespread adoption of threshold based weed management utilizing computer models is the lack of a motivating incentive. The current financial incentive in the form of a reduction in herbicide expenditures is not large enough to counter the costs associated with detailed sampling and the negative impact on large-scale operations. In the past it has been more cost effective to be less precise in weed management in return for expediting the entire process of weed management on a large scale. However, computer models make intensively managing extensive data feasible and cost-effective.

Growers view threshold based weed management as involving added risk without an adequate compensating incentive. Many of the risks are due to biological constraints. Standardization of weed management programs facilitated large-scale operations with a minimum of field-specific decision making and monitoring. Growers of many crops view this approach to be more profitable than intensively managing fewer acres. The development of an information base and appropriate equipment to manage it will eliminate the obstacles to information intensive weed management on a large scale.

Farm operator-land owner relationships involving the management of rented land influences the acceptance of threshold based weed management. A considerable portion of cultivated land in the Midwest is operated by growers who are not the land owners. Maintaining control of rented land is an important economic issue for these operators. Maintaining a sound operator-land owner relationship is therefore an important economic issue to the grower. Many landowners take pride in the appearance of their land even though they are not directly involved in production agriculture. They have tended to view a field free of weeds as an indication of a well-managed farm. Threshold based weed management, which may allow a low weed population to exist, has often not been well accepted by landowners. It has been difficult for a grower to implement threshold based weed management without jeopardizing the relationship with the owner of rented land. Education of landowners with respect to threshold based weed management is essential if a change is to occur.

IV. Conclusions and Future Research

A number of weed management computer models have been developed since 1980. These models are valuable tools for evaluating large amounts of information and ranking weed management strategies. Most of these models are based on herbicide efficacy with a relatively small number based on thresholds. Herbicide efficacy based models represent a refinement in conventional weed management strategies. This coupled with a minimum of required detailed input data has facilitated acceptance of herbicide efficacy based models. Threshold based weed management models have been less widely accepted for several reasons. Threshold based weed management represents a different philosophical approach than many growers are comfortable with. Increased risk is perceived to be associated with threshold based weed management. The lack of an adequate financial incentive to compensate this risk is a major deterrent to widespread adoption.

The potential risk in using threshold based models is associated with sampling errors, accuracy of yield loss predictions, and the future cost of weed seed production. The detailed, time-consuming, and expensive sampling required of current threshold based models is also a major deterrent. This is a limitation for situations where the available operator's time and labor is limited. Development of less demanding sampling procedures would remove this obstacle. Growers are concerned about the consequences of allowing even a subthreshold population of weeds to mature and produce seed. Developing less labor-intensive sampling, improving accuracy of crop loss predictions, and addressing the weed seed production issue would remove much of the risk associated with threshold based models and would improve grower acceptance.

References

Bauer, T.A. and D.A. Mortensen. 1992. A comparison of economic and economic optimum thresholds for two annual weeds in soybeans. *Weed Technol.* 6:228-235.

Bauer, T.A., D.A. Mortensen, G.A. Wicks, T.A. Hayden, and A.R. Martin. 1991. Environmental variability associated with economic thresholds for soybeans. *Weed Sci.* 39:564-569.

Coble, H.D. and D.A. Mortensen. 1992. The threshold concept and its application to weed science. *Weed Technol.* 6:191-195.

Johnson, G.A., D.A. Mortensen, and A.R. Martin. 1992. The influence of spatial weed distribution on amount of herbicide applied. *North Central Weed Sci. Soc. Proc.* 47:10.

King, R.P., D.W. Lybecker, E.E. Schweizer, and R.L. Zimdahl. 1986. Bioeconomic modeling to simulate weed control strategies for continuous corn (*Zea mays*). *Weed Sci.* 34:972-979.

Lindquist, J.L., B.D. Maxwell, D.D. Buhler, and J.L. Gunsolus. 1995. Modeling the population dynamics and economics of velvetleaf (*Abutilon theoprasti*) control in a corn (*Zea mays*) soybean (*Glycine max*) rotation. *Weed Sci.* 43:269-275.

Lindquist, J.L., D.A. Mortensen, S.A. Clay, R. Schmenk, J.J. Kells, K. Howatt, and P. Westra. 1996. Stability of corn-velvetleaf interference relationships. *Weed Sci.* 44:(in press).

Lybecker, D.W., E.E. Schweizer, and R.P. King. 1991. Weed management decisions in corn based on bioeconomic modelling. *Weed Sci.* 39:124-129.

Mortensen, D.A. and H.D. Coble. 1991. Two approaches to weed control decision-aid software. *Weed Technol.* 5:445-452.

Pacala, S.W. and J. Weiner. 1991. Effects of competitive asymmetry on a local density model of plant interference. *J. Theor. Biol.* 149:165-179.

Steckel, L.E., M.S. Defelice, and B.D. Sims. 1990. Integrating reduced rates of postemergence herbicides and cultivation for broadleaf weed control in soybeans (*Glycine max*). *Weed Sci.* 38:541-545.

Swinton, S.M. and R.P. King. 1994. A bioeconomic model for weed management in corn and soybean. *Agric. Syst.* 44:313-335.

Weaver, S.E. 1986. Factors affecting threshold levels and seed production of jimsonweed (*Datura stramonium* L.) in soybeans [Glycine max (L.) Merr.]. *Weed Res.* 26:215-223.

Weaver, S.E. 1990. Size-dependent economic thresholds for velvetleaf (*Abutilon theophrasti* Medic.), cocklebur (*Xanthium strumarium* L.), and jimsonweed (*Datura stramonium* L.) in soybeans [Glycine max (L.) Merr.]. *Weed Sci. Soc. Am. Abstr.* 30:298.

Wilkerson, G.G., S.A. Modena, and H.D. Coble. 1991. HERB: decision model for postemergence weed control in soybean. *Agron. J.* 83:413-417.

Wyse, D.Y., D.A. Mortensen, and G.A. Johnson. 1993. The influence of spatially variable weed populations on crop yield. *North Central Weed Sci. Soc. Proc.* 48:94.

Developing the Next Generation of Weed Management Systems: Economic and Social Challenges

Michael Duffy

I. Introduction

Natural selection determines what types of plants and animals are present in an ecosystem. The plant community, the focus of this paper, goes through a succession until a climax stage is reached. In this final state, the plants maintain an equilibrium until a disruption occurs. Disruption is followed by a new succession and an equilibrium.

The exact date of the first cultivation of crops is lost in antiquity. Regardless when it occurred, this adaptation allowed our species to expand, thrive, and develop.

The cultivation of crops means ecological disruption. This fact cannot be avoided. We replace the multiple species that existed with the single, desired species. The disruption is necessary, so the question becomes how to hold the impact of the disruption to a tolerable level.

Whenever we choose to plant a crop, two things can happen. First, we may create an environment that will be favorable to some plants in addition to the chosen one. Second, plants not present in the previous ecosystem can enter the altered environment and thrive. These two types of plants become what we refer to as weeds. A weed is simply a plant in a place where it is not desired and weeds are simply a fact of life. As long as we cultivate crops, we will have weeds.

Weeds compete with the desired (in this context, agronomic) plants for water, sunlight, nutrients, and space. Their presence can reduce yields. The question becomes how to keep the weeds out or minimize their impact on the yield of the desired crop.

Now, with this very basic review of plant community dynamics fresh in our minds, let's similarly define economics. Economics is the study of how we allocate scarce resources. Our resources are land, labor, capital, and management.

Obviously, how we should allocate our resources depends on our goals. At the farm level, the goal most frequently assumed is profit maximization. It is important to remember that profits can and should include more than just the strictly monetary, market goods and services. Which of the other goods and services we include depends on the goals of the decision-maker and the level of analysis.

For now, we will assume simple profit maximization of the individual farmer. In a later section of this paper, I will return to the issues of the nonmarket goods and services and alternative levels of decision making. We must recognize that there are more costs and benefits than just the ones valued in the market.

The next section will examine some of the past trends in weed management. Then the current situation will be discussed. The fourth section will look at two possible future directions. Finally, I will offer some conclusions and outline the challenges we face.

II. The Past

Farmers have always fought weeds. Throughout history there has been a progression of methods used for weed control (Cochrane, 1979). We progressed from hand, to beasts of burden, to mechanical methods for fighting weeds, to chemical methods. This progression has essentially been driven by the primary energy source used. We progressed from human to animal to fossil fuel energy.

Cultural methods, primarily crop rotations, have also been used for centuries to combat weeds. Different crops grown in successive years decrease the chances for opportunistic plants (Liebmann and Dyck, 1993).

For most of history, cultural or mechanical practices were the only tools available to protect against weeds. The development of herbicides was a major breakthrough. “Herbicides are relatively new and only recently have been used in large quantities for weed control. In 1950 there were only 15 basic herbicides. By 1974 there were more than 180 with about 6000 formulated products.” (Sils, 1982).

The advent of herbicides changed agriculture as much as or more than any single development. They allowed cropping patterns and practices that were previously impossible. “Advances in chemical weed control technology have had far reaching impact on all phases of crop production, including the selection of crops and varieties, seedbed preparations, methods of seeding and seeding rates.” (Sils, 1982).

Herbicide materials have also gone through their own technological evolution. First came the contact herbicides and then the preemergents. Their usage has gone from pounds of active ingredients to ounces and now grams per hectare.

It is interesting to note farmer adoption of herbicides. Contact herbicides were rather quickly assimilated into production practices because their efficacy was apparent. Preemergent herbicide adoption was slower. Klonglan interviewed

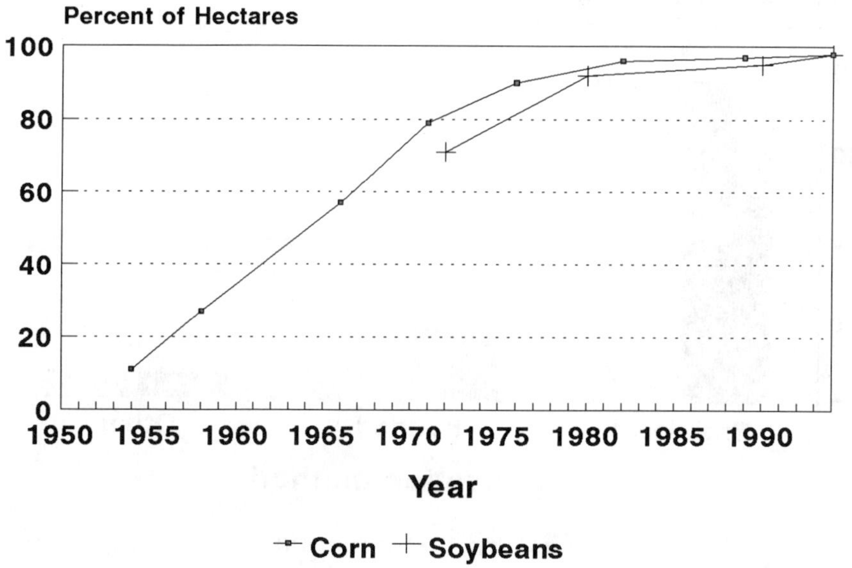

Figure 1. Percent of U.S. hectares treated with herbicides. (Adapted from data obtained from Inputs: Outlook and Situation, USDA-ERS, various years.)

farmers who had used new preemergent products. He asked whether they would use the material again. About 25 percent of the farmers who were satisfied with the level of weed control achieved said they would not use the preemergents again. When queried as to why, the response was that they did not believe that it was the material that prevented weed germination (Klonglan, 1963). Technology had outstripped those farmers' ability to comprehend or believe in it.

III. The Current Situation

Today farmers have four basic options available for weed management: chemical, cultural, mechanical, and biological. Chemical and mechanical methods dominate.

Figure 1 shows the change in percent of corn and soybean hectares treated with herbicides. These percentages are for the U.S., and they illustrate the rapid rise in herbicide use. Note that for the past decade, over 90 percent of the corn and soybean hectares have been treated with herbicides.

Figure 2 presents the predominant methods of herbicide application in Iowa during 1994. The majority of herbicide applications are broadcast. Mechanical cultivations represent another widely used weed management tool. Figure 3 shows the percentage of hectares row cultivated by crop.

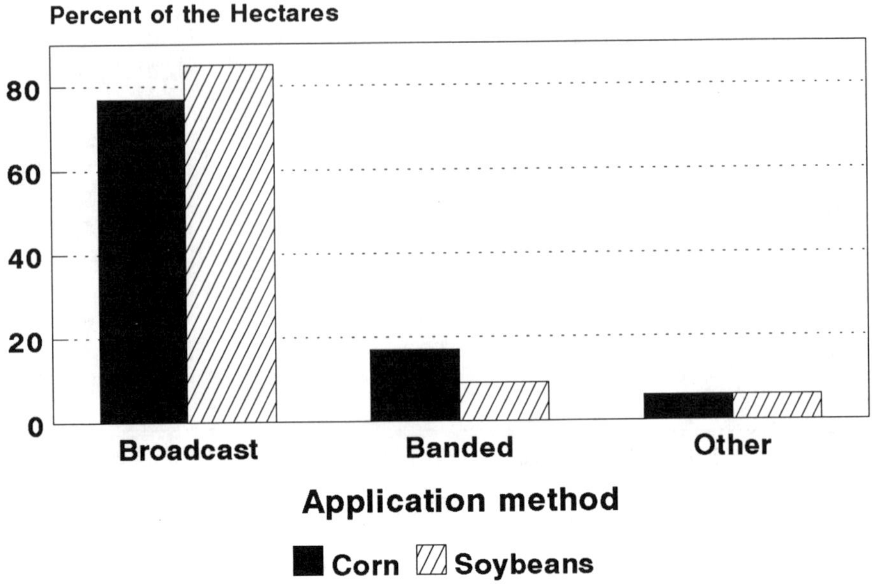

Figure 2. Herbicide application methods in Iowa, 1994. (Adapted from 1994 USDA Crop Production Survey.)

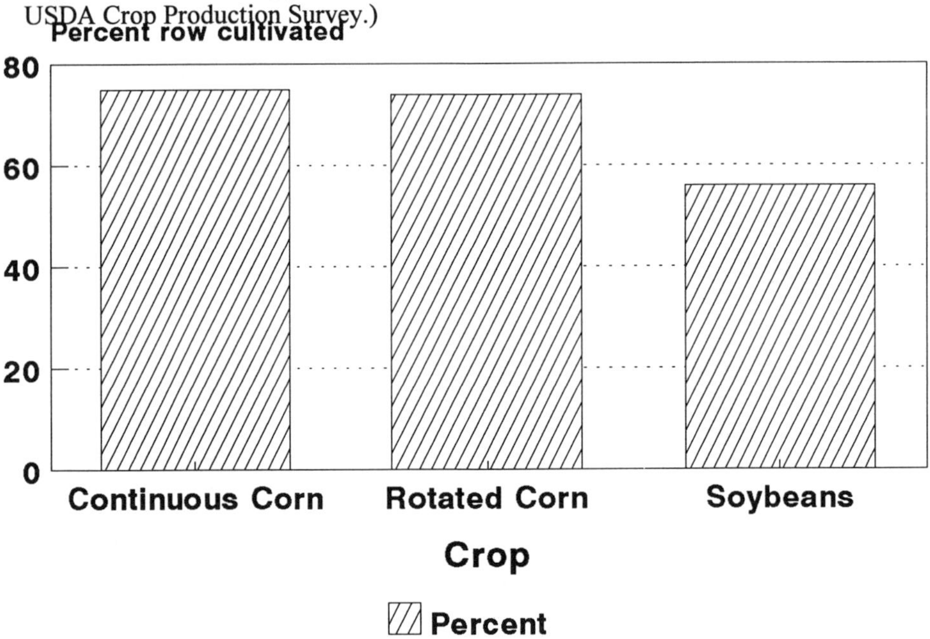

Figure 3. Percent of fields row cultivated in Iowa, 1994. (Adapted from 1994 USDA Crop Production Survey.)

Table 1. Weed management costs per hectare by crop in Iowa, 1994

	Continuous corn	Rotated corn	Soybeans
Herbicides[a]			
Average	$57.27	$59.63	$65.41
Range	0 - $144.59	0 - $195.20	0 - $187.99
Total Weed Costs[b]			
Average	$73.40	$75.06	$81.76
Range	$5.83 - 156.25	$5.83 - $212.69	$13.59 - $207.13

[a]Herbicide costs are for the material and amount used times an average price for each material.
[b]Total weed costs include herbicides, application costs, and row cultivations; machinery costs are from ISU Extension Publication, FM-1712, "Estimated Costs of Crop Production in Iowa, 1995."
(From USDA-NASS, 1994.)

Table 1 presents the direct weed management costs by crop in Iowa for 1994. Herbicide materials make up over three-fourths of the total weed management costs. Note the tremendous variation in costs.

The cross-sectional data presented in Table 1 limit our ability to draw inferences from the weed management practices. Although 1994 was an almost ideal year, with record yields in Iowa for both corn and soybeans, there were still some early season deficit soil water problems and carryover problems from the 1993 floods. These problems may have led to resprays and additional weed management practices. The range in costs shows that farmers vary considerably in their approach to and expenditures on weed management. Analysis of the data revealed no correlation between the herbicide costs and yields.

Primary tillage practices are also used for weed management as well as soil conditioning. However, the use of heavy tillage equipment has diminished over the past several years. Table 2 shows the tillage operations by crop and how they have changed over the last five years.

The change in tillage practices can be attributed to many factors. The most salient factor is the conservation compliance provision of the 1985 Food Security Act. Under this law, farmers with highly erodible land are required to have an approved soil conservation plan, if they are to receive government program benefits. The conservation plan requires farmers to achieve and maintain a minimum level of soil erosion (Napier, 1990).

These conservation plans can include measures such as terraces, grassed waterways, tillage practices, or rotations. Most farmers choose to increase residue cover rather than change cropping sequence or structure.

The mechanical operations changes shown in Table 2 are indicative of the move toward leaving more crop residue. Additionally, a major shift to drilled soybeans has implications for weed management.

Table 2. Common operations used in Iowa by percent of hectares by crop for 1989 and 1994

	Continuous corn		Rotated corn		Soybeans	
	1989	1994	1989	1994	1989	1994
Moldboard plow	23	4	7	1	25	5
Chisel plow	37	28	20	11	30	12
Sub soiler (V ripper)	1	7	15	4	13	--
Soil saver	6	15	2	3	4	16
Field cultivator	50	51	61	61	61	57
Disk	71	50	54	22	73	51
Drag	19	1	14	1	11	2
Rotary hoe	19	--	21	--	17	--
Row cultivator	88	75	88	74	85	56

(Adapted from USDA-NASS, 1994, and Duffy and Thompson, 1991.)

Table 3. Comparison of row planted and drilled soybeans in Iowa, 1994

	Row planted	Drilled
Percent of hectares	72	28
Average yield (kg ha^{-1})	3460	3575
Seed costs ($ ha^{-1})	$35.47	$47.25
Herbicide costs	$63.63	$70.02
Total weed management costs	$80.77	$84.45
Return to management	$181.47	$188.09

(From USDA-NASS, 1994.)

Table 3 presents a comparison of selected characteristics of row-planted versus drilled soybeans in 1994. Note that on the average, drilled soybeans produced 115 kilos more per hectare. The returns, however, were only $6.62 per hectare higher due to higher production costs. Note also that the drilled soybeans had $6.39 per hectare higher herbicide costs but only $3.68 higher total direct weed management costs. This shows the shift from the chemical and mechanical strategies toward more of a single herbicide approach for the drilled soybeans relative to row-planted soybeans.

These trends — decreases in tillage, shift in seeding patterns and others — produce impacts that are difficult to quantify. Decreasing soil erosion, increasing chemical use, decreasing labor, and decreasing direct fuel requirements are all examples of impacts from changes that have occurred in the last few years.

Again, the development of weed management strategies can be characterized by the shift in energy source and reliance on multiple methods. Today we rely heavily on chemicals.

This shift to reliance on herbicides has led to unanticipated yet fundamental problems. One problem is that some plant species have developed resistance to widely used herbicides. A second problem is that some herbicide materials have become so widely used that they are appearing in ground and surface water as well as in rain water (Goolsby et al., 1993; IDNR, 1990).

A third problem is that many farmers find it increasingly difficult to make a living at farming. They have substituted capital for labor and significantly increased output, but they have not significantly increased profits. In the 1950s, net farm income averaged 34 percent of gross farm income. During the 1980s, and thus far in the 1990s, net is only 15 percent of gross income (Duffy, 1995). That means it takes twice as much gross income to earn the same net income. The changes in weed management are one of the reasons for the tightening profit margins.

To summarize, weed management, as we prepare to enter the 21st century, increasingly relies on a single, less-than-ideal management technique. What the future holds is uncertain. At least two alternative directions are available.

IV. The Future

Faced with the problems just discussed, farmers and society are asking, how can we control weeds without the negative side effects? The basis of the question, how to control weeds, lies at the heart of the difference between the two future alternatives.

The concept of "controlling" weeds presumes mastery. It suggests that simply finding the right technological fix will solve the problems. On the other hand, the concept of weed management, as opposed to control, acknowledges that we are a part of a natural system. We must strive to learn how to manage weeds realistically, within the bounds imposed by the natural system.

The current research path seeks to control weeds. This is the so-called industrial approach to agriculture. With this approach we place the emphasis on how to kill the plants we do not like.

If we continue down this path, the challenge will be to continually develop more selective herbicides. We have already seen the progression from pounds to ounces to grams. We will continue to breed the desirable plants for resistance to these herbicides. We will continue to seek ways to maintain monoculture cropping patterns.

This industrial path will mean continual high research and development costs and high input costs. We will continually be struggling to find new products as old ones become ineffective and new environmental concerns are raised.

The alternative path could be characterized by a more sustainable approach. Others have debated what sustainable agriculture means (Lochertz, 1988). Here I only consider three broad characteristics: profitability, environmental soundness, and social responsibility. More specifically, in the context of weeds, a sustainable approach would be characterized by management, not "control."

If the more sustainable approach is taken, the basic premise of our research changes — from one where short-run pecuniary profits are all that matters — to working with nature, seeking a long-run equilibrium. With the sustainable approach we will struggle with understanding the nature of weeds and how to manage them.

The sustainable approach presents challenges in at least four major areas. First, we must gain a better understanding of **basic weed ecology**. What are the important weed species in the fields that most seriously impact the cultivated plants? What are their life cycles? What can be done to discourage their presence? The basic approach must come from an understanding of what we are facing and how best to manage the problems.

The second major challenge is to **look beyond a single year**. We must learn to take a longer, more systems-oriented perspective to the weed problems confronting us. Profits need to be considered over an entire rotation sequence, not just year by year.

The third challenge of a more sustainable approach involves continually **seeking new materials and methods**. We must look for materials, whether at a higher or lower rate, that are efficacious and environmentally benign. We must be cognizant of what happens to the material as it decomposes. We must ask ourselves, is our research going to allow more people to make a living off the land or simply produce more at higher costs, increasing the risks and speeding the loss of family farmers?

We must constantly question and challenge ourselves to find a better way, not just a continuation. We must ask ourselves, what are the true consequences of our research?

Finally, the fourth major challenge in the sustainable approach is to **take a systems perspective**. Savory (1988) described this best with the analogy of different colors of clay. Think of each discipline as a color; for example, weed science is red, economics is green, soil science is blue, and so forth. The current approach is like a rainbow, with each discipline clearly identified by its own jargon, paradigms, and distinctions. The problems we face, however, are more gray (all the colors blended together). We must learn to address the problems from our different perspectives by working together. We must contribute what the red has to say about the gray, what the blue has to say, and so forth.

To be sure, disciplinary rigor and advancement are necessary. But to truly solve the problems of weeds in a broader context, there has to be some examination of the whole. Some of us have to work at the system level if we are to avoid or minimize the unintended problems.

The sustainable approach is demanding and in many ways the more difficult. We have to look at the whole system, over time. A sustainable approach would require a shift in emphasis from how to kill plants to how to live with them.

V. Conclusions

Weeds are an inevitable part of cultivated agriculture. Human activity disrupts the plants present and this creates an imbalance. It is this question of imbalance and how to be the least disruptive that should lead our research endeavors. As Aldo

Leopold noted, "We forget that no species is inherently a pest, and any species may become one." (Leopold, 1943).

Our research must be guided by a sense of trying to increase our comprehension of the complexity of the plants in the fields. Only by understanding life cycles, niches, and attributes of the plants can we truly manage them profitably. We must decide whether we will continue to try to control weeds or merely manage them to fit within our goals.

No matter which direction we choose, we face a struggle. If we choose control, we face a constant struggle to find new more potent and selective materials. We may open a Pandora's box of unanticipated environmental concerns. But if we choose to manage weeds, we must constantly struggle with changing conditions and we must broaden our considerations to include economic conditions and social conditions, as well as environmental conditions.

These are two major challenges facing us. First, we must decide which direction is the most appropriate for us to go. Collectively, we must choose between the industrialized or sustainable approach in the long run.

The second major challenge, regardless of which way we go, is to rid ourselves of the attitude that we can find single, simple answers to complex problems. Each farm and farmer is different. Each is endowed with a different set of the basic resources: land, labor, capital, and management.

The most appropriate combination of those resources depends on the individual. Each farm is unique. What may be appropriate for one may be totally inappropriate for another. Our challenge is to get people to choose the most appropriate mix of resources and to give them as many options as possible. The more options we have available, the more likely we are to find the most appropriate solution.

Economics has, for the most part, fostered a short-run, pecuniary mind-set when it comes to weed management. Future research must look at not only the farmer costs of our decisions but costs to society as well.

Additionally, comprehensive risk management must become a part of research and decision-making. Our current perspective with respect to weed risk management is very limited. What happens to costs when resistance develops, what happens when we rely on a single technique, what happens when we decrease biodiversity and finally, what are the costs beyond the farm gate? Farmers and society face risks surrounding our choice of weed management techniques.

Searching to find the one material that fits every situation, or relying on a single method, may appear to be the right approach. However, such approaches are doomed to failure. We must learn to view natural diversity as an ally and emulate it in our solutions, rather than regarding it as an enemy.

Our challenge is to start thinking about the direction we are heading and where we want to be. We must start thinking about the whole system and not just our little piece of it.

A Catholic prayer goes something like this, “Remember man that you are dust and to dust you shall return.” In other words, we are a part of nature and not apart from it. Our challenge is to decide how we can best manage the natural system to not only meet our goals but allow our children and their children to meet their goals.

References

Cochrane, W.W. 1979. *The Development of American Agriculture, A Historical Analysis*. University of Minnesota Press.

Duffy, M. 1995. Profitability in Farming. Presentation at "Partners and Projects: A Farming Research Exchange." Leopold Center for Sustainable Agriculture, Fifth Annual Conference, Iowa State University, Ames, IA, March 1995.

Duffy, M. and L. Thompson. 1991. The Extent and Nature of Iowa Crop Production Practices, 1989. ISU Extension Publication, FM-1839, January 1991.

Goolsby, D.A., L.L. Boyer, and G.E. Mallard (eds.). 1993. Selected Papers on Agricultural Chemicals in Water Resources of the Midcontinental U.S., U.S. Geological Survey Open-File Report 93-418, Denver, Colorado.

Iowa Department of Natural Resources (IDNR). 1990. The Iowa Statewide Rural Well Water Survey; Water Quality Data: Initial Analysis. Tech Info. Series 19. DNR, Des Moines, Iowa.

Iowa State University Extension Publication, FM-1712. 1995. Estimated Costs of Crop Production in Iowa, 1995. Iowa State University, Ames, Iowa.

Klonglan, G.E. 1963. Role of Free Sample Offer in the Adoption of a Technological Innovation. Unpublished Ph.D. Dissertation, Iowa State University, Ames, IA.

Leopold, A. 1943. "What is a Weed?" p. 306-309. In: S.L. Flader and J.B. Callicott (eds.), *The River of the Mother of God and Other Essays*, 1991. University of Wisconsin Press.

Liebmann, M. and E. Dyck. 1993. Crop Rotations and Intercropping Strategies for Weed Management. *Ecol. Applic.* 3:92-122. February 1993.

Lochertz, W. 1988. Open Questions in Sustainable Agriculture. *Am. J. Alt. Agric.* 3(4). Fall.

Napier, T. 1990. *Implementing the Conservation Title of the Food Security Act of 1985*. Soil and Water Conservation Society, Ankeny, Iowa.

Savory, A. 1988. *Holistic Resource Management*. Island Press, Washington, D.C.

Sils, W. 1982. *Plant Protection*. Iowa State University Press, Ames, Iowa.

USDA, ERS. Various years. Inputs: Outlook and Situation. Washington, D.C.

USDA-NASS. 1994. Cropping Practices Survey. Washington, D.C.

Index